MARIN FLORA

Madroño (*Arbutus Menziesii*) in the tanbark oak-madroño forest
near Mountain Theater, Mount Tamalpais.

MARIN FLORA

Manual of the Flowering Plants and Ferns of Marin County, California

By

JOHN THOMAS HOWELL

PHOTOGRAPHS BY CHARLES T. TOWNSEND

*Second Edition
with supplement*

UNIVERSITY OF CALIFORNIA PRESS

BERKELEY AND LOS ANGELES

1970

UNIVERSITY OF CALIFORNIA PRESS
BERKELEY AND LOS ANGELES
CALIFORNIA

―◇―

UNIVERSITY OF CALIFORNIA PRESS, LTD.
LONDON, ENGLAND

STANDARD BOOK NUMBER
520-00578-3

LIBRARY OF CONGRESS CATALOG CARD NO. 71-100608

PRINTED IN THE UNITED STATES OF AMERICA

Preface to the Second Edition

MARIN FLORA was published December 21, 1949, so in its way, the present publication celebrates the twentieth anniversary of the first edition. A large body of floristic information relating to California has become available in the last twenty years and much that pertains to the plants of Marin County has been assembled in the present Supplement. The extent of this new information is shown by the following analysis of more than 350 notations that make up the Supplement.

Indigenous plants new to the flora of Marin County:

    4 genera
    26 species
    23 subspecies, varieties, or forms
    7 named hybrids

Adventive or naturalized plants new to the flora:

    1 family
    37 genera
    99 species
    7 subspecies, varieties, or forms

Taxa deleted [by union with another taxonomic group]:

    5 genera
    7 species
    8 subspecies, varieties, or forms

In the section of the introduction to MARIN FLORA entitled Numerical Analysis of the Flora (p. 23; and Table 1, p. 24), the following altered totals result from these additions and subtractions: 118 families; 564 genera; 1023 indigenous species; 408 introduced species; 174 subspecies, varieties, and forms; and 7 named hybrids.

Not all changes of plant names that have been proposed in the last two decades have been accepted, mostly in an effort to retain a conservative evaluation of genera. Nevertheless 101 name-changes are listed in the Supplement, changes involving corrections, alteration of status, reidentification of introduced plants, etc. The nomenclature enlisted in any flora is subject to progressive change and that in MARIN FLORA has been and will continue to be no exception!

Distributional notes on plants within Marin County have been kept to a minimum. However, 32 plants are noted in the Supplement which formerly were believed to be endemic or to have a northern or southern Coast Range distributional limit within Marin County but which are known to be more widespread. This figure is offset somewhat by the addition of several plants that are now known to reach a distributional limit in Marin County or to be endemic.

As always, the compilation of a floristic work is accomplished only with the help of many persons—botanists, both professional and amateur, and lay folks who are interested in plants. Although this help is frequently noted in the citation of collections and of botanical literature in the following pages, I wish to name here some of those collaborators without whose interest and help this Supplement would be woefully less complete: R. C. Bacigalupi, T. C. Fuller, V. F. Hesse, Walter and Irja Knight, E. McClintock, L. McHoul, Arthur and Barbara

Menzies, J. Peñalosa, P. Raven, Douglas Ripley, P. Rubtzoff, Malcolm and Laura
Smith, G. H. True; and, for help in preparing the manuscript, Virginia Moore.
To these and all others who have helped I am most grateful.

J. T. H.

California Academy of Sciences
July 18, 1969

Preface

THE METROPOLITAN area about San Francisco Bay is noted for the ease with which one may get from crowded cities into open country. For many decades, Marin County has been considered one of the most scenic parts and its hills and forests have been a constant lure to those who wish to lose themselves in the beauty of a natural scene. In 1938 I turned to the Marin countryside for weekly recreational walks, and, to add a botanical motive to my outings, I began almost immediately to collect data on the rich and diverse flora which was everywhere about me. The elaboration of these simple data, so pleasantly acquired, has resulted in the present work.

My aim has been to prepare a usable tool with which both amateur naturalist and student can become familiar with a remarkable flora lying within ready access of more than a million people, and no effort has been spared to give as accurate and full an account as possible. Although my botanical field studies in Marin County have been pursued more intensively for the past ten years from the California Academy of Sciences, they actually began over twenty years ago while I was still a student of W. L. Jepson. Over this period of years almost every part of the county has been explored, and some places, such as Mount Tamalpais, Point Reyes Peninsula, and Tiburon Peninsula, have been visited many times. Even after this prolonged and detailed survey, plants new to the area are being found, and because so many of the species in the region are characterized by an extremely localized occurrence or restricted range, they will probably continue to be found.

It was originally hoped that the present work would be relatively simple and nontechnical, but such an aim has been only partly realized. In a flora as extensive and as diverse as that of Marin County, there is no way of preparing keys for the identification of the plants without recourse to scientific terminology. The meaning of these botanical terms can be found in the glossary preceding the index at the end of the book. The details in the keys given for the determination of family, genus, species, variety, or form apply only to the plants of Marin County, though much of the data should hold for central coastal California. To save space, descriptions of the different groups, from families to forms, are not given beyond the points needed to identify them in our region. Students desiring a fuller treatment of a plant will find it described in detail in the floras of L. R. Abrams or W. L. Jepson, either under the name used for the plant in this work or under the synonym placed at the end of the ecologic and distributional note that is given for each Marin County plant recognized. Synonymy, beyond this kind supplied for reference purposes, is rarely given.

Systematic work in botany, whether it is floristic or monographic, leans so heavily on what has been done by others that excessive pretensions to novelty and originality can generally be discounted. Of the utmost importance and help in assembling floristic data in the present work was a manuscript catalogue of Marin County plants given to me by the late John W. Stacey. A shorter general list was received from Dorothy Sutliffe and a catalogue of plants found

in the Sausalito Hills was received from Elsie Zeile Lovegrove. The Stacey catalogue approached the proportions of a complete flora, 1,006 species being listed
from many localities. Though its importance and usefulness in the present work
can be realized by the number of times it is referred to for records of occurrence,
it is to be regretted that botanical specimens substantiating these records are
usually lacking. Besides the help from these three lists, many field records and
specimens have been provided by Hans Leschke, whose name also appears frequently with the others at appropriate places in the text. Numerous other records have come from Alice Eastwood—in conversations with her, from her extensive writings, or from her abundant Marin collections preserved in the Herbarium of the California Academy of Sciences. In the present work, the western
America floras of L. R. Abrams, W. L. Jepson, T. H. Kearney and R. H. Peebles,
P. A. Munz, and M. E. Peck, as well as the floristic works of others, have been
consulted constantly, and the works of L. H. Bailey and A. Rehder have been
referred to frequently in connection with the identification of naturalized garden
plants. Wherever possible, account and use have been made of pertinent revi-
sional notes and monographic studies of many students. Most of the American
botanists who are authorities on families or genera represented in the Marin
flora have generously answered questions, identified specimens, and given other
taxonomic or bibliographic help, and to all these I am very grateful. It is from
all these sources that the presently acceptable name for each Marin County
plant, as interpreted in my judgment, has been obtained.

Besides the debt owed to many for botanical assistance, I also wish to express
special gratitude to the following: Rupert C. Barneby, who has read the manuscript and contributed many helpful suggestions; Hans Leschke, who has not only
read the manuscript but has also been a keen and constant collaborator in field
work; Thomas H. Kearney, who has helped to clarify many problems, botanical
and bibliographic; Lewis S. Rose, who has coöperated in field work and has as-
sisted in many ways in the herbarium; Evelyn M. Deasy, who has prepared the
manuscript. A special debt is owed to Charles T. Townsend for his fine photo-
graphic record of the Marin scene and to Malcolm G. Smith who has drawn the
maps. Finally, I wish to acknowledge my obligations to the California Academy
of Sciences, which for nearly a century has been the scientific institution nearest
to Marin County, and to Alice Eastwood, its Curator of Botany, who for over
fifty years has been the one most concerned with Marin County plants.

J. T. H.

California Academy of Sciences
November 22, 1948

Contents

Introduction

WHEN JOHN C. FRÉMONT proposed the name *Chrysopylae,* or Golden Gate, for
the narrow channel between San Francisco Bay and the Pacific Ocean, he could
hardly have known that within a few decades the works of man would have all
but obliterated that which was natural on the southern shore of the channel.
To the north in Marin County, however, beyond the small towns and cities, long
stretches of country are altered only as cattle have grazed and man has cut
timber, built roads, or cleared ground; in spite of its proximity to a world port
and metropolis, the landscape still displays a wildness and natural beauty that
have been but little impaired. For the hills along the north shore have for the
most part been kept as military reservations and the slopes of Mount Tamalpais are
mostly included in federal, state, and municipal preserves. The marshes and hills
along the bay have been little affected by dredging and filling; the coastal bluffs
and ridges of Point Reyes Peninsula are much as they were when Sir Francis
Drake landed in 1579; and the broad open hills at the northern end of the county
must still appear as they did to the Russians, when, more than a hundred years
ago, those colonists looked southward across Bodega Bay from the Sonoma hills
of Russian America to the Marin hills of Alta California. Indeed, throughout
the 529 square miles of the Marin County Peninsula, the scenes of beauty are an
encouragement and an invitation to see and learn more of nature.

Marin County has remained relatively wild and large cities have not been
built along its shores chiefly because of its hilly, even mountainous, topography.
Nearly everywhere hills and wave-cut bluffs rise from the water or tidal marshes,
and only occasionally is there level or gently sloping ground where hills are
more gradual or where narrow valleys and canyons enter the bay or ocean.
Along the Golden Gate the rise from the water is so abrupt that the highest
point in the Sausalito Hills, 960 feet, is attained in less than a mile; the slope
east of Stinson Beach rises 2,000 feet above the ocean in little more than a mile;
and the precipitous bluffs at Point Reyes have an elevation of 600 feet.

Sand beaches are mostly small and narrow and are restricted to the heads of
coves except in several notable places. At the head of both Bolinas Bay and Drakes
Bay remarkable sandspits have developed which almost cut off from the ocean
Bolinas Lagoon and Drakes Estero, and along the ocean north of Point Reyes and
again north of Tomales Bay are long broad beaches and the most extensive
dunes in the county. The Point Reyes beach and dunes have completely cut off
from the ocean Abbotts Lagoon, a freshwater sandbar lake; and at Dillons
Beach, Rodeo Lagoon, and other places, larger or smaller ponds of fresh or
brackish water have been formed behind beaches and dunes. Elsewhere, particularly along San Francisco Bay and at the head of Tomales Bay, broad marsh
lands penetrated by meandering sloughs and tidal channels are a distinctive feature. But nearly everywhere, beyond the marshes, back of the beaches, above the
seacliffs, is the hill country of Marin County.

[1]

The hills are frequently aligned or grouped to form rounded or elongate ridges, separated from one another by passes, canyons, or valleys. Three of these ridges project into the bay as Tiburon, San Quentin, and San Rafael peninsulas, while two circular areas to the south are Angel Island, separated from the mainland by Raccoon Strait, and the Sausalito Hills, cut off from the adjacent hill country by a low pass between the bay and the ocean known as Elk Valley. North of Elk Valley rise the hills that culminate in Mount Tamalpais, 2,610 feet in elevation, the highest and most prominent point in Marin County. Since the slopes of the mountain rise abruptly from the bay and from the ocean, it has a beauty and impressiveness usually found only in peaks much higher, and deep, steep-walled canyons separated by sharp rocky ridges add to the general wildness and interest of the scene.

This country of rugged relief, of which the summit of Mount Tamalpais is the highest part, extends to the north for some distance, and is bounded by Ross and Nicasio valleys and the lower part of Lagunitas Creek. In this area there are the well-defined elevations of Bolinas, Carson, and San Geronimo ridges, and the uplands are dissected by the deep canyons of Lagunitas and Corte Madera creeks and their tributaries. The ruggedness of this country in many parts is such that, if one is off roads and trails, travel may be arduous and a cross-country trip may still be difficult and adventurous. Beyond this rugged Tamalpais area and north to the Sonoma County line, the valley lands are more ample and the slopes of the hills are usually less abrupt. Nowhere does an elevation exceed 2,000 feet, though that height is approached at the top of Big Rock Ridge, elevation 1,905 feet, and Burdell Mountain, a little farther north, 1,560 feet in elevation. These ridges, like Mount Tamalpais, rise from near sea level, but they lack the bold contours of that mountain and are therefore not so impressive.

Geologic Structure and Rocks

On the west, the Tamalpais highland is bounded by a narrow valley which, together with Bolinas Lagoon on the south and Tomales Bay on the north, separates Point Reyes Peninsula from the rest of Marin County. The most prominent topographic feature of this part of the county is Inverness Ridge, the long straight ridge that parallels the narrow valley and the bays at either end. The ridge is not high, the highest points being Mount Wittenberg, 1,403 feet, and Mount Vision, 1,336 feet, but the narrow elongate highland is regular. with only Bear Valley breaking its continuity. On the east front it is quite steep, but on the west it slopes more gradually to the edge of precipitous ocean bluffs, to the marshy borders of Drakes Estero, and to the broad coastal downs above the Point Reyes dunes.

The straight narrow valley which separates the Tamalpais area from Point Reyes Peninsula is noteworthy not only from the physiographer's point of view, but also from the geologist's, since it lies along the San Andreas Fault and represents one of the most pronounced and remarkable fault-trace features in this part of California. This fault, which extends northward from the Gulf of California through the mountains of southern California and the South Coast Ranges, crosses west of the Golden Gate between the mainland and the Farallon Islands and enters Marin County at Bolinas Lagoon. After extending along the fault-trace valley and Tomales Bay in Marin County, it passes across the Bodega roadstead, separates Bodega Head from the mainland in southern Sonoma County, again passes out into the ocean, and reappears on the mainland for the last time in northern Sonoma and southern Mendocino counties, before finally disappearing in the depths of the Pacific just north of Point Arena. It was along the northern part of the San Andreas Fault that the movement took place which, in 1906, caused the earthquake that resulted in the disastrous San Francisco fire.

While the fault line in Marin County is most prominently marked by the rift valley and the steep escarpments of Inverness and Bolinas ridges, there are also other less conspicuous features associated with it, such as trenching, sag ponds, displaced hills and streams, and fault breccia. Although these physiographic features make the area unusually interesting, the chief significance of the fault line is that it separates regions with different geologic histories. East of the fault, the rocks are very old and belong to a series known as the Franciscan Group which is generally believed by geologists to belong to the Jurassic period of geologic history. These rocks are largely sedimentary and consist of shales, sandstones, and conglomerates, together with sometimes local and sometimes extensive occurrences of radiolarian chert. Intruded into these sediments are basic igneous rocks, which, like the outcrops of the chert, may be localized or more extensive. The most conspicuous occurrences of these igneous rocks are in those areas where one of them has been metamorphosed to form serpentine, a very distinctive rock type that is common on Tiburon Peninsula, Mount Tamalpais, Carson Ridge, and in other places in the Franciscan area.

The noteworthy geologic feature of the southern and middle part of Marin County is that there is no trace of younger rock formations anywhere, the only recent sedimentary accumulations being the dunes and beaches along the shore and the limited soil deposits on level ground of valley and coastal flats. This fact indicates that Mount Tamalpais and the surrounding area belong to an ancient upland or mountain mass that has been above sea level as long as any area in western coastal California and much longer than most areas; otherwise sediments of younger formations would be found overlying the ancient Franciscan rocks. This old formation is widespread in the Coast Ranges, but almost everywhere it is found together with younger sedimentary rocks that were deposited on the Franciscan series at a time when that particular area was depressed or below sea level. This is the condition that prevails in Marin County north of the Tamalpais physiographic area, for there the Franciscan rocks are overlain by sediments of Pliocene age and along the ocean there are localized beach deposits of Pleistocene age.

To the west of the San Andreas Fault another very old rock is found, but unlike the sediments and intrusive igneous rocks of the Franciscan formation, this rock is granite. The occurrence of this coarsely crystalline igneous type is rather extensive along Inverness Ridge near Inverness and at the northern end of the peninsula, and there is a localized outcrop near Point Reyes. The granite is not so easy to recognize when it is much weathered, but the granitic ocean cliffs at McClure Beach are distinctive and much of the beauty of the bluffs at Shell Beach comes from the granite of which they are composed. This type of granite, which is commonly known as Montara granite because of its occurrence on Montara Mountain south of San Francisco, has never been found east of the

San Andreas Fault, but to the west it has been found not only in Marin and San Mateo counties, but also on Bodega Head in Sonoma County, Cordell Bank west of Point Reyes, Farallon Islands in San Francisco County, and far to the south on the Monterey Peninsula and in the Santa Lucia Mountains in Monterey County. Wherever it occurs, the granite is generally overlain by much younger sediments. On Point Reyes Peninsula these rocks are of several kinds. Most common and widespread are the pale bedded rocks of shaly character that have been referred to the Monterey series of the Miocene epoch, a formation covering most of the peninsula west of Inverness Ridge. On the west side of Bolinas Lagoon and extending a little to the north along the San Andreas Fault are Pliocene marine sands that overlie the Monterey series in that area, and near Point Reyes is again a localized Pleistocene beach deposit. Then on top of these relatively young formations are the very recent beach and dune deposits, already described.

SOILS

In regions having a desert climate it is generally much easier to examine the geologic structure and study the rocks than in places like Marin County, where, owing to the moderate amount of rainfall, the rocks break down in a relatively short time to form a layer of soil. Most of the area is covered with clayey or sandy loam, depending on the character of the underlying rocks; but in many places the terrain is too steep for the soil to remain in place and landslides are frequent. Thus there are many rocky bluffs along the shoreline, and on the more abrupt slopes of hills and canyons, rock outcrops are frequent and sometimes extensive. The rugged country of the Tamalpais physiographic area is especially rocky, and on the steep slopes of Mount Tamalpais itself the soil layer is usually very thin or even entirely lacking. Outcrops of chert and serpentine—slowly weathering rocks—are generally easily recognized since they are frequently quite devoid of a soil cover; nevertheless, some soil accumulates in level places and mingles with broken fragments of the substratum. Adjacent to dune areas along the coast the soil becomes increasingly sandy as the beach is approached, and a dune-derived soil of considerable depth may develop. Deep soils also collect as an alluvium in the restricted valley bottoms or along the base of hills, and this deep wash soil may merge locally with siltlike deposits on tidal flats, especially along the bay and the maritime esteros.

CLIMATE AND WEATHER

The Marin climate, like that of most of coastal California, is characterized by warm dry summers and cool rainy winters, the same general type of climate that is found in the Mediterranean regions of southern Europe and northern Africa. The Pacific Ocean and the fog, however, have a more moderating effect than the Mediterranean, and the annual range of average temperature is not so great as it is farther inland in California.

The extremes of temperature recorded for specific stations in Marin County may vary by as much as 95° Fahrenheit, while the mean temperatures for January and July may differ by only a fourth to a third as much. Thus the maximum ranges recorded for Mount Tamalpais, Kentfield, and Hamilton Field are 19°–100°, 17°–112°, and 27°–102°, respectively, while the mean temperatures for January and July for the same stations are: Mount Tamalpais, 43.7° (Jan.), 69.0°

1. Redwoods (*Sequoia sempervirens*) in Muir Woods National Monument.

2. Grassy hills at the west end of Mount Tamalpais, looking southeast to Belvedere, Angel Island, and the Berkeley-Oakland hills with San Francisco on the right. *Pseudotsuga taxifolia, Quercus chrysolepis, Q. agrifolia,*

3. Grassland at the west end of Mount Tamalpais. Douglas firs (*Pseudotsuga taxifolia*) and goldcup oaks (*Quercus chrysolepis*) are in the small grove.

4. Oaks (*Quercus agrifolia, Q. lobata,* and *Q. Kelloggii*) in the Greenbrae Hills near Kentfield.

5. Coast live oak (*Quercus agrifolia*) in a grove on Tiburon Peninsula.

6. Buckeye (*Aesculus californica*) and oaks in the Greenbrae Hills near Kentfield. The East Peak of Mount Tamalpais in the background.

7. Buckeye (*Aesculus californica*) overlooking the Greenbrae salt marshes near Kentfield. Bushes of poison oak (*Rhus diversiloba*) in the foreground.

8. Chaparral on the south slopes of Mount Tamalpais, looking northeast toward West Point Inn and East Peak.

9. Lake Lagunitas and East Peak of Mount Tamalpais, with tanbark oak and madroño on the right, redwood across the lake, and chaparral on the ridge. Many herbs can be found on the strand as the water recedes.

10. Coastal grassland on Point Reyes Peninsula overlooking an arm of Drakes Estero. Coastal brush grows on the steeper slopes and wave-cut bluffs, and a small salt marsh occupies the head of the cove.

11. Wild rock garden near Point Reyes, showing *Polypodium californicum* var. *Kaulfussii*, *Echeveria caespitosa*, and *Gaultheria Shallon*.

12. Looking southeast from Shell Beach across Tomales Bay to Black Mountain. Bishop pines (*Pinus muricata*) grow on the slopes above the granitic bluffs.

13. Bishop pine (*Pinus muricata*) on Inverness Ridge west of Inverness. The shrubs in the foreground are coyote brush (*Baccharis pilularis* var. *consanguinea*).

14. Sargent cypress (*Cupressus Sargentii*) on the serpentine above Bootjack Camp, Mount Tamalpais. The shrubs in the foreground are Tamalpais manzanita (*Arctostaphylos montana*).

15. Dillons Beach, looking southwest across the broad dunes and Tomales Bay towards Point Reyes Peninsula. The rocks in the foreground are covered with a fruticulose lichen.

16. Dune grass (*Ammophila arenaria*) on the Bolinas Lagoon sand spit near Stinson Beach.

17. Dunes on Point Reyes Peninsula. The plants shown are: *Ammophila arenaria, Mesembryanthemum chilense, Lupinus arboreus, Baccharis pilularis, Erigeron glaucus, Aplopappus ericoides,* and *Franseria Chamissonis.*

18. Mount Tamalpais and Richardson Bay from the south end of Tiburon Peninsula. In the foreground is grassland

19. Grassland and wind-blown California laurels (*Umbellularia californica*) near the north end of Tiburon Peninsula.

20. Wind-blown California laurel (*Umbellularia californica*) on Tiburon Peninsula.

21. Wind-blown California laurel (*Umbellularia californica*) on the north end of Tiburon Peninsula with Mount Tamalpais in the distance.

22. Mount Tamalpais from the Greenbrae salt marshes. Corte Madera Creek is the tidal slough in the foreground and a slope of the Greenbrae Hills covered with oak woodland is on the right.

23. The beauty of the Marin countryside: Mount Tamalpais from the San Rafael Hills.

24. Lagunitas Creek below Lake Lagunitas. White alder (*Alnus rhombifolia*) is the common tree. Big-leaf maple (*Acer macrophyllum*) is the tree with the rough bark on the right and California laurel (*Umbellularia californica*) is the evergreen tree beyond it. The shrubs are California coffee-berry (*Rhamnus californica*).

(July); Kentfield, 45.8° (Jan.), 65.5° (July); and Hamilton Field, 47.6° (Jan.), 66.3° (July). For these three stations the average annual mean temperature is 55.0°, 56.4°, and 57.8°.

While the average seasonal range of temperature for these three places is about 20° between summer and winter, there are places along the ocean where the range is less than 5°. Thus at Point Reyes, where the minimum and maximum are 27° and 98°, the average temperatures are 49.8° (Jan.), 53.7° (July), and 52.5° (annual). This is a good example of an isothermal climate which is rarely found on land and which is in marked contrast with the continental type of climate. Along the eastern boundary of California, across the Sierra Nevada and beyond the influence of the ocean, the averages for January and July may vary by as much as 45° and differences between minimum and maximum temperatures may approach 140°.

While the average annual mean temperature varies but a few degrees for stations on the coast and in the interior of Marin County, the extremes in annual rainfall are much greater and differ markedly within a short distance. At Point Reyes the average is only 18 inches but at Point Reyes Station near the head of Tomales Bay it is 32 inches. At Kentfield and Hamilton Field the differences are even greater, 45 inches at the former and only 28 inches at the latter. The precipitation on the lee side of the San Rafael Hills and at Black Point is perhaps still less.

The summer fogs that are prevalent along the coast are a departure from the usual character of Mediterranean climate, and the influence of the cool moist air is felt far inland, even beyond the point where the mists are still condensed and visible. Fogs are most frequent and persistent along the immediate coast; and, although high country like the Sausalito Hills, Mount Tamalpais, and Inverness Ridge are effective barriers controlling their inland movement, there is no part of Marin County to which they do not penetrate when most extensive. It is to the fog that Point Reyes owes its equable, albeit chilly, summer temperature—a climate which impressed Sir Francis Drake and the crew of the *Golden Hinde* during their sojourn on the Point Reyes coast in June and July, 1579, as set forth in the notes of Chaplain Francis Fletcher in *The World Encompassed*. "During all which time, notwithstanding it was in the height of summer, and so neere the sunne, yet were wee continually visited with like nipping colds as we had [never] felt before" and by "those thicke mists and most stinking fogges." So impressed was the narrator by this unseemly summer weather that he devoted almost one-sixth of the account concerning the California visit to the cold and wind and to theorizing why the sun, even "in the pride of his heate," could not dissipate the "insufferable sharpnesse." All who have tried to explore the Point Reyes dunes and downs in the midst of a cold summer fog can fully sympathize with the early English visitors and even forgive them the slight exaggerations that color their record.

LIFE ZONES AND PLANT ASSOCIATIONS

There is so much difference in climate between the coastal and interior parts of Marin County that two life zones, as defined by C. Hart Merriam, can be recognized with the aid of plant indicators. On moister slopes and in canyons near the coast the Transition Zone is marked by such trees as *Pinus muricata*,

Pseudotsuga taxifolia, and *Sequoia sempervirens,* and also by many shrubs, among which are *Corylus californica, Ribes Menziesii, Rubus parviflorus, Ceanothus thrysiflorus,* and *Vaccinium ovatum.* In this zone a large variety of herbaceous plants grow on open coastal slopes, in wooded canyons, or in meadows among the hills. This humid coastal belt gives way toward the interior to the Upper Sonoran Zone, which is represented by extensive grassland, usually by scattered oaks, and by chaparral on rocky, exposed ridges. These areas are relatively drier and may actually have less rainfall than the coastal districts, or the areas may be drier because of the steep slopes, rocky soil, windy or sunny exposures, or varying combinations of these features. In this zone the grassland and chaparral are usually evident enough, and the oaks, *Quercus Douglasii* and *Q. lobata,* may serve as useful zone indicators.

Although the life zones are helpful in analyzing the flora of a region in a general way, particularly in an extensive area, the life-zone concept in Marin County is not so useful because conditions arising from the interaction of climate, topography, and substratum are extremely and locally varied. Within a short distance, conditions may change so completely that two entirely different groups of plants which are neighbors as far as space is concerned may be quite unrelated in their requirements of soil and moisture. Grassland may end abruptly on the edge of brush, forest may pass into chaparral, and dunes and salt marshes may adjoin one another. Varied physical conditions in a restricted area produce diverse expressions in the vegetation, and these can be best understood and appreciated by a consideration of the plant associations of the county.

As is true in the defining of life zones, it is not always easy to limit or define a plant association; but in a general way this can be done satisfactorily enough, if due regard is given to the fact that nearly every plant association blends or merges gradually with one or more other associations which it may adjoin. Sometimes the intermediate area is narrow and scarcely noticeable, but at other times it may be quite extensive and may itself have a characteristic appearance. Gradual, rather than abrupt, change from one set of physical and biological conditions to another is largely responsible for this blending: it is in this intermediate zone where the plant communities are making their floristic readjustments as a new set of conditions is approached.

The associations and the belt where they blend may be thought of as something more or less permanent and static, but usually this interrelationship of plant communities is a moving changeable thing that reflects in a vital way the response of the plants that make up the associations. Thus, in a landscape which constantly changes owing to geologic or other processes, new conditions are arising to which the plant communities respond; the development of the plant communities themselves creates conditions that modify the basic stability of the community as an organized entity; and natural or unnatural accidents (that may be catastrophic in relation to the plant association concerned) may occur at any time.

So, while the plant communities are usually definite enough, they are not always stable, and intergradation from one to another is to be expected. In Marin County the following twelve associations can generally be distinguished: redwood forest, tanbark oak–madroño woodland, oak-buckeye woodland, Douglas fir forest, bishop pine forest, chaparral, coastal brush, grassland (sometimes closely related to one or another of these listed associations), streambank and lakeshore groves or thickets, freshwater marsh, saltwater marsh, and dunes. The photographs used to illustrate this work have been selected with the idea of depicting these plant associations in various aspects of the natural scene in Marin County. Not only do Charles Townsend's photographs portray to the botanist critical views of the vegetation types but to the amateur they reveal scenes of beauty in the several plant associations.

THE REDWOOD FOREST

The redwood (*Sequoia sempervirens*) in its finest development forms one of the noblest forests in the world, and, though Marin County does not boast trees comparable to the largest in Humboldt and Del Norte counties, it has in Muir Woods a beautiful and representative primeval grove (see plate 1). This occupies the floor and lower slopes of a narrow canyon on the southern side of Mount Tamalpais where the tall columnar trunks rise above thick masses of huckleberry (*Vaccinium ovatum*), and the western azalea (*Rhododendron occidentale*) forms open graceful thickets along the stream. The pawnbroker bush (*Euonymus occidentalis*) is occasional on brushy slopes together with the California hazel (*Corylus californica*), wood rose (*Rosa gymnocarpa*), thimble berry (*Rubus parviflorus*), blue blossom (*Ceanothus thyrsiflorus*), and other shade- or moisture-loving shrubs. The attractive herbaceous ground cover is flowery in the spring but remains fresh and green throughout the year; it includes the redwood sorrel (*Oxalis oregana*), redwood violet (*Viola sempervirens*), and ginger (*Asarum caudatum*), while other herbs that are lovely in flower or in leaf are the vanilla grass (*Hierochloe occidentalis*), fairy lantern (*Disporum Smithii*), trillium (*Trillium ovatum*), fetid adders-tongue (*Scoliopus Bigelovii*), clintonia (*Clintonia Andrewsiana*), anemone (*Anemone quinquefolia* var. *Grayi*), *Adenocaulon bicolor*, and others. The California sword fern (*Polystichum munitum*) is abundant and especially attractive on steep slopes, and in wet ground along the streambank the lady fern (*Athyrium Filix-femina*) raises its ample fronds above the water.

Mingling with the redwoods in Muir Woods are both California laurels (*Umbellularia californica*) and tanbark oaks (*Lithocarpus densiflorus*), but it is in other parts of Marin County, where the redwoods were cut for lumber fifty or a hundred years ago, that the laurels and tanbark oaks form a more conspicuous part of the forest. In such places, as in Mill Valley, Corte Madera, and Lagunitas Canyon, the redwoods have grown rapidly as crown sprouts from the stumps of the old trees and the forest might be taken for the original growth except for the occasional stumps of forest giants that are still visible and tell of a glory that has gone. The people of Marin County and California should be perennially grateful to William Kent, who gave Muir Woods to the nation to be preserved and enjoyed for all time.

In Marin County the redwood is restricted to the long-emergent Tamalpais area where the rocks are all of Franciscan age and recent sediments are entirely lacking. No redwoods are known on Point Reyes Peninsula west of the San Andreas Fault and they do not occur in the northern part of the county, which is extensively overlain by Pliocene sediments. Certainly it is evident that the distribution of the redwood is intimately connected with the geologic history of the region.

The Tanbark Oak–Madroño Forest

The redwood forest is best developed in canyons where water is plentiful but, like most plant associations, it does not stop abruptly but follows the narrowing stream courses to slopes high on the side of Mount Tamalpais, where pygmy groves of stunted individuals merge with the shrubbery of the chaparral. Or more commonly the redwood forest ascends the canyonsides and blends gradually with the tanbark oak–madroño forest that is characteristic of the borderland of the redwood forest and that is almost always found along its drier margins (frontis., pl. 9). This woodland is made up largely of broad-leaved evergreen trees. Besides the tanbark oak (*Lithocarpus densiflorus*) and the madroño (*Arbutus Menziesii*), there are several other trees that occur generally and may even be abundant: coast live oak (*Quercus agrifolia*), canyon live oak (*Q. chrysolepis*), chinquapin (*Castanopsis chrysophylla*), and California laurel (*Umbellularia californica*). The Douglas fir (*Pseudotsuga taxifolia*) is frequently a characteristic member of this forest, as for example at the west end of the Tamalpais ridge, where its massive lichen-covered trunks and broad wind-blown crowns are a picturesque feature of the landscape. Around springs and along streams the wax-myrtle (*Myrica californica*) forms arborescent thickets and the California nutmeg (*Torreya californica*) occurs occasionally as a slender tree along streams or as a tall shrub on the edge of the chaparral. In Marin County it is in this redwood borderland that two of the woody climbers of the region are found, Dutchman's pipe (*Aristolochia californica*), which is uncommon, and California honeysuckle (*Lonicera hispidula* var. *vacillans*), which is abundant; and here, too, are found the Scouler willow (*Salix Scouleriana*) and big-leaf maple (*Acer macrophyllum*), usually in springy places or along watercourses. The following shrubs are found commonly in the tanbark oak–madroño forest but none of them is restricted to it:

Whipplea modesta	*Ceanothus thyrsiflorus*
Rosa gymnocarpa	*Arctostaphylos virgata*
Rhus diversiloba	*Rhododendron occidentale*
Rhamnus californica	*Gaultheria Shallon*

The following herbs are found on rock outcrops in shallow soil:

Polypodium californicum	*Sedum spathulifolium*
Selaginella Wallacei	*Romanzoffia californica*
Delphinium nudicaule	

On moist flats or in wet places around springs and along streams are:

Woodwardia fimbriata	*Circaea pacifica*
Carex Bolanderi	*Aralia californica*
Tellima grandiflora	*Synthyris reniformis* var. *cordata*
Lathyrus Torreyi	*Galium triflorum*
Viola glabella	*Petasites palmatus*

Even more numerous are the perennial herbs of the drier slopes of the open woodland:

Dryopteris arguta	*Nemophila parviflora*
Hystrix californica	*Cynoglossum grande*
Melica Harfordii	*Satureja Douglasii*
Carex globosa	*Pedicularis densiflora*
Dentaria integrifolia	*Castilleja latifolia* var. *rubra*
Polygala californica	*Madia madioides*
Sanicula laciniata	*Hieracium albiflorum*
Trientalis latifolia	

THE OAK–BUCKEYE FOREST

Farther inland there is a more extensive belt that is drier than the tanbark oak–madroño woodland but this area is neither so exposed as the ridges on Mount Tamalpais which are covered with chaparral nor so dry as the hills and valleys still farther north and east which are covered with grassland. This intermediate belt is occupied by a relatively dense oak woodland in which the California buckeye (*Aesculus californica*) is a common and conspicuous feature (plates 4, 6, 23). In contrast with the more humid tanbark oak–madroño forest, most of the distinctive trees of this oak-buckeye woodland are deciduous, and besides the buckeye, include the California black oak (*Quercus Kelloggii*), Garry oak (*Q. Garryana*), and the valley oak (*Q. lobata*). Two of the evergreen trees of the tanbark oak–madroño forest are also common—the California laurel and the coast live oak,—but the tanbark oak is generally absent and the madroño is only occasional. Although this oak woodland has rather an extensive development in Marin County, it forms a much more important plant association farther north in the Coast Ranges where the Garry oak is more common. Inasmuch as it lies along the edge of the tanbark oak–madroño forest, most of its shrubby and herbaceous species are found either in that woodland or in the coastal brush formation to which the oak woodland is also closely allied. A few of the more common shrubs and woody climbers found in the oak-buckeye woods are:

Ribes californicum	*Sambucus coerulea*
Holodiscus discolor	*Symphoricarpos rivularis*
Rhamnus californica	*Lonicera hispidula* var. *vacillans*
Rhus diversiloba	*Baccharis pilularis* var. *consanguinea*
Mimulus aurantiacus	*Artemisia californica*

THE DOUGLAS FIR FOREST

A wooded area much more localized than the oak woodland is the Douglas fir forest on Point Reyes Peninsula, a weak southern representation of another northern forest association. In the Black Forest on Inverness Ridge, however, the stand of Douglas fir resembles in density and uniformity the forest groves of this tree that occur far to the north in western Oregon and Washington; the size and aspect of the trees, too, are reminiscent of those northern woods. Typically this sort of woodland is made up not only of Douglas fir but also of coast hemlock (*Tsuga heterophylla*), Sitka spruce (*Picea sitchensis*), lowland fir (*Abies grandis*), and canoe cedar (*Thuja plicata*), a mixed forest that in California extends southward only to northern Sonoma County. Even without this variety of fine trees, however, the Douglas fir woodland on Inverness Ridge is more closely related to that forest than to any other and so undoubtedly it should be considered a southern outpost. It is interesting and perhaps significant that both in Marin County and in northern Sonoma County, these related coastal forests lie west of the San Andreas Fault. As in all the woods in Marin County, the California laurel occurs to a limited extent in the Black Forest, and such shrubs and woody climbers as poison oak (*Rhus diversiloba*), California coffeeberry (*Rhamnus californica*), blue blossom (*Ceanothus thyrsiflorus*), sticky monkey flower (*Mimulus aurantiacus*), and California honeysuckle (*Lonicera hispidula* var. *vacillans*) are occasional or common.

THE BISHOP PINE FOREST

On Inverness Ridge there is a second localized woodland, the bishop pine forest, but unlike that of the Douglas fir which is scarcely representative of the northern forests to which it is related, the bishop pine woodland is highly developed and in every way is typical of the closed-cone pine forests of coastal California (plates 12, 13). Even in its limited extent it resembles other groves of the bishop pine (*Pinus muricata*) or of the related Monterey pine (*P. radiata*), both of which generally occur in small isolated groves from northern California south into Lower California. On Inverness Ridge, the bishop pine grows chiefly on the granitic part at the north end of the ridge, whereas the Douglas fir forest has its best development on the rocks of the Monterey series that overlie the granite at the south end of the ridge. The pine is also found to a very limited extent on Franciscan rocks in the Carson country north of Mount Tamalpais, but there it is associated with the chaparral and does not form the forest association that has so high a development near Inverness. Just as the Marin redwoods are restricted to the Franciscan area of Tamalpais and vicinity, so it would seem that the bishop pine in Marin County was once restricted to the granite west of the San Andreas Fault and has spread eastward only in relatively recent time.

Growing with the pines on Inverness Ridge are also trees that occur in the other forest associations of Marin County, such as the coast live oak, wax-myrtle, laurel, buckeye, madroño, and others. Shrubs form a dense understory beneath the trees and for the most part are the species that are found in the tanbark oak–madroño woodland. There are, however, several additional species which are related to chaparral types and not to those of the woodland, such as chinquapin (*Castanopsis chrysophylla* var. *minor*), chamise (*Adenostoma fasciculatum*), ceanothus (*Ceanothus gloriosus* var. *porrectus*), and manzanita (*Arctostaphylos Cushingiana*). These and other species produce a low shrub formation which under the pines on the higher rocky ridges closely resembles certain aspects of the chaparral. Both floristically and ecologically, and perhaps also historically, the two shrub formations are related; certainly it would seem that some of the important elements of the chaparral have been derived in the course of its evolution from shrubs originally associated with the closed-cone pine forests.

THE CHAPARRAL

It is the chaparral that gives to Mount Tamalpais its distinctive texture, the same effect that is produced on many of the lower mountains in California by this dense and uniform covering of shrubs. From a distance, there is a velvety quality that characterizes it and gives depth to the blues and purples that pervade the slopes; from near at hand there is still that seeming smoothness and a lawnlike quality that belie the tough and rugged character of the plant cover. Up steep slopes, over rolling summits, and across broad flats spreads the unbroken array of shrubs, dense, erect, stiff—the pile in the fabric of the mountain's mantle (plate 8).

California's climate has undoubtedly been the chief factor in the evolution of this distinctive plant association. Otherwise why should there be the maquis of the Mediterranean slopes of southern Europe, the only other plant formation in the world with a physiognomy like that of the California chaparral? Surely it must have been the Mediterranean climate, common to the two regions, which has been the selecting agent that has brought together two unrelated groups of shrubs to grow in similar form under similar conditions so far apart. In many places in the world there are soils devoid of humus or rocky slopes devoid of soils, but in very few places is there a Mediterranean type of climate. While the two plant associations are so remarkably alike in overall effect, it is not strange that floristically they are entirely different: what the floral provinces of the Mediterranean had to offer in the development of the maquis is in no specific detail similar to what came into the chaparral from the plateaus of Mexico, from the islands of Tertiary California, or from the boreal regions of America. Is it little wonder that the brooms of southern Europe are finding California so agreeable a home? The pity is that the Californians are putting up so poor a fight against the inroads of these aggressive Old-World weeds.

Although shrubs of many families have come together to form the chaparral, it is the oak, rose, pea, buckthorn, and heather families that are most abundantly represented both by species and by individuals. One of the remarkable features of the formation is that plants with such diverse hereditary histories should come to live together so intimately under such specialized conditions and, from a Lamarckian point of view, with such unity of purpose. Although the general effect is uniform throughout the formation, close observation reveals a distributional pattern of species even within the formation. Thus certain slopes (where the soil may be more acid) are rich in various members of the heather family (species of *Arctostophylos, Schizococcus, Gaultheria,* and *Vaccinium*); another sterile slope will be dominated by chamise (*Adenostoma fasciculatum*); and steep shale and sandstone ridges, where little else grows, provide the chosen home of *Arctostaphylos canescens*. The serpentine outcrops are the most distinctive of all and support many species of shrubs and herbs not found on the adjacent sedimentary rocks (plate 14). Here is found the Sargent cypress (*Cupressus Sargentii*) which usually attains the size of a low tree, although sometimes it matures as a shrub and bears cones when only a few feet tall. Like the redwood, the Sargent cypress in Marin County is restricted to the Franciscan rocks of the Tamalpais area.

The chaparral has its best development in Marin County on the higher slopes of Mount Tamalpais and northward through the Carson country, but there are localized occurrences on Angel Island, San Rafael Hills, Big Rock Ridge, and elsewhere. The chaparral of the serpentine is restricted to Mount Tamalpais and the Carson country; the serpentine of Tiburon Peninsula with its distinctive flora will be described in the section devoted to grassland. The following lists, although partial, give an idea of the richness and diversity of the chaparral flora of Marin County.

SHRUBS

Castanopsis chrysophylla var. *minor*
Quercus Wislizeni var. *frutescens*
Dendromecon rigida
Adenostoma fasciculatum
Cercocarpus betuloides
Photinia arbutifolia
Rosa spithamea var. *sonomensis*
Pickeringia montana

Lupinus Douglasii var. *fallax*
Lotus scoparius
Ceanothus foliosus
Ceanothus sorediatus
Ceanothus ramulosus
Ceanothus gloriosus var. *exaltatus*
Ceanothus Masonii
Helianthemum scoparium var. *vulgare*
Garrya elliptica

Garrya Fremontii
Arctostaphylos canescens
Arctostaphylos Cushingiana
Arctostaphylos glandulosa
Schizococcus sensitivus
Eriodictyon californicum
Lepechinia calycina
Castilleja foliolosa
Aplopappus arborescens

HERBS (PERENNIAL AND ANNUAL)

Xerophyllum tenax
Zigadenus Fremontii
Calandrinia Breweri
Silene multinervia
Papaver californicum
Lotus junceus
Sidalcea Hickmanii
Hypericum concinnum
Navarretia mellita
Phacelia suaveolens
Cryptantha muricata
Cryptantha Torreyana var. pumila
Orobanche fasciculata var. franciscana
Orobanche bulbosa
Boschniakia strobilacea
Campanula angustiflora
Madia exigua

WOODY PLANTS ON THE SERPENTINE

Cupressus Sargentii
Quercus durata

Ceanothus Jepsonii
Arctostaphylos montana

HERBACEOUS PLANTS ON THE SERPENTINE

Cheilanthes siliquosa
Scribneria Bolanderi
Calamagrostis purpurascens var. ophitidis
Zigadenus fontanus
Allium falcifolium
Brodiaea peduncularis
Carex serratodens
Carex debiliformis
Eriogonum vimineum var. caninum
Arenaria Douglasii
Silene californica
Parnassia californica
Aquilegia eximia
Streptanthus batrachopus
Streptanthus glandulosus var. pulchellus
Linum micranthum
Viola ocellata
Sanicula tuberosa
Navarretia rosulata
Monardella neglecta
Stachys pycnantha
Castilleja stenantha
Helenium Bigelovii
Calycadenia multiglandulosa var. cephalotes
Cirsium Vaseyi

THE COASTAL BRUSH

Besides the chaparral, there is a second shrub formation in Marin County which has been listed above as coastal brush, and, though the two are alike in the shrubby habit of the plants that characterize them, their plants are mostly unrelated and the aspects of the two are quite different. While the chaparral is restricted to exposed and relatively dry slopes of the Upper Sonoran Zone, the more widespread coastal brush is found on slopes that are much moister, and, according to Merriam's scheme, would belong to the Transition Zone. As the coastal brush is defined here, it is perhaps more extended than might be expected, for it includes the low wind-pruned plants of maritime bluffs and mesas as well as the dense rank thickets or "soft chaparral" of interior hills and canyons (plates 10, 15). Although the two habitats are quite unrelated and the plants in each frequently assume markedly different growth forms, most of the shrubby species in the two are the same, and only a few of the perennial herbs are restricted to one habitat or the other. Because this floristic interrelationship is so apparent and also because the extremes of the two pass from one to the other so frequently and gradually, they are here regarded as part of the same association.

In the hills and canyons, the plants on the moister, shadier, northern slopes differ somewhat from those on the drier, sunny, southern slopes. Poison oak (Rhus diversiloba) is an all-too-abundant plant in both places, but in the former, rampant plants form tall dense thickets, and in the latter, the plants are lower and more bushy. California blackberry (*Rubus ursinus*) is a woody climbing or trailing plant found in both moister and drier habitats, but it is much more vigorous in the former. On the moister slopes, currant (*Ribes sanguineum* var. *glutinosum*), ninebark (*Physocarpus capitatus*), and purple nightshade (*Solanum Xanti*) are more common, while on the drier slopes, sticky monkey flower (*Mimulus aurantiacus*) and California sagebrush (*Artemisia californica*) are characteristic. The following lists include the shrubs and herbs found more or less generally throughout the coastal brush formation:

SHRUBS

Salix lasiolepis
Corylus californica
Ribes sanguineum var. glutinosum
Ribes Menziesii var. leptosmum
Rubus parviflorus var. velutinus
Rubus spectabilis var. franciscanus
Osmaronia cerasiformis
Holodiscus discolor
Physocarpus capitatus
Amelanchier pallida
Photinia arbutifolia
Lupinus rivularis
Rhus diversiloba
Rhamnus californica
Ceanothus thyrsiflorus
Dirca occidentalis
Solanum Xanti
Sambucus coerulea
Symphoricarpos rivularis
Lonicera hispidula var. vacillans
Mimulus aurantiacus
Baccharis pilularis var. consanguinea
Eriophyllum confertiflorum
Artemisia californica

HERBS (PERENNIAL AND ANNUAL)

Pityrogramma triangularis
Adiantum Jordani
Agrostis diegoensis
Calamagrostis rubescens
Luzula multiflora
Trillium chloropetalum
Smilacina amplexicaulis
Urtica californica
Actaea arguta
Thalictrum polycarpum
Cardamine oligosperma
Heuchera micrantha
Lithophragma affine
Potentilla glandulosa

Lupinus latifolius
Psoralea physodes
Vicia gigantea
Lathyrus Bolanderi
Oxalis pilosa
Godetia amoena
Sanicula crassicaulis
Ligusticum apiifolium
Lomatium californicum
Heracleum maximum
Navarretia squarrosa
Nemophila heterophylla
Phacelia californica
Phacelia nemoralis
Phacelia malvaefolia
Cryptantha micromeres
Stachys rigida var. quercetorum
Scrophularia californica
Castilleja affinis
Castilleja latifolia var. rubra
Castilleja latifolia var. Wightii
Galium Nuttallii
Galium californicum
Marah oregonus
Marah fabaceus
Grindelia hirsutula
Solidago lepida var. elongata
Aster chilensis
Aster radulinus
Gnaphalium ramosissimum
Gnaphalium californicum
Gnaphalium chilense
Anaphalis margaritacea
Wyethia glabra
Madia gracilis
Layia gaillardioides
Eriophyllum lanatum var. arachnoideum
Achillea borealis var. californica
Artemisia Douglasiana
Agoseris plebeia
Rafinesquia californica

On the ocean bluffs, the growth is usually not so dense, the shrubs are lower and more compact, and the herbaceous or suffrutescent perennials are much more conspicuous (plates 10, 15). The dense and rounded mounds of *Eriophyllum staechadifolium* var. *artemisiaefolium*, one of the few shrubs in the brush

formation restricted to the coastal zone, are conspicuous and attractive even when
they are not covered with golden flower heads. Most of the shrubs are depressed
and windblown, and some have assumed a low or creeping habit. Several of
these have been distinguished from their inland relatives by name, such as:
Ceanothus thyrsiflorus var. *repens, Monardella villosa* var. *franciscana,* and typical *Baccharis pilularis.* Among the herbs the following are nearly or quite restricted to the coastal slopes:

Allium dichlamydeum   *Arabis blepharophylla*
Habenaria Greenei   *Angelica Hendersonii*
Cerastium arvense   *Plantago juncoides* var. *californica*
Silene pacifica   *Erigeron glaucus*
Delphinium californicum

The coastal brush assòciation, although a very real thing in any view that may
be taken of the plant formations of Marin County, is perhaps the least definite
in its boundaries. In the midst of the brush, islands of California laurel and
coast live oak may occur, and these trees are almost always present along the
stream courses in the bottom of brushy canyons. From this close relation between
the trees and shrubs, it is but a step to the arboreous plant associations in which
members of the coastal brush frequently form a shrubby story under the trees.
The separation between the Transition Zone brush and the Upper Sonoran
chaparral is usually more definite, but even here blending is frequent, as in those
places where poison oak, blue blossom, and coyote brush (*Baccharis pilularis*
var. *consanguinea*) grow with the chaparral shrubs. Also the boundary between
the brush and grassland is often not very definite. This may be due either to a
gradation in the controlling factors of soil and water or to the evident spread of
the brush into the grassland. The coyote brush is common as a colonizer on
grassy slopes and it may either give rise to a pure society or, more frequently,
it may be the forerunner of a mixed society containing blackberry, poison oak,
and other shrubs. Another commonly observed type of colonization occurs where
the spread of the shrubs is anticipated by dense growths of bracken, rush,
blackberry, potentilla, yerba buena, and other herbaceous or suffrutescent perennials in the midst of grassland or along the margin of established shrubbery.
This change from grassland to shrub formation may be a natural development
or succession, but the advance of the shrubbery is hastened by overgrazing in
the grassy areas.

THE GRASSLAND

The grassland itself is usually definite enough as a plant association, but there
are many types of grassland, and in an area even as limited as Marin County
three of four kinds can be recognized; their differentiation is attributable to the
differing locations, amounts of moisture, character of the soil, and other factors
affecting the development. When, as frequently happens in this region, the
grassland is more closely allied to an adjacent area of brush or forest than to
some other grassy area, the grassland types should properly be described in connection with that related formation. Here, however, the different kinds of Marin
grassland are considered together but the several types are recognized under the
following names: coastal grassland, hill and valley grassland, mountain meadow,
serpentine grassland, and vernal pools.

COASTAL GRASSLAND

The coastal grassland is closely allied to the dunes on the one hand and to hill grassland and coastal brush on the other (plate 10). It is different from all of these, however, and has a distinctive development on Point Reyes Peninsula behind the dunes. There are two phases: the drier mesa-like flats that are covered with low annuals and perennials, and the wetter swales that support rank growths of grasses, sedges, and rushes in the form of large tussocks. The following list includes a number of the plants found in the coastal grassland:

Deschampsia holciformis
Calamagrostis nutkaensis
Festuca californica
Juncus effusus
Carex obnupta
Camassia Quamash var. linearis
Brodiaea terrestris
Calochortus Tolmiei
Maianthemum dilatatum
Sisyrinchium bellum
Iris Douglasiana
Habenaria leucostachys
Spiranthes Romanzoffiana
Polygonum bistortoides
Montia sibirica
Stellaria littoralis

Ranunculus orthorhynchus
Delphinium decorum
Sidalcea rhizomata
Viola adunca
Godetia amoena
Lilaeopsis occidentalis
Veronica scutellata
Castilleja Leschkeana
Orthocarpus erianthus var. roseus
Orthocarpus floribundus
Plantago hirtella var. Galeottiana
Campanula californica
Corethrogyne californica
Baeria hirsutula
Cirsium Andrewsii

HILL AND VALLEY GRASSLAND

Along its inner border the narrow coastal belt passes into the hill and valley grassland that includes most of the grassy areas of Marin County (plates 2, 3, 18, 19). A more detailed consideration would perhaps recognize several grassland types instead of the one given here, but many of the species are found throughout the grassy areas from coastal hills to interior slopes and valleys; so they are treated together. However, there are differences between the open coastal slopes having a more turflike plant development and the hills and valleys along the eastern margin, where annual species are more common and where there is frequently a scattered occurrence of valley oaks (Quercus lobata). Then the grassland of the bald hills and glades high up on Mount Tamalpais is again different, and still different are small grassy flats in the chaparral. It seems likely that the coastal grassland is related to the coastal brush, the grassland of the glades and bald hills belongs to the tanbark oak–madroño woodland, and that of the innermost hills and valleys is the nearest approach in Marin County to the grassland of the interior valleys of California. In the following list, however, the characteristic plants of these grassy areas are all grouped together:

Pteridium aquilinum var. pubescens
Stipa lepida
Stipa pulchra
Poa scabrella
Festuca idahoensis
Festuca dertonensis
Avena barbata
Avena fatua
Briza minor
Sitanion jubatum

Hordeum brachyantherum
Hordeum leporinum
Hordeum Hystrix
Lolium multiflorum
Danthonia californica
Bromus mollis
Bromus rigidus
Elymus triticoides
Carex tumulicola
Juncus patens

Juncus occidentalis
Calochortus luteus
Brodiaea pulchella
Brodiaea elegans
Sisyrinchium bellum
Iris longipetala
Rumex Acetosella
Delphinium hesperium
Ranunculus californicus
Platystemon californicus
Eschscholzia californica
Lepidium nitidum
Tillaea erecta
Acaena californica
Fragaria californica
Lupinus nanus
Lupinus bicolor
Lupinus densiflorus
Lupinus albicaulis
Sidalcea malvaeflora
Lotus micranthus
Lotus subpinnatus
Lotus humistratus
Viola pedunculata
Godetia amoena
Oenothera ovata
Sanicula arctopoides
Sanicula bipinnatifida
Daucus pusillus
Lomatium utriculatum
Convolvulus subacaulis
Phlox gracilis
Linanthus acicularis
Linanthus parviflorus

Linanthus androsaceus
Nemophila Menziesii
Cryptantha flaccida
Plagiobothrys nothofulvus
Amsinckia intermedia
Trichostema lanceolatum
Stachys ajugoides
Orthocarpus attenuatus
Orthocarpus densiflorus
Orthocarpus faucibarbatus
Plantago erecta
Marah fabaceus
Grindelia hirsutula var. brevisquama
Chrysopsis Bolanderi
Solidago californica
Pentachaeta bellidiflora
Pentachaeta alsinoides
Lessingia hololeuca
Micropus californicus
Filago californica
Stylocline amphibola
Hemizonia luzulaefolia
Hemizonia congesta
Madia sativa
Madia capitata
Layia platyglossa
Layia chrysanthemoides
Lagophylla ramosissima
Achyrachaena mollis
Baeria chrysostoma
Soliva daucifolia
Cirsium quercetorum
Microseris Douglasii
Agoseris hirsuta

MOUNTAIN MEADOW

The wet meadows high on the slopes of Mount Tamalpais are not extensive but are interesting because they contain a few plants showing a relationship to the flora of the high North Coast Ranges and the Sierra Nevada. Among the noteworthy plants found in these Tamalpais meadows are:

Carex Cusickii
Eleocharis Engelmannii
Eleocharis rostellata
Juncus Kelloggii
Zigadenus fontanus
Habenaria leucostachys
Spiranthes porrifolia

Rumex salicifolius
Potentilla Micheneri
Hypericum Scouleri
Epilobium Halleanum
Prunella vulgaris var. lanceolata
Heterocodon rariflorum
Cirsium Vaseyi

SERPENTINE GRASSLAND

Whereas the serpentine on Mount Tamalpais is mostly covered by woody plants, the serpentine of Tiburon Peninsula is covered by grasses and other herbaceous plants (plate 18). Here are found two of the most restricted endemics of California, Streptanthus niger and Castilleja neglecta, while two other plants, Allium lacunosum and Linum congestum, are not known elsewhere in Marin County. These plants, with perhaps the exception of the Castilleja, indicate a floristic relation between the Tiburon grassland and similar serpentine grassland in the South Coast Ranges. Other plants characteristic of this area are:

Calamagrostis purpurascens var. ophitidis
Zigadenus fontanus
Brodiaea peduncularis
Carex serratodens
Eriogonum vimineum var. caninum
Montia spathulata

Arenaria Douglasii
Monardella neglecta
Stachys pycnantha
Orthocarpus lithospermoides
Helenium Bigelovii
Calycadenia multiglandulosa var. cephalotes

VERNAL POOLS

A localized but distinctive group of spring and summer plants grows in low places in the valleys where water stands during the rainy season. Such pools are infrequent in hilly Marin County but where they occur an interesting plant society can be found that is related to the vernal pool flora so richly developed in the Sacramento and San Joaquin valleys. The following plants have been found growing on the desiccated beds of Marin pools:

Alopecurus Howellii
Pleuropogon californicus
Ranunculus pusillus
Elatine brachysperma
Boisduvalia glabella
Eryngium oblanceolatum
Eryngium aristulatum
Navarretia intertexta
Allocarya californica

Allocarya bracteata
Gratiola ebracteata
Veronica peregrina var. xalapensis
Downingia concolor
Gnaphalium palustre
Lasthenia glabrata
Lasthenia glaberrima
Blennosperma nanum

STREAMBANK PLANTS

A typical streambank association is rarely developed in Marin County because there are only a few creeks that carry water through the summer. Along the intermittent streams, coast live oaks and laurels are found both in the coastal brush and in the grassland, while along the larger perennial or nearly perennial streams are willows, alders, maples, ashes, and shrubby dogwoods. Near the coast and in the redwood groves the red alder (Alnus rubra) is found, but along streams farther inland the white alder (A. rhombifolia) takes its place (plate 24); otherwise there is no marked difference in this riparian group of plants from the coast to the interior. A list of the streambank trees, shrubs, and herbs follows:

Salix lasiandra
Salix laevigata
Salix lasiolepis
Salix Hindsiana
Salix Coulteri
Alnus rhombifolia
Alnus rubra
Acer macrophyllum
Acer Negundo var. californicum
Cornus californica
Fraxinus oregona
Woodwardia fimbriata
Glyceria leptostachya
Carex nudata

Scirpus microcarpus
Eleocharis macrostachya
Juncus effusus
Lilium pardalinum
Epipactis gigantea
Boykinia elata
Psoralea macrostachya
Aralia californica
Mimulus cardinalis
Mimulus guttatus
Holozonia filipes
Helenium puberulum
Petasites palmatus

LAKESHORE PLANTS

Around the ponds and coastal lagoons the woody plants are mostly willows, alders, and dogwoods; and frequently they form a very rank and dense growth. The artificial reservoir lakes on the north side of Mount Tamalpais are too recent to have developed a woody strand flora, but along the margin and on the desiccated lake bottom an interesting group of annual herbs can be found in summer and autumn as the water is withdrawn (plate 9). Several of these plants are not known from other stations in Marin County, and some are rare any-where in California. Among the plants found in this man-made lacustrine habitat are:

Amaranthus californicus	*Ammannia coccinea*
Cypselea humifusa	*Lythrum Hyssopifolia*
Mollugo verticillata	*Lindernia anagallidea*
Spergularia rubra	*Limosella aquatica*
Tillaea aquatica	*Aster exilis*
Rorippa curvisiliqua	*Xanthium calvum*
Euphorbia serpyllifolia	*Gnaphalium palustre*
Elatine brachysperma	

FRESHWATER MARSHES

The number of freshwater ponds, lagoons, and reservoirs in Marin County is not very great, but they are extensive enough to support a greater variety of aquatic plants than they actually do. The flora of California is not noted for an abundant occurrence of aquatics, although it includes a general and diversified representation of them; so it is perhaps not surprising that their number is rather limited in Marin County. The Olema Marsh near the south end of Tomales Bay is perhaps the best freshwater marsh area in the county, though the laguna in Chileno Valley develops an interesting flora during those periods when it is not drained. More species would probably become established in Phoenix, Lagunitas, and Alpine lakes, the reservoirs on the north side of Mount Tamalpais, if there were not such extremes between high and low water. Perhaps the coastal sandbar lagoons become a little brackish in the autumn, which, if even only occasionally true, would restrict the number of species. Even so, in Marin County, there is an interesting variety of submerged or floating aquatic plants and a somewhat larger number of marsh herbs that grow along muddy strands or on shallow bottoms. A few of the more characteristic are listed here, divided into two groups by their manner of growth:

AQUATIC HERBS

Azolla filiculoides
Isoetes Howellii
Zannichellia palustris
Potamogeton nodosus
Potamogeton pectinatus
Najas guadalupensis
Lemna minor
Lemna minima
Spirodela polyrhiza
Ceratophyllum demersum
Ranunculus Lobbii
Ranunculus aquatilis var. *capillaceus*
Callitriche hermaphroditica

Callitriche palustris
Callitriche Bolanderi
Hippuris vulgaris
Myriophyllum exalbescens
Myriophyllum hippurioides
Utricularia vulgaris

TERRESTRIAL HERBS

Typha angustifolia
Typha domingensis
Typha glauca
Typha latifolia
Sparganium Greenei
Triglochin striata

Lilaea subulata
Alisma Plantago-aquatica var. *triviale*
Damasonium californicum
Sagittaria latifolia
Glyceria pauciflora
Beckmannia Syzigachne
Carex exsiccata
Scirpus californicus
Scirpus validus
Eleocharis macrostachya
Polygonum coccineum

Polygonum punctatum
Polygonum hydropiperoides
Nuphar polysepalum
Potentilla Egedii var. *grandis*
Ludwigia palustris
Cicuta Douglasii
Cicuta Bolanderi
Oenanthe sarmentosa
Hydrocotyle ranunculoides
Bidens laevis

SALT MARSHES

In contrast to the rather restricted representation of the streambank and fresh-water marsh associations in Marin County, there is a notable development of the salt marsh association in rather extensive areas on both the bay and ocean sides of the peninsula (plates 10, 22). The marshes bordering or adjacent to Point Reyes Peninsula on the ocean side have nearly the same flora as that found in the marshes bordering San Francisco Bay south of San Rafael Peninsula, but northward along San Pablo Bay there is a slight but discernible floristic difference that probably reflects the decreased salinity of the tidal waters. Only two species, *Puccinellia grandis* and *Grindelia stricta,* are found in the ocean marshes and not in the bay marshes; on the other hand *Salicornia Bigelovii* and *Grindelia humilis* are found in the bay marshes but not in the ocean marshes. In the bayside marshes, *Cordylanthus maritimus* is found to the south of San Rafael Peninsula while to the north there is not only the rare *Cordylanthus mollis* but also *Scirpus acutus* and *Glaux maritima.* In these more northern marshes, *Rosa californica* grows on the edge of the marsh area, and a form of *Achillea borealis* var. *californica* occurs commonly in the marshes.

Within the salt marshes there is a noticeable difference in the distribution of species from the outer lower margin that is covered daily by the tides to the innermost part that is reached only by the highest tides. Cordgrass, *Spartina foliosa,* is the conspicuous colonizer on the lowest tidal flats where it sometimes develops an extensive pure society before soil is built up. It then gives way to the usual salt marsh flora that includes the following: *Salicornia virginica, Frankenia grandifolia, Limonium commune* var. *californicum, Plantago juncoides,* and *Jaumea carnosa.* The *Salicornia* is the most widespread plant in the marshes. Associated with it along its lower border is *Triglochin concinna,* while the related *T. maritima* is more nearly restricted to the upper, less saline borders. *Rumex occidentalis* var. *procerus* and *Grindelia humilis* are also more often found in the upper reaches of the marshes where they frequently grow on the edge of tidal channels; in the channels themselves, particularly where the water is not too saline, *Ruppia maritima* forms tangled masses.

Just beyond the marshes are two other belts or zones which, though not a part of the salt marshes, can perhaps be mentioned here. The lower one is found only on the Pacific side of Marin County and corresponds to the littoral zone of the bays and rocky coast. On the silty shallows of the bays is found eel grass, *Zostera marina,* while on more exposed rocky coasts two species of surf grass, *Phyllospadix Torreyi* and *P. Scouleri,* are occasional or common. The broad ribbon like leaves of the eel grass and the narrow cordlike leaves of the surf grass are familiar and abundant in the littoral debris that strews the beaches following a severe storm.

Beyond the upper edge of the salt marshes is a saline or subsaline belt that may be steep and narrow, or nearly flat and much more extensive. This area is best developed along the bay marshes and is probably more closely related to the grassland than to the marshes. Salt grass, *Distichlis spicata*, and various species of *Spergularia* are especially characteristic of this intermediate belt, though various other salt-tolerant native and introduced species also occur here. A rare pepper grass, *Lepidium oxycarpum*, is one of the plants that has been found on these flats, and *Monerma cylindrica* and *Parapholis incurva*, two queer exotic grasses from the Old World, are rather common.

All in all, the oozy, smelly, insectiferous salt marshes are anything but monotonous and hold for the field biologist a world of life and interest. Although excursions in these areas are caviare to the general public (and unfortunately, even to the general biologist), repeated visits to them reward the student, professional and amateur, with ecologic and systematic data that can be had nowhere else. How tragic is the present-day tendency to "reclaim" so perfect and balanced a creation by destroying it!

DUNES AND BEACHES

In common with the silted sediments in the marshes and the alluvial soil of the valley bottoms, the dunes and beaches are among the most recent geologic deposits; and likewise, in common with the plant associations covering the Marin marshes and valleys, the dune and beach flora of the county is well developed (plates 15, 16, 17). Although the present dunes may be very, very young geologically and may seem very transitory, they represent a geologic feature as old as any shoreline, and any thoughtful survey of the various plants adapted for so specialized a habitat indicates the age-long process by which plants have been prepared to fill this niche in the natural scene. There are no evolutionary or genetic misfits on the ocean dunes: natural selection has seen to that.

Most of the dune plants are low-growing annual or perennial herbs, although some of the latter may be more or less woody near the ground. There are, however, several definitely woody plants, but these too are frequently low-growing or even prostrate, so in habit they blend with the herbaceous and suffrutescent perennials. Especially noteworthy among the shrubs are three species of *Lupinus*, *L. albifrons*, *L. arboreus*, and *L. rivularis*, that are especially abundant and widespread on the Point Reyes dunes where they produce such a fine wildflower spectacle when in bloom. Then there are two shrubby members of the Sunflower Family, *Aplopappus ericoides* and *Baccharis pilularis*, the latter differing from the common var. *consanguinea* of the coastal brush association in its prostrate habit and smaller leaves. Farther inland, on flats that perhaps belong to the coastal grassland, are two prostrate shrubs that are among the most attractive in Marin County: glory mat, *Ceanothus gloriosus*, and kinnikinnick, *Arctostaphylos Uva-ursi*.

From the beach sand where the highest surf all but washes and from the mobile dune where a thousand rootlets literally strive to hold their ground, the herbaceous plants are distributed inland to the more stable dune hills, sandy flats, sand-choked swales, and ponds. Just as in the several zones in the salt marshes, different kinds of plants are found in these different places: *Cakile edentula* var. *californica* and *C. maritima* grow on the outermost beaches; *Abronia latifolia, Cymopterus littoralis, Franseria Chamissonis,* and many other plants grow on higher, more stable beaches; *Poa Douglasii, Elymus mollis,* and *Ammophila arenaria* grow on the least stable dunes; *Eschscholzia californica, Cirsium occidentale,* and many more grow with the shrubs on fixed dunes and stable flats; still farther inland, where the dunes blend with the coastal grassland, are *Poa confinis, Agrostis Blasdalei, Chorizanthe villosa,* and *Fragaria chiloensis;* and in wet places along the beaches and behind the dunes have been found such plants as *Triglochin striata, Sagina crassicaulis,* and *Lilaeopsis occidentalis.* Certainly there are many kinds of places on the dunes where plants live, and a detailed exploration may be as pleasant and as profitable to the student as his visits to the salt marshes.

The following are other Marin County dune plants not already mentioned:

Agropyron arenicola
Elymus vancouverensis
Juncus Leseurii
Polygonum Paronychia
Rumex salicifolius var. *crassus*
Eriogonum latifolium
Chorizanthe cuspidata
Atriplex leucophylla
Atriplex californica
Abronia umbellata
Mesembryanthemum chilense
Cardionema ramosissimum
Erysimum concinnum
Lupinus Layneae
Lupinus variicolor
Lotus eriophorus
Astragalus Nuttallii

Lathyrus littoralis
Oenothera contorta var. *strigulosa*
Oenothera micrantha
Oenothera cheiranthifolia
Convolvulus Soldanella
Gilia Chamissonis
Gilia millefoliata
Phacelia distans
Cryptantha leiocarpa
Amsinckia spectabilis
Grindelia arenicola
Layia carnosa
Baeria uliginosa
Achillea borealis var. *arenicola*
Tanacetum camphoratum
Artemisia pycnocephala
Agoseris apargioides

THE EFFECT OF FIRE

In this survey of the plant associations in Marin County, the influence of earth water, and air on the distribution of the different species has been stressed from various points of view, and now only a brief account of the fourth of Empedocles' elements, fire, remains to complete a consideration according to his classic enumeration. The summer-dried hills and mountains of California are unusually susceptible to devastating visitations of this element and in the form of brush and forest fires it has periodically swept over the valleys and lower slopes for ages. The element of fire has, therefore, became a factor, at least of secondary importance, in the evolutionary development of the plant associations where it has been most prevalent, and in such associations, only the plants that are able to recover or reproduce between recurring fires have survived and become members of those particular associations. This is especially true of the chaparral, for the shrubs that are found in it reappear promptly after a burn either as vigorous crown sprouts or root sprouts or as fast-growing seedlings. Many of the crown sprouts and some of the seedlings develop fast enough to flower and fruit during the first or second season of growth following a fire. And not only is the brushy cover quickly replaced but it comes back with the renewed vigor of youth. Chaparral fires, if they are not too frequent and too severe, actually rejuvenate the formation, and the plant community, that might otherwise become decadent and pass into some other type of formation, such as grassland or forest, depending on the character of the soil and climate, renews its dominant hold on the area concerned.

What is true of the recovery of the chaparral is, in somewhat less degree, also true of the coastal brush association and some of the trees of the woodland associations. In these plant communities there are both trees and shrubs that are also able to regenerate through crown sprouts or root sprouts, and here, too, the growth is rapid and vigorous. The redwood is very tenacious of life, and even a badly burned and charred trunk will begin to replace its crown with green tufted branchlets within a few weeks after a fire. Other trees, like the oaks and madroños, if severely burned, are killed to the ground and must be regenerated by crown sprouts.

After the extensive Marin County burn in 1945, the following trees and shrubs were observed to sprout from crowns or underground stems:

Sequoia sempervirens
Torreya californica
Corylus californica
Castanopsis chrysophylla
Quercus agrifolia
Quercus durata
Quercus Wislizeni var. frutescens
Quercus chrysolepis
Lithocarpus densiflorus
Myrica californica
Umbellularia californica
Holodiscus discolor
Adenostoma fasciculatum
Rosa spithamea var. sonomensis
Cercocarpus betuloides
Photinia arbutifolia

Amelanchier pallida
Amorpha californica
Pickeringia montana
Rhus diversiloba
Aesculus californica
Rhamnus californica
Garrya elliptica
Rhododendron occidentale
Vaccinium ovatum
Arctostaphylos Cushingiana
Arctostaphylos glandulosa
Arbutus Menziesii
Eriodictyon californicum
Symphoricarpos rivularis
Baccharis pilularis var. consanguinea

Some of the trees are completely killed by a fire, such as the bishop pine, Sargent cypress, and Douglas fir. The first two generally carry many cones that remain closed until they are opened by the fire that kills the tree, and the seeds, released over the ashes of the burn, may give rise to a grove of seedlings where only one individual stood before. The Douglas fir, on the other hand, is not only killed by the fire, but its replacement must first await the development of a brushy cover in the protection of which the young fir seedlings can grow. Carl Purdy used to say that the blue blossom thickets are the nurseries of the Douglas fir. So, while fire can be so destructive to the Douglas fir, it can serve as an instrument of propagation and dispersal for the bishop pine and Sargent cypress.

Besides the shrubs and trees that are benefited by properly spaced burns, there is a group of annual herbs that are also benefited and that respond as fireweeds. Some of these plants are seen only after fires, while others may be found occasionally here and there in the brush and chaparral between fires, but it is only after a burn that any of them occurs in abundance. At such a time the blackened slopes are carpeted with a lush, flowery growth where before only shrubs of the chaparral had monopolized all of the space. Among the most abundant of the fireweeds following the 1945 Marin County burn was *Phacelia suaveolens*, a rare species that had not been found in the Tamalpais area for nearly forty years; *Silene multinervia* which was abundant for the first time in three decades; and *Papaver californicum* which had not been seen since the 1929 Tamalpais fire. Other rare or little-collected herbs that were common or occasional after the 1945 fire were: *Montia gypsophiloides* var. *exigua, Calandrinia Breweri, Antirrhinum Hookerianum, Mimulus modestus, M. Rattanii,* and *Campanula angustiflora.* This page "in Nature's infinite book," telling of what happens after a fire in the California chaparral, is one to be perused with interest and profit by the systematic botanist, professional or amateur.

NUMERICAL ANALYSIS OF THE FLORA

So numerous and varied are all the different places and conditions where plants grow in Marin County, that it is not surprising that a large and representative section of the California flora is found in the 529 square miles of this relatively restricted area. In this work, 1,313 species and 144 named varieties and forms are recognized as occurring in 117 families and 528 genera. Of the species, 1,004 are indigenous and 309 are introduced. It is of interest to compare these figures with those given for all of California by Jepson in *A Manual of the Flowering Plants of California* in 1925, which are the most recent figures available for the entire state. There the total of 4,019 species is divided between 3,727 natives and 292 alien immigrants in 142 families. According to these figures, it would appear that 32.5 per cent of the California flora is represented in Marin County, but this is not a fair estimate, as can be seen from the fact that there are more introduced plants recognized here for Marin County than Jepson recognized for all of California. Likewise the specific concept in the present work is less conservative and less inclusive than it was in Jepson's Manual, so that here many plants are accepted as species that there were either treated as varieties or even ignored. The California flora, enumerated by the norm of species accepted here, would undoubtedly approach 5,000 species; so it is probably safe enough to conclude that the flora of Marin County represents approximately 25 per cent of the flora of the entire state.

The accompanying tables give an analysis of the Marin County flora, the first according to the chief groups of vascular plants, the second according to the ten largest families in the county.

GEOGRAPHIC DISTRIBUTION OF PLANTS IN MARIN COUNTY

A local flora such as this is prepared primarily as a list of the plants growing in a given area, but secondary deductions of equal interest, can be obtained from the list if data are also included for analytic studies of the ecologic and geographic distribution of the plants. Already an account of the plant formations based on the ecology of the region has been given; now the flora will be examined geographically in an attempt to show something of its relationship to the area that it occupies.

Although most of the native plants within Marin County have a wide and generalized distribution, an examination of the 1,004 indigenous species shows that about 300 of them are more restricted and that more than 250 of these have a distribution chiefly oriented around three centers: Mount Tamalpais, Point Reyes Peninsula, and the Sonoma borderland. While Mount Tamalpais itself

TABLE 1

VASCULAR PLANTS IN MARIN COUNTY

Major groups	Families	Genera	Species	Named varieties and forms
Pteridophyta	7	18	32	5
Gymnospermae	4	6	6	—
Monocotyledoneae	18	105	299 (70)[a]	18
Apetalae (Salicaceae to Aizoaceae)	14	31	91 (24)	11
Polypetalae (Portulacaceae to Garryaceae)	48	181	460 (125)	63
Gamopetalae (Ericaceae to Compositae)	26	187	425 (90)	47
Total	117	528	1313 (309)	144

[a] The figures in parentheses give the number of introduced species.

claims 73 species found nowhere else in Marin County, the number is swelled to 136 when all the Tamalpais area of Franciscan rocks is included. On Point Reyes Peninsula are 61 species found nowhere else in Marin County; and along the Sonoma borderland are 31 species, besides 21 more, common to Point Reyes Peninsula and the Dillons Beach area. Much smaller numbers of restricted species are found elsewhere; for example, there are only 10 in the Sausalito Hills-Angel Island area and 6 on Tiburon Peninsula. In this segment of the county flora characterized by restricted distributions, there are only 12 species found both in the Tamalpais area and on Point Reyes Peninsula, and 12 species common to Tiburon and Tamalpais, but occurring nowhere else in the county.

The noteworthy fact disclosed by this analysis of what may be called the restricted element of the indigenous Marin flora is that almost all the species are found in one of three areas with very different geologic histories: the long-emergent Tamalpais area of Franciscan age and devoid of Tertiary rocks; the Point Reyes Peninsula of granite overlain largely by Miocene sediments and separated from the rest of Marin County by the San Andreas Fault; and the Sonoma borderland of Pliocene sediments overlying the Franciscan. Of more than passing interest is the very small number of species common to the Tamalpais and Point Reyes areas, which is an impressive floristic reflection of

TABLE 2

TEN LARGEST FAMILIES IN MARIN COUNTY

Families	Genera	Species	Named varieties and forms
Compositae	82	183 (45)[a]	20
Gramineae	55	151 (65)	10
Leguminosae	19	107 (37)	25
Scrophulariaceae	19	55 (8)	11
Cyperaceae	5	55 (2)	—
Cruciferae	20	44 (18)	2
Umbelliferae	28	42 (9)	4
Liliaceae	16	36 (1)	5
Rosaceae	16	33 (5)	5
Caryophyllaceae	10	31 (11)	1

[a] The figures in parentheses give the number of introduced species.

The seven largest genera, with the number of species in each, are: *Carex,* 35; *Trifolium,* 27; *Lupinus,* 16; *Agrostis,* 15; and *Festuca, Juncus,* and *Lotus,* each with 14.

the very different geologic histories of the two areas. The foregoing data concerning the restricted indigenous plants can be summarized as follows:

Tamalpais group of species 136
   Plants on Mount Tamalpais, 73
   Plants elsewhere in the Tamalpais Franciscan area, 63
Point Reyes Peninsula group of species 61
Sonoma borderland group of species 52
   Plants along northern border, 21
   Plants in the Tomales–Dillons Beach area, 10
   Plants common to Point Reyes Peninsula and Dillons Beach area, 21
Sausalito and Angel Island area 10
Tiburon Peninsula 6
Tiburon and Tamalpais area ....., ... 12
Tamalpais area and Point Reyes Peninsula 12

Marin County Flora and the Flora of the Coast Ranges

In the present work, the distributional relationship of the indigenous species and varieties to the rest of the Coast Range flora has been noted especially, and lists have been prepared of those plants for which Marin County is the northern or southern distributional limit in the Coast Ranges. According to these lists, 97 entities reach a southern limit in Marin County and only 34 extend northward from the South Coast Ranges to reach their northern limit in the county. The difference between these two figures would seem to indicate that the Golden Gate plays an important role in limiting the northward and southward distribution of plants of the Coast Ranges, but an examination of the lists discloses that this impressive physiographic barrier has not exercised as great an influence as might be assumed. If the 97 plants reaching their southern limit in Marin County are grouped according to the general locality where they stop, the Tamalpais area and Point Reyes Peninsula immediately emerge as the centers that are truly significant, the 97 plants aligning themselves as follows: south to the Tamalpais area, 51; south to Point Reyes Peninsula, 34; south to the Sonoma borderland, 7; south to the Golden Gate, 5. Likewise in an analysis of the 34 plants noted as extending north to Marin County, the figures are similar though not so impressive: north to the Tamalpais area, 12; north to Point Reyes Peninsula, 11; north to the southern border (i.e., to Sausalito Hills, Angel Island, and Tiburon Peninsula), 7; north to the Sonoma borderland, 4.

Just as the Tamalpais and Point Reyes areas stood out as being important centers when the distribution of localized plants was considered, now these two areas figure prominently in the much more important matter of plant distribution in the central Coast Ranges. In the earlier analysis, the so-called Sonoma borderland also stood out as important, but in the second it is no more significant than the Golden Gate appears to be. The reason for this change of status is clear enough. The Sonoma borderland area is the narrow southern fringe of a floral province not otherwise represented in Marin County, and, though its plants have a narrow distribution in the county, they are not critical in the broader phytogeographic view because they are widely distributed in the Coast Ranges both north and south of the Tamalpais area.

One further detail on the relative unimportance of the Golden Gate as a floristic barrier may be had by comparing the number of species reaching a northern limit in Marin County with the number reaching a northern limit in

San Francisco. From a detailed and critical list of San Francisco plants published by Katharine Brandegee in 1892, it can be deduced that there are only about 11 species for which San Francisco is the northern limit in the Coast Ranges. This figure is of the same order of size as the number of southern plants that reach Mount Tamalpais (12) and Point Reyes Peninsula (11). These numbers represent approximately the rate at which northward ranging species drop from the flora in this region as consecutive geologic areas of commensurate size and importance are reached.

As an effective barrier, the Golden Gate and San Francisco Bay are important now, but all floristic evidence indicates the establishment of the flora in the region long before the formation of this waterway, which, geologically speaking, is of recent origin. The rise of water along the shore of the bay is so recent, in fact, that Indian shell mounds are in places partly submerged. This inundation of valley lands, where San Francisco Bay now is, may have occurred when the bay block of the earth's crust was down-faulted, thus allowing the ocean waters to enter through the already existing Golden Gate gorge. More recently the theory has been advanced that the original San Francisco Valley has been flooded by a rise of ocean water caused by the melting of the continental ice sheets of the Pleistocene epoch. Probably both processes have contributed to the formation of this major physiographic feature of the central California coast: there is abundant geologic evidence for the differential movement of earth blocks in the San Francisco Bay region, and there is mathematical evidence that the water from the continental ice increased the depth of the ocean by a hundred feet or more. In whatever manner San Francisco Bay was formed, the plant migrations of the Coast Ranges had already taken place.

Since the waterway consisting of the Golden Gate and San Francisco Bay can be practically ignored as a distributional barrier, a comparison of the total number of plants reaching northern or southern limits in Marin County can be made directly without reference to it. Such a comparison shows the southward movement to have been much stronger than the northward, and this is as might be expected because in this county the flora is predominantly northern in its relationship. Almost all this northern influence is coastal, with only the slightest suggestion of a tenuous montane boreal relationship in such plants as *Cheilanthes gracillima*, *Eleocharis Engelmannii*, and *Epilobium Halleanum* on Mount Tamalpais. In connection with the northern and southern limits of distribution in the Marin area, it is of interest to compare the results of detailed floristic studies made on Mount Diablo by Mary L. Bowerman and in the Mount Hamilton Range by Helen K. Sharsmith. Since it has been shown that San Francisco Bay is of no importance phytogeographically, it is of little moment that these areas lie east and south, respectively, of the bay; but it is important to note that they are in the eastern or inner Coast Ranges while Marin County is in the outer. According to these workers, 30 species reach a northern limit on Mount Diablo and 40 species in the Mount Hamilton Range, and in each, only 6 species are recorded as reaching a southern limit. When these figures are compared with those already given for Marin County, they are seen to be reversed in value and disclose also a reversal of direction in the chief plant migrations in the two adjacent but unlike regions of the Coast Ranges. The southern elements, that are relatively weak in the Marin flora, become the dominant elements in the hotter and drier inner Coast Ranges, and there the northern derivatives, so numerous in Marin County, are much reduced in number. It is interesting that in this reduced northern part of the Mount Hamilton flora, there are also meager but discernible north montane elements, just as there are on Mount Tamalpais. Whereas the southern elements in the inner-range flora can be traced to several geologic floras austral in development, the northern elements in the outer Coast Ranges come from such diverse geologic floras of boreal derivation as those related to the redwood, Douglas fir, and closed-cone pine forests. From these observations, the three chief directional sources of the present-day flora of the Coast Ranges can be recognized: the south-interior elements that extend northward in the drier easternmost ranges and are weakly represented on the coast; the north coastal elements that range southward near the coast and become rapidly weaker toward the interior; and the north montane elements that range southward in appreciable amount to the summits of the high North Coast Ranges and that are feebly represented on such peaks as Mount Tamalpais and Mount Hamilton, but more strongly on the higher Santa Lucia Mountains in Monterey County.

ENDEMIC PLANTS

Closely related to these matters of floristic distribution, and yet apart from them, are the endemic plants, those plants restricted to Marin County. In the county, 10 species, 8 varieties, and 1 form are herein recognized as endemic, and they may be listed according to the geographic (and generally geologic) areas where they occur:

MOUNT TAMALPAIS FRANCISCAN AREA
Montia spathulata var. *rosulata*
Streptanthus glandulosus var. *pulchellus*
Streptanthus batrachopus
Ribes Menziesii var. *Victoris*
Lupinus Douglasii var. *fallax*
Astragalus Gambellianus var. *Elmeri*
Lessingia ramulosa var. *micradenia*
Cirsium Vaseyi

POINT REYES PENINSULA
Agrostis aristiglumis
Lupinus Layneae
Sidalcea rhizomata
Arctostaphylos Cushingiana f. *repens*

Castilleja Leschkeana
Blennosperma nanum var. *robustum*

TIBURON
Streptanthus niger
Castilleja neglecta

MISCELLANEOUS
Melica Geyeri var. *aristulata*
(Sausalito and Tamalpais area)
Arctostaphylos montana
(Tamalpais area and Point Reyes Peninsula)
Arctostaphylos virgata
(Tamalpais area and Point Reyes Peninsula)

Again it is of interest to compare the number of Marin County endemics with those recognized in the local floras of Mount Diablo and the Mount Hamilton Range. Dr. Bowerman lists 6 endemic species on Mount Diablo and Dr. Sharsmith gives 3 species and 1 variety for the Mount Hamilton Range. The sizes of the floras and areas where these occur are: 630 plants in 55 square miles on Mount Diablo, and 761 plants in 1500 square miles in the Mount Hamilton Range. On Mount Tamalpais (not including the Carson country) there are 796 species and 55 varieties in about 36 square miles.

NATURALIZED PLANTS

While matters relative to the composition and relationship of the Marin County flora are being considered, it is necessary to add a few remarks about the all-too-abundant members of the vegetation that have been introduced from other places. It has already been noted that more weedy plants are accepted in this work than Jepson recognized for the entire state of California. Whereas Jepson recognized as an alien member of the flora only those species "which are really established and have a true competitive status," here the view is not so conservative and any plant reproducing itself without cultivation is included, whether the plant is aggressively spreading or seems only passively and locally established. The inclusion of such exotics as Australian species of *Acacia* and *Eucalyptus* may be considered extreme, but they must be included if this work realizes its purpose of treating all uncultivated Marin County plants. "The waifs of today may become the weeds of tomorrow," and in the past, even "promising" waifs have been too frequently ignored. Fleeting fugitives from cultivation, however, are not accepted unless they show promise of becoming floristically fixed or else are known as weeds in other parts of California.

Also not treated in this work are those native conifers which have been planted on Mount Tamalpais in an attempt to increase the forest cover of its slopes. Among these are *Pinus radiata, P. muricata, P. attenuata, P. Coulteri,* and *Chamaecyparis Lawsoniana.* A few California species have been included as weeds, for, although they are native in other parts of California, they are not believed to be indigenous in Marin County. Such plants as *Salix melanopsis, Mentzelia laevicaulis, Chrysopsis oregona, Heterotheca grandiflora,* and perhaps *Grindelia camporum* belong to this interesting group of introductions.

Finally it should be added that the importance of the weed flora in many parts of Marin County cannot be realized even by the 1:3 ratio it bears to the native population. In all too many districts the natives are being replaced by immigrants, and there are some places where, alas, even now the native flora is all but vestigial. One of the most important functions served by a local work such as this is that the detailed floristic account, now set down in fact, will serve a generation hence as a historic record of a flora that may have become a fleeting memory. The preparation of local floras for valley and foothill areas cannot be urged too strongly before it is too late, for between weeds and civilization the native plants in the lower hills and valleys of California are disappearing all too quickly and completely.

The situation is, however, not entirely without hope, for, if federal and state reserves and monuments are properly developed and administered, they can become plant-life sanctuaries even if the area concerned has been set aside primarily for historical or cultural reasons. Already the importance of preserving the redwoods and Monterey cypresses has been recognized. A fine example of the bishop pine forest is included in the Shell Beach State Park on Tomales Bay. Larger reservations like these are needed for groves and forests, but there are many small areas needing protection in order to preserve samples of valley and hill flora. It is to be hoped that the idea of small botanical monuments may be developed so that in the near future characteristic or outstanding remnants of our lowland flora may be preserved by such a means. One can scarcely foretell how precious a half dozen acres of marsh or meadow will be a century from now.

EARLY BOTANICAL EXPLORERS

Although there have been many changes in the flora of Marin County owing to weeds and civilization, much of the flora is as it was more than a hundred years ago, when, still a provincial frontier of Spain and, later, of Mexico, the region was first visited by botanists. Apparently the first person primarily interested in natural history to visit the region of San Francisco Bay was Langsdorff, who, as naturalist on the expedition of Count Resanov, was at the Presidio of San Francisco in the spring of 1806. That was before the missions at San Rafael and Sonoma were established, so instead of going north to explore, he went south to Mission San Jose on the southeast shore of San Francisco Bay.

The first man prompted by scientific curiosity to land on the shores of present-day Marin County was Chamisso, who, with Eschscholtz, was a naturalist on the first Kotzebue expedition, which spent the month of October, 1816, at the Presidio of San Francisco. Chamisso must have yearned to get away from the dunes and downs of the San Francisco Peninsula and on to the high steep hills and mountains across the water, although, as he writes, "the year was already old, and the country, which in the spring months (as Langsdorff has seen it) blooms like a flower-garden, presented now to the botanist only a dry, arid field." His chance came, and, though he probably collected only specimens of the red radiolarian chert (which he misidentified), he must be counted the first of the long line of botanists destined to explore those hills. "On the ninth of October, some Spaniards were sent to the northern shore to catch horses with a *riata*, . . . and I seized the opportunity also to look about me on that side. The red-brown rocks there, . . . as may be seen in the mineralogical museum at Berlin, are siliceous schist. . . ."

Although Eschscholtz apparently did not go to the "north shore" in 1816, it is interesting that he was probably its next scientific visitor, when, in the fall of 1824, as naturalist on Kotzebue's second expedition, he was again at the then Mexican Presidio of San Francisco. At that time, he visited the recently estab-lished missions at San Rafael and Sonoma and thence made the overland trip to the Russian settlement at Bodega on the Sonoma coast. There is no record that he collected any specimens during this short northern trip, but it was from his botanical collections made on the second Kotzebue expedition that such well-known plants as yellow sand-verbena (*Abronia latifolia*), blue blossom (*Ceanothus thyrsiflorus*), and coffee-berry (*Rhamnus californica*) were named and described.

During the rest of the Mexican period and before the discovery of gold, only two other botanists are known to have traversed the Marin countryside, both in the employ of the Royal Horticultural Society of London but neither an Englishman: the well-remembered Scotsman, David Douglas, and the German, Karl Theodor Hartweg. Douglas was at Mission San Rafael on July 27, 1831, on his way to Sonoma, but there is no record that he collected anything in Marin County (though he did collect the type of *Pogogyne parviflora* Benth. in Sonoma County). According to a memoir by F. R. S. Balfour, Douglas climbed Mount Tamalpais in the fall of 1833, but this record of what may be a first ascent seems doubtful unless further information substantiates it.

To Karl Theodor Hartweg goes the credit for making in 1846 and 1847 the first collections definitely known to come from our area, though his fame in California rests chiefly on his explorations in the Sacramento Valley and in the Sierra Nevada, from both of which regions he was the first botanist to obtain extensive collections. In Marin County, however, he did collect at Corte Madera one plant described over three decades later as new to science, *Fritillaria lanceolata* var. *gracilis* Wats.

In less than a decade from that quiet day when Hartweg collected his fritillary near the outermost bounds of provincial Mexico, California was to be briefly independent, to become the southwestern frontier of the United States, and to stage a gold rush with world-encircling reverberations. In these boisterous times, Marin County was designated one of the first twenty-seven counties of California, the federal transcontinental surveys reached the Pacific coast, and the California Academy of Natural Sciences was founded in San Francisco.

John M. Bigelow, botanist on the Pacific railroad survey under Lieutenant A. W. Whipple, came to the San Francisco Bay region in the spring of 1854, and from April 16 to 20 he made a botanical collecting trip through the redwoods north of Mount Tamalpais to the ocean on Point Reyes Peninsula. Never before had these parts been visited by a botanist, and his collection, with respect to the novelties in it, is the richest ever made in the region. *Scoliopus* and *Whipplea,* two genera of frequent occurrence in the woods of Marin County, were based on Bigelow's collection, and no fewer than twenty species and varieties were described by Torrey and others in the botanical reports of the expedition and elsewhere. From Marin County, Bigelow went on to other rich fields in Sonoma and Napa counties, and thence to the foothills and middle slopes of the Sierra Nevada, but it is not likely that anywhere in all his travels, in or out of California, did he ever enjoy an excursion so rich as the one from Corte de Madera and Rancho San Geronimo to Punta de los Reyes.

Even as Bigelow was making his memorable Marin excursion, the infant Academy of Sciences was celebrating its first anniversary in San Francisco. Founded in April, 1853, by a small group of men devoted to the promotion of scientific studies, this little institution on the fringe of western civilization soon won the recognition and respect of students everywhere, and for California and the West it quickly became the chief center of research in the natural sciences. While western America and the Pacific Ocean were early assumed as the academy's field of interest, the unexplored country in the vicinity was not overlooked, and, from the very earliest days, Marin County, just across the bay, furnished the scientists of the academy a profitable field for exploration and study. The botanical riches of this area were early recognized, and, among the first plants described in the Proceedings of the Academy by Albert Kellogg, one of its founders and original curators, was a new species of snapdragon which was reported at the meeting of January 29, 1855, thus: "Dr. Kellogg exhibited a specimen and drawing of a Linariad considered new—it was found by Dr. Andrews near Punta de los Reyes"; and the formal description of *Antirrhinum vexillo-calyculatum* followed. This, apparently, is the first plant ever described from Marin County. It is an indication of the vision, zeal, and industry of these early-day Californians that the first plant named from Marin County was described not in St. Petersburg, London, or Boston, but on home ground in San Francisco.

LATER BOTANICAL EXPLORERS

In the years that followed, the botanists at the academy, more than any others in California, have taken a special interest in the flora of Marin County, and most of the plants that have been described or reported from there have been collected or identified by those connected with that institution. So besides Albert Kellogg, there should be mentioned H. H. Behr, H. N. Bolander, Mary Layne

Curran (later Mrs. Katharine Brandegee), T. S. Brandegee, and E. L. Greene, the principal workers on Marin plants up to about 1890. In 1860 the California Geological Survey was organized under the direction of Josiah Dwight Whitney, and although William H. Brewer, its chief botanist and erstwhile secretary to the academy did little work in Marin County, he did make three brief visits there in 1862. Henry N. Bolander made repeated trips there and about a half dozen new species were described from his collections, chiefly while he was collecting for the survey.

In 1885, E. L. Greene became Instructor in Botany at the University of California and, not only did he take into his new work his deep interest in the Marin flora, but he imparted this interest to his students, Victor K. Chesnut, Elmer Drew, Charles A. Michener, Frederick T. Bioletti, and Willis Linn Jepson. For each of these enthusiastic young neophytes Greene named a species from the slopes of Tamalpais. Altogether, during both his academy and university years, Greene described about 28 plants from Marin County, the largest number described from there by any botanist. Interest in the Marin County flora at the university was carried on by W. L. Jepson and later by J. Burtt Davy, H. M. Hall, H. L. Mason, and others, down to the present day. All through the years, various collectors and botanists have visited Marin County, seeking either specimens or recreation. No fewer than fourteen persons, not named here, have collected one or more specimens that were subsequently described as new to science, and many other visitors through their collections have added rare or critical specimens to local or distant herbaria.

In 1892 Alice Eastwood came to the California Academy of Sciences where, as joint curator of botany with Katharine Brandegee, she soon became interested in the flora of Mount Tamalpais and of Marin County generally. Year after year, weekly trips afoot over miles of trail gave her an intimate and extensive knowledge that no one else has had of Marin plants as they grow. It was this critical field knowledge that enabled her to distinguish the different species of manzanita (*Arctostaphylos*) where before her work all had been conflicting opinion and confusion, and to demark as generically distinct those related plants with friable fruits (*Schizococcus*). More specimens of Marin County plants have probably been collected and distributed by Miss Eastwood through the academy than by anyone else; and after Greene, more plants from the county have been described by her than by any other botanist. John Milton Bigelow, Edward Lee Greene, and Alice Eastwood are the botanists who have done most to make the Marin flora known to science and the world.

It is on this ample, century-old foundation, to which these botanists have contributed so much, that the present work rests.

L'envoi

Marin County may be likened to that fabulous siren of antiquity of whom it was written: "Age cannot wither her, nor custom stale her infinite variety." Her scenes are fair to see, her sounds are sweet to hear, her airs are fresh to breathe. Her hills and mountains look down on cathedral groves and over boundless waters. Her voices are myriad for him who listens: the crash of surf on cliffs and beaches; the roar of gale through fir and pine; the trill of wren-tits from chaparral and brush; the sounds of many creatures in the marshes; the silence of the sunshine and the mist. Her seasons can be told by smell: the tonic perfume of a thousand flowers in spring; the fruity fragrance of manzanita in the summer sun; the pungent odor of tarweed drenched in autumn dew; the earthy smell of age-old redwood humus in the winter. The feel of summer fog on sunburned cheek; the silky smoothness of madroño stem; the taste of fountains in the serpentine—how sensuous are her manifold lures!

> How often have I loitered o'er thy green . . .
> How often have I paused on every charm.

KEY TO THE PRINCIPAL GROUPS OF FAMILIES

a. Plants not producing seeds (*Pteridophyta,* the ferns and fern allies) ....
....Group 1, p. 37
a. Plants producing seeds (*Spermatophyta,* the seed plants) ....b
b. Plants without flowers, the ovules borne on a scale and not in a closed
   receptacle or ovary; leaves scalelike or needle-like (*Gymnospermae,* the
   conifers and related plants) ....Group 2, p. 37
b. Plants with flowers, the ovules borne in a closed receptacle or ovary;
   leaves rarely needle-like or scalelike (*Angiospermae,* the flowering
   plants) c
c. Cotyledon usually 1; flower parts commonly in 3s; veins of the leaves
   generaly parallel and extending the length of the leaf (*Monocotyle-
   doneae,* the monocotyledons) ....Group 3, p. 38
c. Cotyledons 2; flower parts usually in 4s and 5s, rarely 3s; veins of the
   leaves generally not parallel (*Dicotyledoneae,* the dicotyledons) ....d
d. Flowers without a perianth or, if a perianth is present, the segments
   alike in color and texture ....e
d. Flowers with a perianth, the segments in 2 series usually well differentiated in color and texture, the outer segments, the sepals, forming
   the calyx, the inner, the petals, forming the corolla. (In some families
   of Groups 9 and 12, the calyx is reduced or obsolete and the flowers
   then appear to belong to Group 4 or 5) ....f
e. Trees, shrubs, and woody climbers ....Group 4, p. 39
e. Herbs, sometimes woody at the base ....Group 5, p. 40
f. Petals distinct or rarely united at the very base ....g
f. Petals conspicuously united ....j
g. Ovary inferior or partly inferior ....Group 9, p. 44
g. Ovary superior ....h
h. Plants woody ....Group 6, p. 41
h. Plants herbaceous i
i. Flowers irregular ....Group 7, p. 42
i. Flowers regular or nearly so ....Group 8, p. 43
j. Stamens more than 5 ....Group 10, p. 45
j. Stamens 2–5 ....k
k. Ovary superior ....Group 11, p. 46
k. Ovary inferior ....Group 12, p. 47

[35]

(Ferns and fern allies)

a. Leaves (fronds) coiled in vernation and usually numerous, large, more than 2.5 cm. long, with definite stalks (stipes) and blades. Sporangia borne on the lower surface of leaves.—True ferns . . . . .*Polypodiaceae,* p. 49

a. Leaves not coiled in vernation (except in *Pilularia*), small and scalelike and mostly less than 1 cm. long, or if longer, either sheathlike or grasslike, or, in *Botrychium,* large and ample but usually only two, one fertile and one sterile arising from a common stalk.—Fern allies .b

b. Leaves ample and rather fernlike, usually 2, very unlike, one fertile and one sterile. Sporangia borne along the edge of the numerous narrow subdivisions of the fertile blade; stems entirely subterranean . . . .*Ophioglossaceae,* p. 53

b. Leaves numerous, scalelike, sheathlike, or grasslike c

c. Leaves grasslike . . . . .d

c. Leaves not grasslike . . . . .e

d. Stem a subterranean corm, bearing numerous leaves in a close tuft; sporangia in a cavity at the base of leaves . . . .*Isoetaceae,* p. 53

d. Stem elongate and slender, at or just below the surface of the ground, bearing 1 to few leaves at each node; sporangia in round or ovoid short-stalked sporocarps . . . .*Marsileaceae,* p. 54

e. Plants annual, floating on the surface of water, sometimes becoming terrestrial on muddy strands; stems less than 2.5 cm. long, covered with imbricate leaves; sporangia in sporocarps on the lower side of stems . . . .*Salviniaceae,* p. 54

e. Plants perennial, terrestrial; stems mostly more than 2.5 cm. long; sporangia in conelike spikes . . . . .f

f. Aerial stems jointed, the main stem hollow; leaves forming sheaths at the nodes; sporangia borne in conspicuous conelike spikes at the end of stems . . . .*Equisetaceae,* p. 54

f. Stems solid, creeping; leaves small, numerous, imbricate, the plants mosslike; sporangia in terminal 4-angled spikes of slightly modified leaves . . . .*Selaginellaceae,* p. 55

GROUP 2

(Gymnosperms)

a. Foliage leaves scalelike, opposite. Cone globose, the scales radiating from the center with broad flattish tops . . . .*Cupressaceae,* p. 57

a. Foliage leaves elongate, spreading, the leaves or leaf bundles alternate .b

b. Bases of foliage leaves not decurrent along stem. Cone elongate, the scales imbricate, attached at one end to the central axis . . . .*Pinaceae,* p. 57

b. Bases of foliage leaves decurrent along the stem c

c. Leaves falling singly, not attached to short branchlets; fruit fleshy, 1-seeded . . . .*Taxaceae,* p. 55

c. Leaves mostly falling attached to short branchlets; fruit a woody, globose, many-seeded cone .... *Taxodiaceae,* p. 56

GROUP 3

(Monocotyledons)

a. Plants terrestrial, or if growing in water, the leaves and culms aerial .b
a. Plants aquatic, stems and leaves floating or submerged (or the plants sometimes stranded as the waters recede) ....n
b. Perianth present, the segments in 2 series, generally conspicuous and petal-like, fleshy, never scalelike c
b. Perianth none, or if present, scalelike or bristle-like, not fleshy ....h
c. Perianth segments greenish, not petal-like. Carpels 6 (6 fertile or 3 fertile and 3 sterile), at maturity the fertile carpels separating from a central axis ....*Juncaginaceae,* p. 60
c. Perianth segments petal-like, or the 3 outer smaller or green ....d
d. Carpels numerous, distinct, in fruit becoming achenes. Leaves basal; outer perianth segments green, the inner white ....*Alismaceae,* p. 61
d. Carpels 3, united, in fruit becoming a capsule or berry ....e
e. Ovary superior ....f
e. Ovary inferior ....g
f. Stems prostrate, rooting at the nodes, very leafy; outer perianth segments green, inner white ....*Commelinaceae,* p. 98
f. Stems (i.e., aerial) erect, not rooting at the nodes; outer perianth segments mostly petal-like (except in *Trillium* and *Calochortus*) ....
....*Liliaceae,* p. 100
g. Perianth regular ....*Iridaceae,* p. 108
g. Perianth irregular ....*Orchidaceae,* p. 109
h. Inflorescences dimorphic, one type largely subterranean and consisting of single sessile pistillate flowers, the other type scapose and consisting of a number of staminate, pistillate, and perfect flowers ....
....*Lilaeaceae,* p. 60
h. Inflorescences all alike or essentially so, not dimorphic i
i. Flowers unisexual, without conspicuous bractlets, the staminate and pistillate flowers separate but on the same axis ....j
i. Flowers perfect or unisexual, subtended by conspicuous bracts or bractlets l
j. Flowers in globose heads ....*Sparganiaceae,* p. 59
j. Flowers in dense elongate spikes ....k
k. Leaves with broad blades; spike (spadix) subtended by a white petal-like bract (spathe) ....*Araceae,* p. 97
k. Leaves linear; spike subtended by a leaflike bract ....*Typhaceae,* p. 58
l. Fruit a capsule; perianth conspicuous, of 6 scales in 2 series ..*Juncaceae,* p. 98
l. Fruit an achene; perianth none or inconspicuous and bristle-like .....m
m. Culms usually round and hollow; flowers subtended by 2 bractlets ....
....*Gramineae,* p. 62
m. Culms solid, frequently 3-angled; flowers subtended by 1 bractlet ......
....*Cyperaceae,* p. 88

n. Plants marine, on rocky reefs or in sheltered bays ....*Zosteraceae*, p. 60

n. Plants in fresh or brackish water of lakes, streams, or tidal sloughs ....o

o. Plants floating, small (2–6 mm. in diameter), leafless .....*Lemnaceae*, p. 97

o. Plants partly or entirely submerged, the leaves and flowers sometimes floating ....p

p. Leaves without stipules, opposite below, whorled above, 1-nerved; perianth present, petaloid ....*Hydrocharitaceae*, p. 61

p. Leaves with stipules or with a broadened sheathing base, alternate or opposite, 1- to many-nerved; perianth, if present, not petaloid .....q

q. Leaves entire, with stipules; flowers in sessile or stalked clusters ......
....*Potamogetonaceae*, p. 59

q. Leaves minutely spinulose-toothed, with broadened base; flowers solitary in the leaf axils ....*Najadaceae*, p. 60

GROUP 4

(Woody apetalous dicotyledons)

a. Plants parasitic on branches of trees and shrubs. Leaves opposite, evergreen; fruit a 1-seeded berry ....*Loranthaceae*, p. 117

a. Plants not parasitic ....b

b. Calyx (fused together with the corolla) falling as a cap or lid when the flower opens ....*Eucalyptus*, p. 196

b. Calyx, if present, not falling as a lid or cap when the flower opens ...c

c. Leaves alternate ....d

c. Leaves opposite. Plants dioecious or in *Clematis* the flowers polygamous ....n

d. Leaves resinous-dotted, especially on the lower surface. Leaves evergreen, without stipules; flowers in short unisexual catkins ....
....*Myricaceae*, p. 116

d. Leaves not resinous-dotted ....e

e. Flowers, staminate or both staminate and pistillate, in catkins. Leaves with stipules, deciduous except in several species of *Quercus* ......f

e. Flowers not in catkins i

f. Both staminate and pistillate flowers in catkins ....g

f. Staminate flowers in catkins, pistillate flowers at the base of the catkin or 1 to several in small clusters. Plants monoecious; fruit a nut ....h

g. Plants dioecious; fruit a many-seeded capsule ....*Salicaceae*, p. 111

g. Plants monoecious; fruit a small winged nutlet ....*Betulaceae*, p. 112

h. Pistillate flower subtended by 2 bractlets that enlarge to form a tubular, calyx-like cover for the fruit ....*Corylaceae*, p. 113

h. Pistillate flowers subtended by an involucre that enlarges to form a spiny bur or scaly cup ....*Fagaceae*, p. 113

i. Stems twining; leaves cordate; flowers very irregular ......*Aristolochia*, p. 118

i. Stems not twining; leaves generally rounded or cuneate; flowers regular or nearly so ....j

j. Leaves evergreen. Fruit fleshy ....k

j. Leaves deciduous l

k. Leaves without stipules, aromatic; flowers perfect; trees ....*Lauraceae*, p. 140

k. Leaves with stipules, not aromatic; flowers mostly unisexual; shrubs ...
....*Rhamnus,* p. 188
l. Trees with unisexual flowers. Fruit a samara ....*Ulmaceae,* p. 117
l. Shrubs with perfect flowers ....m
m. Leaves toothed; fruit an achene with a long plumose tail ....
....*Cercocarpus,* p. 159
m. Leaves entire; fruit a greenish drupe ....*Thymelaeaceae,* p. 196
n. Leaves simple. Shrubs with flowers in catkins ....*Garryaceae,* p. 210
n. Leaves compound ....o
o. Woody vines; sepals creamy white, petal-like; fruit an achene with a
   plumose tail ....*Clematis,* p. 136
o. Trees; sepals not petal-like; fruit a samara ....p
p. Fruit a double samara; leaflets generally 3, coarsely and irregularly serrate or lobed ....*Aceraceae,* p. 187
p. Fruit a simple samara; leaflets 5–7, finely and evenly serrate ..*Oleaceae,* p. 217

GROUP 5

(Herbaceous apetalous dicotyledons)

a. Plants parasitic on pines. Leaves opposite, reduced to connate scales;
   fruit a 1-seeded berry ....*Loranthaceae,* p. 117
a. Plants not parasitic ....b
b. Leaves in whorls, without stipules. Plants aquatic c
b. Leaves alternate or opposite, or if rarely in whorls, then with stipules
   ....d
c. Ovary superior; achene beaked by the persistent style; stems jointed
   ....*Ceratophyllaceae,* p. 136
c. Ovary inferior; fruit without a beak; stems not noticeably jointed ....
   ....*Haloragaceae,* p. 201
d. Fruits 1-seeded ....e
d. Fruits 2- to many-seeded ....n
e. Fruit indehiscent, an achene ....f
e. Fruit dehiscent (sometimes tardily so), a utricle or capsule ....j
f. Pistils more than 1. Leaves, at least above the base, compound ....
   ....*Ranunculaceae,* p. 136
f. Pistil 1 ....g
g. Flowers unisexual; herbs with stinging hairs. Leaves simple, with
   stipules ....*Urticaceae,* p. 116
g. Flowers perfect (or sometimes unisexual in *Sanguisorba*); herbs without
   stinging hairs ....h
h. Leaves with stipules, pinnately compound or in *Alchemilla* simple and
   palmately lobed ....*Rosaceae,* p. 155
h. Leaves simple, not palmately lobed, with or without stipules i
i. Flowers in heads, showy; achene enclosed by the hardened persistent
   base of the calyx tube ....*Nyctaginaceae,* p. 127
i. Flowers generally on short slender pedicels, inconspicuous if subsessile;
   achene not enclosed by base of calyx tube ....*Polygonaceae,* p. 118

j. Leaves opposite, with conspicuous scarious stipules ...*Caryophyllaceae*, p. 131

j. Leaves alternate or opposite, without stipules ....k

k. Fruit a capsule; leaves with forked or stellate hairs l

k. Fruit a utricle; leaves without forked or stellate hairs ....m

l. Flowers perfect; fruit round, strongly flattened ....*Athysanus*, p. 149

l. Flowers unisexual; fruit ovoid, turgid ....*Eremocarpus*, p. 184

m. Bracts of the inflorescence herbaceous or lacking; filaments free ......
....*Chenopodiaceae*, p. 123

m. Bracts of the inflorescence present, dry and scarious, not herbaceous;
   filaments united at the base ....*Amaranthaceae*, p. 126

n. Flowers without a calyx, consisting of single stamens or pistils ....o

n. Flowers with a calyx ....p

o. Stamens and pistils (i.e., flowers) clustered within an involucre that resembles a single flower; plants with milky juice ....*Euphorbia*, p. 184

o. Stamens and pistils (flowers) solitary or clustered in the axils of leaves,
   not in a flower-like involucre; aquatic or marsh plants without milky
   juice ....*Callitrichaceae*, p. 185

p. Leaves alternate ....q

p. Leaves opposite or verticillate ....u

q. Leaves simple, entire or undulate ....r

q. Leaves pinnately divided to bipinnatifid or compound ....s

r. Leaves cordate; calyx 3-lobed ....*Asarum*, p. 118

r. Leaves not cordate; calyx 4-lobed ....*Tetragonia*, p. 128

s. Fruit a red or white berry ....*Actaea*, p. 136

s. Fruit a casule ....t

t. Ovary superior; seeds 2 ....*Cruciferae*, p. 142

t. Ovary inferior; seeds numerous ....*Datiscaceae*, p. 195

u. Ovary inferior. Plants in water or on wet ground; style and stigma 1
....*Ludwigia*, p. 197

u. Ovary superior v

v. Perennial; style and stigma 1 ....*Glaux*, p. 217

v. Annual; styles or stigmas more than 1 ....w

w. Styles, 3–5, distinct ....*Caryophyllaceae*, p. 131

w. Style 1, stigmas 2 or 3 ....*Aizoaceae*, p. 127

GROUP 6

(Polypetalous dicotyledons with woody stems and superior ovary)

a. Leaves opposite ....b

a. Leaves alternate ....f

b. Leaves palmately compound ....*Hippocastanaceae*, p. 188

b. Leaves simple c

c. Leaves with several palmate veins from the base ....*Aceraceae*, p. 187

c. Leaves with one midvein from the base ....d

d. Flowers large and showy; stamens numerous ....*Cistaceae*, p. 194

d. Flowers 2 cm. or less in diameter; stamens less than 10 ....e

e. Calyx broad and shallow; tall shrub of moist woods ......*Euonymus*, p. 187

e. Calyx narrow and tubular; low suffrutescent plant of salt marshes
....*Frankenia,* p. 194
f. Fruit a samara. Leaves large, pinnately compound, with a disagreeable
odor when bruised ....*Simaroubaceae,* p. 184
f. Fruit not a samara ....g
g. Corolla irregular or reduced to 1 petal in *Amorpha* ....*Leguminosae,* p. 161
g. Corolla regular, not reduced to 1 petal ....h
h. Sepals 5, dimorphic, the 3 inner ovate, the 2 outer linear and bractlike
....*Cistaceae,* p. 194
h. Sepals or calyx segments regular or irregular but not dimorphic i
i. Stamens 10 or more ....j
i. Stamens less than 10 ....o
j. Sepals 2; petals 4 ....*Papaveraceae,* p. 140
j. Sepals or calyx segments 5; petals 5 ....k
k. Pistils several to many, developing into follicles, achenes, drupes, or
drupelets ....*Rosaceae,* p. 155
k. Pistil 1 l
l. Styles several; stamens united by their filaments ....*Malvaceae,* p. 191
l. Style 1; stamens distinct or united slightly at the base of the filaments ....m
m. Fruit indehiscent, an achene or drupe; leaves broad and deciduous or
heather-like and evergreen ....*Rosaceae,* p. 155
m. Fruit a capsule; leaves ample and evergreen ....n
n. Capsule 5-celled; leaves simple ....*Ledum,* p. 215
n. Capsule 1-celled; leaves compound or reduced to parallel-veined phyllodes ....*Leguminosae,* p. 161
o. Leaves evergreen ....p
o. Leaves deciduous ....q
p. Leaves compound ....*Berberidaceae,* p. 139
p. Leaves simple ....*Rhamnaceae,* p. 188
q. Leaves simple, palmately veined from the base ....*Vitaceae,* p. 191
q. Leaves compound, the leaflets not palmately veined ...*Anacardiaceae,* p. 186

GROUP 7

(Herbaceous polypetalous dicotyledons with irregular
flowers and superior ovary)

a. Flowers saccate or spurred (the spur adnate to the pedicel and inconspicuous in *Pelargonium*) ....b
a. Flowers not saccate or spurred ....f
b. Sepal spurred c
b. Petals saccate or spurred. Pistil 1, ovary 1-celled ....e
c. Pistils several (generally 3), becoming follicles; petals 4; leaves palmately
parted or divided ....*Delphinium,* p. 138
c. Pistil 1, the fruit not a follicle; petals 5; leaves not palmately parted or
divided ....d
d. Spur conspicuous; ovary 3-celled; leaves peltate ....*Tropaeolaceae,* p. 183

d. Spur inconspicuous and adnate to the pedicel; ovary 5-celled; leaves round-reniform, palmately veined ....*Pelargonium,* p. 180

e. Petals 4, 2 saccate at base; sepals 2 ....*Fumariaceae,* p. 141

e. Petals 5, 1 spurred at base; sepals 5 ....*Violaceae,* p. 194

f. Stamens 8–40, distinct, inserted on one side of the flower; stigmas 3–6, terminating the beaks of the ovary ....*Resedaceae,* p. 150

f. Stamens distinct or generally united, 8–10; stigma 1, generally terminating a slender style ....g

g. Sepals 5, more or less petal-like; petals 3, the lower keeled and beaked; ovary 2-celled ....*Polygalaceae,* p. 184

g. Sepals 5, not petal-like; petals 5, the two lower united by their lower edge and forming a keel; ovary 1-celled ....*Leguminosae,* p. 161

GROUP 8

(Herbaceous polypetalous dicotyledons with regular flowers and superior ovary)

a. Pistils more than 1, distinct or a little united at the very base ......b

a. Pistil 1 ....e

b. Stamens many; fruit an achene or follicle c

b. Stamens 5 or 10; fruit a follicle ....d

c. Stamens hypogynous, attached to the receptable below the pistils ....*Ranunculaceae,* p. 136

c. Stamens perigynous, attached to the upper part of the concave receptacle or hypanthium, or the receptacle disklike and the stamens attached on the outer part ....*Rosaceae,* p. 155

d. Pistils 3–5 ....*Crassulaceae,* p. 150

d. Pistils 2 ....*Saxifragaceae,* p. 151

e. Sepals 2 or 3 ....f

e. Sepals 4–12 ....h

f. Leaves opposite, with stipules; stems creeping, rooting at the nodes ....*Elatinaceae,* p. 194

f. Leaves alternate or opposite, without stipules; stems erect ....g

g. Fruits longer than wide; sepals falling early; petals 4 or 6 *Papaveraceae,* p. 140

g. Fruits about as wide as long; sepals not falling early; petals 5 ....
....*Portulacaceae,* p. 128

h. Stamens numerous, more than 10 i

h. Stamens 2–10 ....n

i. Flowers solitary on stalks arising from the base of the plant; sepals 5–12; petals 10–20 ....j

i. Flowers generally clustered, not arising from the base of the plant; sepals mostly 4 or 5; petals 4–10 ....k

j. Leaves linear, clustered on the crown of a thickened root; petals white to rose; fruit a 1-celled capsule ....*Lewisia,* p. 130

j. Leaves broadly ovate, cordate, arising from rootstocks; petals inconspicuous; fruit 10–25-celled, somewhat fleshy ....*Nymphaeaceae,* p. 135

k. Leaves compound; fruit a berry ....*Actaea,* p. 136

k. Leaves simple; fruit dry, generally a capsule l

l. Stamens monadelphous, the filaments united to form a tube; leaves palmately veined ....*Malvaceae*, p. 191

l. Stamens distinct or united in bundles, not monadelphous; leaves pinnately veined ....m

m. Leaves opposite; styles 3 ....*Guttiferae*, p. 193

m. Leaves alternate; style 1 ....*Cistaceae*, p. 194

n. Ovary 1-celled ....o

n. Ovary 2–5-celled or the pistil consisting of 5 carpels united to a central axis ....u

o. Plants saprophytic, whitish or buff, leaves without chorophyll ....
....*Pityopus*, p. 216

o. Plants with green leaves ....p

p. Leaves opposite ....q

p. Leaves alternate or basal ....r

q. Placentae central; styles 3–5, distinct, rarely united; sepals distinct (or united in *Silene*) ....*Caryophyllaceae*, p. 131

q. Placentae basal; style 1, 2- or 3-cleft; sepals united except at the tip
....*Frankeniaceae*, p. 194

r. Leaves compound ....*Vancouveria*, p. 139

r. Leaves simple ....s

s. Sepals and petals 4; stamens 6; style 1 ....*Cruciferae*, p. 142

s. Sepals and petals 5, the sepals distinct or united; stamens 5 or 10; styles or stigmas 2–5 ....t

t. Styles or stigmas 2–4; fruit a several- to many-seeded capsule ....
....*Saxifragaceae*, p. 151

t. Styles 5; fruit a 1-seeded achene or utricle ....*Plumbaginaceae*, p. 217

u. Style 1 v

u. Styles 2–5 ....z

v. Calyx tubular, the sepals united almost to the tips ....*Lythraceae*, p. 196

v. Calyx not tubular, the sepals distinct or united only near the base ...w

w. Sepals and petals 4 ....*Cruciferae*, p. 142

w. Sepals and petals 5 x

x. Fruit a 5-celled, many-seeded capsule; leaves reduced and scale-like
....*Pyrola*, p. 216

x. Fruit consisting of 5, 1- or several-seeded nutlets, each developing from a carpel; leaves well developed, deeply divided or pinnately compound ....y

y. Nutlet 1-seeded, not spiny; leaves divided ....*Limnanthaceae*, p. 186

y. Nutlet several-seeded, bony and spiny; leaves pinnate ..*Zygophyllaceae*, p. 184

z. Fruit consisting of 5, 1-seeded, tailed carpels ....*Geraniaceae*, p. 180

z. Fruit a capsule ....A

A. Leaves simple ....*Linaceae*, p. 183

A. Leaves 3-foliolate ....*Oxalidaceae*, p. 182

GROUP 9

(Polypetalous dicotyledons with inferior or partly inferior ovary)

a. Petals numerous; leaves fleshy, triangular in cross section ....
....*Mesembryanthemum*, p. 127

a. Petals 2–6; foliage leaves herbaceous or coriaceous (or more or less fleshy in *Portulaca*), not triangular-thickened ....b
b. Trees and shrubs (*Whipplea* a trailing undershrub) c
b. Herbs (*Zauschneria* in *Onagraceae* sometimes suffrutescent) ....h
c. Stamens more than 5 ....d
c. Stamens 4 or 5 ....f
d. Stamens numerous; leaves aromatic. Fruit a woody capsule .*Eucalyptus*, p. 196
d. Stamens 8 to about 20; leaves not aromatic ....e
e. Leaves alternate, with stipules; fruit a fleshy pome ....*Rosaceae*, p. 155
e. Leaves opposite, without stipules; fruit a capsule ....*Whipplea*, p. 153
f. Fruit a partly inferior capsule; petals hooded and long-clawed ....
....*Ceanothus*, p. 188
f. Fruit fleshy, entirely inferior; petals neither hooded nor long-clawed ..g
g. Leaves alternate; petals and stamens 5; fruit a berry ....*Ribes*, p. 153
g. Leaves opposite; petals and stamens 4; fruit a drupe ....*Cornaceae*, p. 210
h. Flowers in umbels i
h. Flowers not in umbels ....j
i. Styles 2; fruit not fleshy ....*Umbelliferae*, p. 202
i. Styles 5; fruit a berry ....*Araliaceae*, p. 202
j. Leaves roundish in outline, palmately veined and lobed ..*Saxifragaceae*, p. 151
j. Leaves neither roundish in outline nor palmately lobed ....k
k. Ovary partly inferior; fruit a circumscissile capsule; sepals 2 ....
....*Portulaca*, p. 128
k. Ovary entirely inferior; fruit not a circumscissile capsule; sepals 2–5, sometimes very small or obsolete l
l. Stamens very numerous; leaves with harsh hairs ....*Loasaceae*, p. 195
l. Stamens 2–10; leaves without harsh hairs ....m
m. Aquatic herbs with at least the submerged leaves in whorls and pinnately divided into narrow or capillary divisions ....*Myriophyllum*, p. 202
m. Plants usually not aquatic or if aquatic then the leaves not whorled or capillary-divided ....*Onagraceae*, p. 197

GROUP 10

(Gamopetalous dicotyledons with stamens more than 5)

a. Flower regular ....b
a. Flower irregular ....g
b. Stamens numerous c
b. Stamens 6–10 ....d
c. Filaments distinct or united at the base; style simple ....*Leguminosae*, p. 161
c. Filaments united into a tube; style branches several ....*Malvaceae*, p. 191
d. Pistils 5, becoming follicles in fruit; leaves generally fleshy-thickened
....*Crassulaceae*, p. 150
d. Pistil 1, becoming a capsule or berry; leaves not fleshy-thickened ...e
e. Styles 5; leaves 3-foliolate ....*Oxalidaceae*, p. 182
e. Style 1; leaves simple ....f
f. Foliage leaves whorled at the top of a slender stem; corolla rotate, pink ....*Trientalis*, p. 217

f. Foliage leaves not whorled; corolla tubular or urnshaped ...*Ericaceae*, p. 211

g. Sepals 2, distinct; petals 4, the 2 outer saccate at the base; stamens 6
....*Fumariaceae*, p. 141

g. Sepals 5, or, if reduced, more or less united; petals 3 or 5; stamens
8–10 ....h

h. Petals 5, the two lower partly joined to form a keel; stamens 10 ......
....*Leguminosae*, p. 161

h. Petals 3; stamens 8 ....*Polygalaceae*, p. 184

GROUP 11

(Gamopetalous dicotyledons with stamens 2 to 5 and superior ovary)

a. Stamens 5, free from the corolla ....b

a. Stamens 2–5, attached to the corolla tube c

b. Plants shrubby, the stems tall and leafy; corolla 3–5 cm. long, the segments united to above the middle ....*Rhododendron*, p. 215

b. Plants herbaceous, the stems short with the leaves in a basal tuft;
corolla 1 cm. long or less, the segments distinct nearly to the base
....*Plumbaginaceae*, p. 217

c. Stamens as many as the corolla lobes and attached opposite the lobes
....*Primulaceae*, p. 216

c. Stamens as many as the corolla lobes or fewer, alternating with the
lobes ....d

d. Pistils 2, united above by a common style or stigma; herbs with milky
juice ....e

d. Pistil 1; herbs without milky juice ....f

e. Stamens distinct, not united to the pistil ....*Apocynaceae*, p. 218

e. Stamens united and also more or less united to the pistil ....
....*Asclepiadaceae*, p. 219

f. Fruit at maturity breaking into 4, 1-seeded nutlets, the rupture sometimes retarded ....g

f. Fruit at maturity not breaking into 4, 1-seeded nutlets i

g. Corolla regular; stamens 5; stems terete; leaves alternate or opposite
....*Boraginaceae*, p. 229

g. Corolla generally more or less irregular; stamens 4; stems 4-sided; leaves
opposite ....h

h. Style apical, arising from the summit of the grooved or entire ovary;
herbage not fragrant ....*Verbenaceae*, p. 232

h. Style arising from a depression between the 4 lobes of the ovary; herbage generally fragrant ....*Labiatae*, p. 233

i. Style 3-lobed, -cleft, or -parted. Ovary usually 3-celled, rarely 1-celled
....*Polemoniaceae*, p. 222

i. Style simple or 2-lobed or -cleft, or sometimes 2 styles distinct to the
base ....j

j. Ovary 1-celled (or in the *Hydrophyllaceae* becoming 2-celled by the
meeting of the placentae in the center) ....k

j. Ovary 2-celled (in *Limosella* in the *Scrophulariaceae*, 2-celled only
below the middle) ....n

k. Corolla regular 1

k. Corolla irregular, 2-lipped ....m

l. Calyx toothed to deeply lobed, not divided nearly to the base; leaves entire, glabrous ....*Gentianaceae,* p. 218

l. Calyx divided nearly to the base; leaves generally toothed, lobed, or divided, hairy or glandular (except in *Romanzoffia*) ....
....*Hydrophyllaceae,* p. 226

m. Stamens 4; parasitic plants without green leaves ......*Orobanchaceae,* p. 249

m. Stamens 2; aquatic plants with green leaves ....*Lentibulariaceae,* p. 249

n. Stamens with anthers 5 ....o

n. Stamens with anthers 2 or 4 ....q

o. Corolla a little irregular, rotate; filaments woolly. Inflorescence spicate or racemose ....*Verbascum,* p. 239

o. Corolla regular, bowlshaped, funnelform, salverform, or subrotate; filaments glabrous or hairy, not woolly ....p

p. Calyx deeply divided nearly to the base; stems frequently twining or trailing; fruit a capsule ....*Convolvulaceae,* p. 219

p. Calyx toothed or shallowly lobed; stems generally erect, rarely trailing; fruit a capsule or berry ....*Solanaceae,* p. 237

q. Corolla more or less irregular; capsule not circumscissile ....
....*Scrophulariaceae,* p. 238

q. Corolla regular; capsule circumscissile ....*Plantaginaceae,* p. 250

GROUP 12

(Gamopetalous dicotyledons with stamens 2–5 and inferior ovary)

a. Flowers few to many in a head subtended by involucral bracts, or rarely the heads only 1-flowered; fruit an achene ....b

a. Flowers not in a head; fruit not an achene (or in the *Valerianaceae* 1-seeded and achene-like) c

b. Stamens free. Flowers perfect ....*Dipsacaceae,* p. 255

b. Stamens united by the anthers, or rarely if the anthers are free, then all or some of the flowers in each head unisexual ......*Compositae,* p. 258

c. Leaves opposite or whorled ....d

c. Leaves alternate ....f

d. Fruit consisting of 2, 1-seeded carpels; leaves whorled .....*Rubiaceae,* p. 252

d. Fruit fleshy and a berry or berry-like drupe, or dry and achene-like; leaves opposite ....e

e. Fruit fleshy, 2- to several-seeded; plants woody ....*Caprifoliaceae,* p. 254

e. Fruit dry, 1-seeded; plants herbaceous ....*Valerianaceae,* p. 255

f. Calyx limb obsolete; corolla regular, rotate; leaves palmately veined and lobed ....*Cucurbitaceae,* p. 257

f. Calyx limb evident or conspicuous; corolla not rotate; leaves not palmate ....g

g. Corolla regular; stamens distinct ....*Campanulaceae,* p. 257

g. Corolla irregular; stamens united by the anthers ....*Lobeliaceae,* p. 258

POLYPODIACEAE. FERN FAMILY

a. Sori dorsal, not on the margin of fronds ....b

a. Sori marginal or submarginal, covered by the more or less revolute leaf edge h

b. Indusium present, sometimes inconspicuous. Fronds 1–3-pinnate, not golden beneath c

b. Indusium none ....g

c. Sori oblong or linear, more than 1 mm. long, parallel to the midrib of pinnules. Fronds evergreen ....5. *Woodwardia*

c. Sori at maturity nearly round or appearing so (if oblongish in *Dryopteris* mostly less than 1 mm. long and oblique to the midrib of pinnules) ......d

d. Stipe scaly only near the base or sometimes with a few scattered scales above; indusium ciliate, inconspicuous or fragile; fronds annual ....e

d. Stipe scaly up to the blade, the rachis also more or less scaly; indusium conspicuous, not fragile; fronds generally evergreen (except perhaps in *Dryopteris dilatata*) ....f

e. Blades 1- or 2-pinnate, mostly less than 3 dm. long; indusium attached at base, hoodlike ....1. *Cystopteris*

e. Blades 2- or 3-pinnate, generally more than 3 dm. long; indusium attached along one side ....2. *Athyrium*

f. Indusium round-reniform; blades 2- or 3-pinnate ....3. *Dryopteris*

f. Indusium peltate; blades 1- or 2-pinnate ....4. *Polystichum*

g. Blades not golden beneath, pinnately parted nearly or quite to the rachis, the segments serrate; sori round ....6. *Polypodium*

g. Blades golden or yellowish beneath, pinnate, the lowest pair of pinnae again pinnate; sori oblong or linear along veins. ....7. *Pityrogramma*

h. Fronds pinnate or pinnatisect, dimorphic, the segments of the fertile frond narrowly linear, of the sterile frond oblong. ....8. *Blechnum*

h. Fronds 2- or 3-pinnate, the fertile and sterile nearly or quite alike (or somewhat dimorphic in *Cheilanthes siliquosa*) i

i. Segments finely and palmately veined (sometimes obliquely so), without a distinct midvein; reflexed margin usually not continuous ....9. *Adiantum*

i. Segments with a distinct, pinnately branched midvein or the venation obscured by hairs or scales; reflexed margin mostly continuous (except in *Cheilanthes californica*) ....j

j. Plants large and coarse, fronds arising singly from stout elongate rootstocks, mostly annual, usually more than 3 dm. long, blades pubescent below and sometimes above ....10. *Pteridium*

j. Plants small, fronds clustered (or arising singly from slender elongate rootstocks in *Pellaea andromedaefolia*), evergreen, mostly less than 3 dm. long, or if longer, the blades glabrous ....k

k. Indusium present, or if lacking, the lower surface of the segments hairy or scaly ....11. *Cheilanthes*

k. Indusium lacking, the sporangia covered by the reflexed leaf margin; blades glabrous ....12. *Pellaea*

[49]

1. Cystopteris

1. **C. fragilis** (L.) Bernh. BRITTLE FERN. Occasional on rocky stream banks in shade or partial shade: Mill Valley, acc. Stacey; Cataract Gulch, Mount Tamalpais; Lagunitas Canyon; Little Carson; San Anselmo Canyon.—*Filix fragilis* (L.) Gilib.

2. Athyrium

1. **A. Filix-femina** (L.) Roth. COMMON LADY FERN. Occasional along streams in wooded canyons or among sedges and rushes in open swamps near the coast, from Sausalito and Mount Tamalpais to Point Reyes Peninsula and Tomales.

Most of the Marin County plants of this variable species belong to var. *sitchense* Rupr., but the peculiar form from the open coastal marshes, in which the pinnules are strongly revolute, is var. *sitchense* Rupr. forma *strictum* (Gilbert) Butters.

3. Dryopteris

a. Fronds 2-pinnate, the pinnules sometimes pinnatifid . . . . .1. *D. arguta*
a. Fronds 3-pinnate . . . . .2. *D. dilatata*

1. **D. arguta** (Kaulf.) Watt. COASTAL WOOD FERN. Abundant and widespread on wooded canyonsides throughout our region except in dense redwood forests, from Sausalito, Tiburon, and Mount Tamalpais to San Rafael Hills, Tomales, and Point Reyes Peninsula.—*Aspidium rigidum* Sw. var. *argutum* (Kaulf.) D. C. Eat.

2. **D. dilatata** (Hoffm.) Gray. SPREADING WOOD FERN. Rare and local in moist coastal woods: Steep Ravine, Mount Tamalpais, *Leschke;* Bear Valley, Point Reyes Peninsula.—*Aspidium spinulosum* (Muell.) Sw. var. *dilatatum* Hoffm.

The spreading wood fern may be distinguished from the lady fern, which it resembles superficially, by the broad base of the leaf blade.

4. Polystichum. SHIELD FERN

a. Pinnae simple, serrate, auriculate along upper margin at base . . . . .b
a. Pinnae more or less pinnately or bipinnately parted c
b. Fronds elongate, the pinnae spreading, not directed forward and not imbricate
. . . . .1. *P. munitum*
b. Fronds shorter, the pinnae crowded, directed forward and obliquely imbricate
. . . . .1*a. P. munitum* var. *imbricans*
c. Pinnae pinnately lobed or divided . . . . .2. *P. californicum*
c. Pinnae pinnately parted to pinnate, the lowest pinnules again cleft or divided
. . . . .3. *P. Dudleyi*

1. **P. munitum** (Kaulf.) Presl. WESTERN SWORD FERN. Wooded hills and canyons in shade or partial shade, widespread and common, especially luxuriant in coastal redwood canyons: Angel Island; Sausalito; Tiburon; Mount Tamalpais; San Rafael Hills; Point Reyes Peninsula; Tomales.

1*a.* **P. munitum** (Kaulf.) Presl var. **imbricans** (D. C. Eat.) Maxon. ROCK SWORD FERN. Confined to exposed rocks about the summit of Mount Tamalpais, as along the Northside Trail. This variety usually occurs at much greater elevations and the Tamalpais plant may be only a form of the species growing under unfavorable conditions.

2. **P. californicum** (D. C. Eat.) Diels. CALIFORNIA SHIELD FERN. Deep shaded canyons in wet places adjacent to cascades and streams, rare: Big Carson; Fairfax

Hills; Mount Tamalpais (Swede George Gulch, *Leschke*, above Phoenix Lake); hills southwest of Mill Valley, *Cannon.—P. aculeatum* (Sw.) Roth var. *californicum* (D. C. Eat.) Jeps.

In Big Carson Canyon near the falls and on Mount Tamalpais above Phoenix Lake, plants have been seen which are intermediate between *P. californicum* and *P. munitum* and it is suspected that they may be of hybrid origin.

3. **P. Dudleyi** Maxon. DUDLEY SHIELD FERN. Rare in deep shade on moist slopes of Mount Tamalpais: canyon west of Ross; above Muir Woods, *Leschke*. These are the only Marin County stations known for this rare fern, which is otherwise confined to canyons of the Santa Cruz Mountains. Under the name *Aspidium aculeatum*, the Dudley shield fern was reported from Marin County by M. E. Parsons in 1891.—*P. aculeatum* (Sw.) Roth var. *Dudleyi* (Maxon) Jeps.

5. Woodwardia

1. **W. fimbriata** Smith. WESTERN CHAIN FERN. Common around springs, in rocky streambeds, or on moist alluvial flats, in shade or sun: Sausalito; Tiburon Peninsula; Mount Tamalpais; Dillons Beach; Point Reyes Peninsula.—*W. Chamissoi* Brack.; *W. radicans* of authors for the plant of western North America.

Some of the largest and most luxuriant specimens of the western Woodwardia that have been seen form ferny thickets in deep shade at Camp Hogan on the north side of Mount Tamalpais, where the tall clustered fronds well exceed the height of a man. Maxon has given the maximum length of fronds as 3 meters.

6. Polypodium. POLYPODY

a. Blades coriaceous or leathery; rhizome glaucous . . . . .1. *P. Scouleri*
a. Blades not leathery; rhizome not glaucous . . . . .b
b. Segments oblong, acute or obtuse . . . . .2. *P. californicum*
b. Segments deltoid-linear, elongate, attenuate-acute . . . . .3. *P. Glycyrrhiza*

1. **P. Scouleri** Hook. & Grev. LEATHER FERN. Occasional near the coast, in crevices of exposed rocks or in moss on the trunks of trees: Sausalito; Steep Ravine, Mount Tamalpais; Point Reyes; Dillons Beach.

2. **P. californicum** Kaulf. CALIFORNIA POLYPODY. Common and widespread on rocks in sheltered shaded canyons or on open exposed hills, varying greatly in size, shape, and texture of blade: Angel Island, Sausalito, and Tiburon to Mount Tamalpais, San Rafael, and Tomales. The more coriaceous specimens with shorter broader blades from maritime habitats, as from exposed rocks at Point Reyes and Tomales, are referable to var. *Kaulfussii* D. C. Eat. (plate 11).

3. **P. Glycyrrhiza** D. C. Eat. LICORICE FERN. Occasional plants on mossy rocks and logs in deep moist canyons near streams and cascades: Mount Tamalpais (Bootjack Canyon above Muir Woods, Cataract Gulch); Lagunitas Canyon; Big Carson.

7. Pityrogramma

1. **P. triangularis** (Kaulf.) Maxon. GOLDBACK FERN. Common on rocky summer-dried slopes in partial shade of shrubs or trees: Angel Island; Sausalito; Tiburon; Mount Tamalpais; San Rafael Hills; Tomales; Point Reyes Peninsula.—*Gymnogramma triangulare* Kaulf.

Unusually large specimens in the San Rafael Hills become as much as 18 inches tall.

8. Blechnum

1. **B. Spicant** (L.) Roth. DEER FERN. Occasional in wet sheltered places near springs or swamps: Sausalito; Tennessee Valley; Bolinas Ridge; Bear Valley and Ledum Swamp, Point Reyes Peninsula.—*Struthiopteris Spicant* (L.) Weis; *Lomaria Spicant* (L.) Desv.

9. Adiantum. MAIDENHAIR FERN

a. Stipe and rachis simple, the stipe not forked ....1. *A. Jordani*
a. Stipe forked, the blade with two equal spreading divisions ....2. *A. pedatum*

1. **A. Jordani** C. Müll. CALIFORNIA MAIDENHAIR. Common on shaded rocky slopes under bushes or trees: Sausalito, Tiburon, and Mount Tamalpais (Blithe-dale Canyon, Bootjack, Cataract Gulch, Phoenix Lake) to San Rafael Hills and Tomales.—*A. emarginatum* D. C. Eat., not Bory.

California maidenhair is especially abundant and lovely above Fairfax in San Anselmo Canyon where rocky walls above the stream are covered by its delicate fronds from late winter to the end of spring.

2. **A. pedatum** L. var. **aleuticum** Rupr. WESTERN FIVEFINGER. Rare in moist shaded canyons: Sausalito; above Muir Woods, *Leschke,* and Steep Ravine, Mount Tamalpais; Big Carson, *Leschke;* Bolinas Ridge; Lagunitas Canyon, acc. M. E. Parsons; Inverness, acc. Stacey.

10. Pteridium. BRAKE FERN; BRACKEN

1. **P. aquilinum** (L.) Kuhn var. **pubescens** Underw. WESTERN BRACKEN. On grassy hills, in brushy canyons, and in open woods, probably the most abundant and widespread fern in Marin County: Angel Island, Sausalito, and Tiburon to San Rafael, Tomales, and Point Reyes Peninsula.—*Pteris aquilina* L. var. *lanuginosa* Bong.

On dry windswept hills the fronds are scarcely more than 6 inches tall, almost hidden among grasses; but in alluvial soil on moist canyon bottoms the fronds may become 4 to 6 feet tall as they do on Point Reyes Peninsula.

11. Cheilanthes. LACE FERN; LIP FERN

a. Blades glabrous; indusium present ....b
a. Blades scaly or hairy below; indusium inconspicuous or lacking c
b. Sori interrupted, between pairs of marginal teeth. ....1. *C. californica*
b. Sori contiguous and confluent on the margin of entire or subentire pinnules
....2. *C. siliquosa*
c. Fronds mostly 2-pinnate, segments oblongish, tomentose beneath ....
....3. *C. gracillima*
c. Fronds mostly 3-pinnate, segments roundish, scaly beneath ..4. *C. intertexta*

1. **C. californica** (Hook.) Mett. CALIFORNIA LACE FERN. Rocky slopes of canyons, rare: above Muir Woods and near Mountain Theater, Mount Tamalpais; Fairfax Hills, *Menzies;* Carson country.

2. **C. siliquosa** Maxon. SERPENTINE FERN. Locally common in serpentine areas around rocks and in protected spots: Mount Tamalpais (West Peak, Laurel Dell, Rifle Camp, Matt Davis Trail); Carson Ridge; San Geronimo Ridge.—*Pellaea densa* (Brack.) Hook.

Around serpentine rocks on Tiburon Peninsula, a form grows in which the fronds are more dissected and the sori are not continuous. Since the sori are chiefly along the sides of the segments rather than in sinuses formed by marginal teeth, the Tiburon plant is referred to *C. siliquosa* rather than to *C. californica*.

Cheilanthes siliquosa is widely distributed in western North America but only in central California is it nearly restricted to serpentine. On Mount Tamalpais it has been seen outside the serpentine areas at only a single station, where it grew in the crevices of sandstone above Cascade Canyon on the south side of the mountain.

3. **C. gracillima** D. C. Eat. LACE FERN. Occasional in protected crevices of exposed rocks along the crest of Mount Tamalpais (Corte Madera Ridge, East Peak, Fern Canyon, Northside Trail, Rock Spring). This fern, which is characteristic of rocky ledges at middle elevations in the Sierra Nevada, is not uncommon on the higher peaks of the North Coast Ranges, but to the south specimens have been seen only from the Santa Lucia Mountains.

4. **C. intertexta** (Maxon) Maxon. COASTAL LIP FERN. Rare in crevices of protecting rocks: Mount Tamalpais (Matt Davis Trail, Mountain Theater, Throckmorton Ridge, Northside Trail); Carson country; San Rafael Hills.—*C. Covillei* Maxon var. *intertexta* Maxon.

12. **Pellaea.** CLIFF BRAKE

a. Segments roundish or elliptic, obtuse . . . . .1. *P. andromedaefolia*
a. Segments narrowly oblong or linear-oblong, mucronate . . . . . .2. *P. mucronata*

1. **P. andromedaefolia** (Kaulf.) Fée. COFFEE FERN. Occasional on dry rocky slopes under brush: Tiburon; south side of Mount Tamalpais; Carson country; San Rafael Hills; Tomales.

2. **P. mucronata** (D. C. Eat.) D. C. Eat. BIRDFOOT FERN. Occasional on hot rocky slopes in chaparral or opens in oak woodland: Throckmorton and Corte Madera ridges, Mount Tamalpais; San Rafael Hills; Carson country; Black Point. —*P. Ornithopus* Hook.

OPHIOGLOSSACEAE. ADDER'S TONGUE FAMILY
Botrychium. GRAPEFERN

1. **B. silaifolium** Presl. LEATHERY GRAPEFERN. In Marin County known only on the north slope of Mount Tamalpais at Willow Meadow, where it grows in partial shade in deep moist soil, and on Point Reyes Peninsula near the radio station, where it borders a swale among sedges and rushes.

ISOETACEAE. QUILLWORT FAMILY
Isoetes. QUILLWORT

a. Plants terrestrial; corm 3-lobed . . . . .1. *I. Nuttallii*
a. Plants amphibious; corm 2-lobed . . . . .2. *I. Howellii*

1. **I. Nuttallii** A. Br. Wet or moist soil in grassy meadows or along ephemeral streamlets: north side of Mount Tamalpais (Rock Spring, Potrero Meadow, Lagunitas Meadows); Little Carson Canyon; Corte Madera; San Rafael.

2. **I. Howellii** Engelm. In shallow ponds which dry during the summer: Willow Meadow and Hidden Lake, Mount Tamalpais; small pond south of Olema,

MARSILEACEAE. Marsilea Family
Pilularia

1. **P. americana** A. Br. The American pillwort has been reported from Marin County but I have seen no specimens from there. It is found occasionally in the interior valleys of California where, in drying rainpools, its many short, slender, grasslike leaves sometimes form limited turflike patches.

SALVINIACEAE. Water Fern Family
Azolla. Water Fern

1. **A. filiculoides** Lamk. American Water Fern; Mosquito Fern. Rare in Marin County on quiet pools: Rodeo Lagoon; Sausalito; Lily Lake; Olema Marshes; near Aurora School; near Abbotts Lagoon, Point Reyes Peninsula.

EQUISETACEAE. Horsetail Family
Equisetum. Horsetail; Scouring Rush

a. Aerial stems annual, of two kinds, the fertile unbranched and whitish or brownish, the sterile green, bearing numerous whorled branches . . . . b
a. Aerial stems of one kind, green, rarely branching, annual or perennial c
b. Sterile stems slender, generally 5 mm. or less in diameter; branches 3- or 4-ribbed, the ribs not longitudinally grooved, bearing one row of low siliceous tubercles; spikes 1.5–3 cm. long . . . . 1. *E. arvense*
b. Sterile stems stouter and taller, 5–20 mm. in diameter; branches 4–6-ribbed, the ribs longitudinally grooved, bearing two rows of prominent siliceous tubercles; spikes generally more than 3 cm. long . . . . 2. *E. Telmateia*
c. Aerial stems annual, slender, the ribs smooth or with fine transverse siliceous bands; spikes obtuse . . . . 3. *E. kansanum*
c. Aerial stems perennial, frequently tall and robust, the ribs with siliceous bands or tubercles; spikes tipped with a short sharp point . . . . d
d. Ribs of stems with siliceous tubercles generally arranged in two rows
. . . . 4a. *E. hyemale* var. *californicum*
d. Ribs of stems with prominent transverse siliceous bands
. . . . 4b. *E. hyemale* var. *robustum*

1. **E. arvense** L. Common Horsetail. Occasional in moist or wet soil on the edge of thickets and along streams, sometimes ruderal: Rodeo Lagoon, Tiburon, and Mount Tamalpais (Laurel Dell, Lagunitas Creek) to the Big Carson, Tomales, and Point Reyes Peninsula (Bolinas, Inverness, Ledum Swamp).

Generally the branches are simple but occasionally plants are found with secondary branches (var. *ramulosum* Rupr.) : Mill Valley; Dillons Beach; Point Reyes Peninsula.

2. **E. Telmateia** Ehrh. var. **Braunii** (Milde) Milde. Giant Horsetail. Common and widespread in deep moist soil in woodlands and in brushy places near streams: Angel Island; Sausalito; Tiburon; Mount Tamalpais; Tomales; Inverness Ridge.

Although the green horsetails are almost always sterile, green branching stems bearing fruiting spikes have been found in Lagunitas Canyon.

3. **E. kansanum** Schaffn. Summer Scouring Rush. In Marin County known only

from Lucas Valley (*Leschke*) and from Little Carson Canyon, where it grew on the brushy side of a steep dry gully.

4*a*. **E. hyemale** L. var. **californicum** Milde. CALIFORNIA SCOURING RUSH. In Marin County definitely known only on moist coastal flats under red alders near Muir Beach and along a stream near Tomales, but probably occurring elsewhere in similar situations near the coast.

4*b*. **E. hyemale** L. var. **robustum** (A. Br.) A. A. Eat. GIANT SCOURING RUSH. Plants apparently referable to this variety have been found on sandy downs of Point Reyes Peninsula near the wireless station and on the edge of woods near Bolinas.—*E. praealtum* Raf.

SELAGINELLACEAE. MOSS FERN FAMILY

Selaginella

1. **S. Wallacei** Hieron. WALLACE MOSS FERN. Widely distributed but not common, on shaded or sunny rocks near the limits of redwood groves: ridge east of Muir Woods; Mount Tamalpais (Bootjack, Nora Trail, Cataract Gulch, Northside Trail, Lake Lagunitas, Alpine Lake); Carson country; Fairfax Hills; San Rafael. The Wallace moss fern, ranging southward from British Columbia and Montana, reaches its southernmost station in Marin County.

TAXACEAE. YEW FAMILY

a. Leaves sharply prickle-pointed; fruit greenish, drupelike ....1. *Torreya*
a. Leaves not prickle-pointed; seed attached to a reddish fleshy cup ..2. *Taxus*

1. Torreya

1. **T. californica** Torr. CALIFORNIA NUTMEG. Well-developed specimens of the California nutmeg in Marin County become medium-sized trees up to 60 feet tall with a straight clean trunk 4 feet or more in circumference and covered with a rather thin, shreddy, grayish brown bark. On Mount Tamalpais large trees are rare; a few grow on the Ocean View Trail above Muir Woods, and others are found in the deep well-watered canyons on the north side of the mountain. In Cataract Gulch a large fallen tree has a length of 86 feet and a trunk diameter of 3 feet at the ground and 1 foot 10 inches at 3 feet above the ground. Smaller trees and shrubby forms of the species, however, are rather common on shaded slopes in the belt between redwood pockets and the chaparral. North of Mount Tamalpais in Marin County the California nutmeg has been seen on Bolinas Ridge and along Lagunitas Creek near Camp Taylor.

Besides the endemic California tree, five other torreyas are known, one from the southeastern United States and four from China and Japan. These present-day species of *Torreya*, which are now so widely separated, are probably derived from forms that were part of a widespread circumpolar flora of high latitudes before the Pleistocene. With the advent of the Ice Age this extensive flora migrated far to the south in Europe, Asia, and North America, where its modern representatives survive as isolated relicts. Two other gymnosperms which today occur in Marin County and which have probably been derived from that ancient boreal flora are the redwood and Sargent cypress.

The common name, California nutmeg, is derived from the nutmeg-like appearance of the inside of the seed.

2. **Taxus.** YEW

1. **T. brevifolia** Nutt. WESTERN YEW. In Marin County known only from the ridge between Olema and Bear Valley, where it was discovered in 1932 by Robert H. Menzies. Widely distributed in western North America, the western yew occurs as far north as Alaska, but south of Marin County it is known only from the Santa Cruz Mountains.

TAXODIACEAE. REDWOOD FAMILY

1. **Sequoia**

1. **S. sempervirens** (Lamb.) Endl. REDWOOD (see plate 1). In Marin County the redwood occupies a broad, island-like area with Mount Tamalpais in the south central part. To the south the trees range only to Frank Valley below Muir Woods with a wide gap beyond to Kings Mountain in San Mateo County. To the north there are scattered groves and pockets as far as Nicasio; beyond that point none occurs until the Russian River country in Sonoma County is reached. To the east redwood pockets occur in the San Rafael Hills, with only a limited occurrence of the tree in the Oakland Hills across San Francisco Bay to the southeast. To the west the San Andreas Fault marks the distributional limit. The absence of the redwood from such well-marked geographic areas as the Sausalito Hills, Tiburon Peninsula, and Point Reyes Peninsula has undoubtedly resulted from the different geologic history of these areas or from their inaccessibility when the redwood was dispersed through the region. Other variations in the floras of restricted districts of Marin County provide further evidence of different geologic and floristic histories of contiguous parts.

The variation in size of redwoods on Mount Tamalpais is extreme and ranges from the impressively large trees in Muir Woods to stunted shrubby trees in the chaparral on exposed mountain slopes. Generally these dwarfs are found along shallow drainage channels where the deep redwood canyons of the lower slopes finger out against the higher ridges. Sometimes on dry elevated flats these stunted redwoods develop pygmy groves in which individuals with slender canelike trunks are so numerous that they form almost impenetrable thickets. The ability of the redwood to stump-sprout following fire or injury undoubtedly has much to do with the development of these dense thickets in the chaparral.

Regeneration from the stump of a redwood may give rise to a number of nearly equal-sized individuals and, after the stump of the parent tree has disappeared, these trees form a larger or smaller redwood circle, a feature not uncommon in the canyons in Marin County. The large circle near the mill in Mill Valley must be very ancient since the present trees represent at least the second regeneration around the base of the original forest giant.

Branchlets of the redwood are of two kinds, those that remain part of the tree for years and form the trunk and branches, and those that are deciduous after a few seasons. The leaves of these two types of branches are generally quite unlike. On the short deciduous branchlets, the leaves spread horizontally to either side and the branchlets are frequently grouped to form flattened leafy sprays. On the persistent branchlets the leaves may spread more or less if in a shaded or protected position, but in exposed places, as about the crown of the tree, the leaves are scalelike and appressed to the branchlet. So unlike in appearance are the extreme forms of these two kinds of branchlets that they are not infrequently mistaken by the casual observer for branchlets from trees of different kinds.

PINACEAE. PINE FAMILY

a. Foliage leaves two in a cluster, the base of the cluster enveloped by scarious bractlike leaves ....1. *Pinus*

a. Foliage leaves solitary without accessory bracts ....2. *Pseudotsuga*

1. Pinus. PINE

1. **P. muricata** D. Don. BISHOP PINE; PRICKLE-CONE PINE (plates 12, 13). In Marin County, the bishop pine forms picturesque groves on the granitic ridge west of Inverness on Point Reyes Peninsula and it also occurs sparsely in the chaparral of the Carson country, as on Pine Mountain and on San Geronimo Ridge. North of Marin County, the species is found along the California coast in Sonoma, Mendocino, and Humboldt counties, and southward it occurs in a highly disrupted range as far as Cedros Island and the adjacent mainland of middle Lower California. In favorable localities, the gray-barked trunks grow tall and are either branchless below a broad flattened crown in close forests or bear branches nearly to the ground in open places. On exposed ridges or in the chaparral, the trees are compact or dwarfed and mature individuals may bear cones when only 5 or 6 feet tall.

The bishop pine belongs to a small group of California pines in which the cones may remain closed and attached to the trunk and branches for many years. Only after fire has swept through a grove of these closed-cone pines do the cones generally open and thus plant a new forest in the ashes of the old. For this reason the trees in a given part of a bishop pine forest are mostly uniform in size and age.

Not only is the bishop pine one of the closed-cone pines characteristic of the pine forest of the California coast today, but from the fossil record it is known to have been part of a similar plant formation since the Pleistocene.

2. Pseudotsuga

1. **P. taxifolia** (Lamb.) Britt. DOUGLAS FIR (plates 2, 3). In Marin County the Douglas fir is a characteristic part of the woodland bordering redwood groves and pockets, its usual associates being tanbark oak, California laurel, and madroño. It is common and widespread on Mount Tamalpais and in the hill country to the north but it is not known from the Sausalito, Tiburon, or San Rafael hills. In the Black Forest on Inverness Ridge southwest of Olema, Douglas fir is the dominant tree and the forest is comparable in development to the fir forests in Oregon and Washington. At the west end of Mount Tamalpais, fir saplings that were probably seeded during the relatively wet years late in the 1930's are very numerous in a border between the grassland and the oak-madroño groves, the saplings frequently being abundant enough to form dense evergreen thickets. With the aging of these firs, this Tamalpais woodland will probably become quite like the Black Forest of Inverness Ridge, with the Douglas fir the dominant forest tree.—*P. mucronata* (Raf.) Sudw.

CUPRESSACEAE. CYPRESS FAMILY

1. Cupressus. CYPRESS

1. **C. Sargentii** Jeps. SARGENT CYPRESS (plate 14). This conifer, like the bishop pine, is an indicator of another plant island found in Marin County, but whereas

the pine is nearly confined to the area west of the San Andreas Fault, the cypress is known only to the east of the fault. To the north the nearest station where the cypress again occurs is in Sonoma County near Occidental, and to the southeast a grove of the cypress grows on Cedar Mountain beyond Livermore in Alameda County. The Sargent cypress is a California endemic, ranging from Mendocino County on the north to San Luis Obispo County on the south, and it occurs chiefly on serpentine.

In Marin County, the Sargent cypress presents two distinct aspects. On Carson Ridge it forms a remarkable pygmy forest which matures when only a few feet tall, the numerous stunted individuals forming distinctive open groves or dense thickets. On the northwestern slope of Mount Tamalpais, where it is not uncommon, the cypress becomes a well-developed tree up to 50 feet tall, commonly with one or more broad spreading branches near the ground. These gray-green trees blend with the gray-green rock of the serpentine barrens to form a picturesque and memorable part of the Mount Tamalpais scene.

TYPHACEAE. CAT-TAIL FAMILY

1. **Typha.** CAT-TAIL

a. Spike usually continuous with the staminate and pistillate parts adjoining; compound pedicels in mature pistillate spikes bristle-like, 1.5–3.5 mm. long ....4. *T. latifolia*

a. Spike usually interrupted with the staminate part separated from the pistillate; compound pedicels in mature pistillate spikes papillate to papillate-bristly, 0.5–1.2 mm. long ....b

b. Pistillate spikes light brown at maturity; bracts numerous, with ovate, apiculate blades. Stigmas linear ....2. *T. domingensis*

b. Pistillate spikes dark brown at maturity; bracts few or numerous, with spatulate, blunt blades c

c. Stigmas linear, sometimes early deciduous; bracts numerous; compound pedicels papillate, 0.5–0.7 mm. long ....1. *T. angustifolia*

c. Stigmas lanceolate-linear, usually persistent; bracts very few; compound pedicels papillate to bristly, 0.6–1.2 mm. long ....3. *T. glauca*

1. **T. angustifolia** L. Locally common in fresh or brackish water: Tiburon; Mill Valley; Phoenix Lake, Mount Tamalpais. Narrow-leaved cat-tails have also been seen at Corte Madera and in the San Rafael Hills, but they may belong to the following species.

2. **T. domingensis** Pers. In Marin County definitely known only from Phoenix Lake, Mount Tamalpais, and from marshes bordering the bay at Burdell but perhaps more widely distributed.—*T. truxillensis* H.B.K.; *T. angustifolia* of most California references.

3. **T. glauca** Godron. In Marin County known only from Phoenix Lake, Mount Tamalpais, and from specimens collected by Hans Leschke in low ground east of San Rafael. These collections were identified by Neil Hotchkiss who has supplied the data by which the four species are distinguished in the foregoing key. According to him, Marin County is the only area in the West where the four are known to occur.

4. **T. latifolia** L. Common in fresh or salt water of lakes and marshes: Rodeo Lagoon; Tiburon; Mill Valley; Mount Tamalpais (Willow Meadow, Hidden

Lake, Phoenix Lake); Stinson Beach; Lily Lake; Burdell; Olema Marshes; Drakes Estero, Point Reyes Peninsula.

SPARGANIACEAE. BUR-REED FAMILY

1. Sparganium. BUR-REED

1. **S. Greenei** Morong. Fresh water marshes: Rodeo Lagoon; Stinson Beach; Olema, the type locality; the laguna, Chileno Valley.—*S. eurycarpum* Engelm. var. *Greenei* (Morong) Graebner.

POTAMOGETONACEAE. PONDWEED FAMILY

a. Leaves opposite (or appearing whorled); flowers represented by single stamens and pistils subcapitately congested in the axils of leaves; peduncle very short or none. Stipules free from blades and generally evanescent 3. *Zannichellia*
a. Leaves alternate (or the uppermost opposite); flowers perfect, on elongate peduncles . . . . .b
b. Stipules free from the leaf blades or if partly attached, the end prolonged as a free hyaline scale; peduncle straight .1. *Potamogeton*
b. Stipules attached to the leaf blades, the free upper part forming 2 short rounded auricles; peduncles spirally coiled in fruit . . . . .2. *Ruppia*

1. Potamogeton. PONDWEED

a. Leaves both submerged and floating, the blades of the floating leaves elliptic
. . . . .1. *P. nodosus*
a. Leaves submerged, linear . . . . .b
b. Stipules partly adnate to the leaf blades, free only at the upper end
. . . . .2. *P. pectinatus*
b. Stipules free from the leaf blades c
c. Peduncles filiform, more than 1 cm. long; leaves usually with a pair of small glands at base. . . . . .3. *P. panormitanus*
c. Peduncles a little clavate, usually 1 cm. long or less; leaves without glands at base . . . . .4. *P. foliosus*

1. **P. nodosus** Poir. Ponds and lakes: Phoenix Lake, Mount Tamalpais; Lily Lake; Point Reyes Station; east of Aurora School; laguna in Chileno Valley; Dillons Beach.— *P. americanus* C. & S.

2. **P. pectinatus** L. In quiet pools and lakes: Phoenix Lake and Lake Lagunitas, Mount Tamalpais; Inverness.

3. **P. panormitanus** Biv. var. **minor** Biv. In Marin County known only from Phoenix Lake, Mount Tamalpais, *Jussel*. Var. *major* Biv. with leaves 1–3 mm. wide (instead of 1 mm. wide or less) is to be expected since it has been found in San Francisco.

4. **P. foliosus** Raf. var. **macellus** Fern. Fresh or slightly brackish water of ponds and streams: Stinson Beach; Dillons Beach; near Abbotts Lagoon, Point Reyes Peninsula; San Antonio Creek.

2. Ruppia

1. **R. maritima** L. DITCH GRASS; WIGEON GRASS. Brackish and saline water, especially common in tidal sloughs of salt marshes: Tiburon; Escalle; Black Point;

Muir Beach; Drakes Estero; Inverness. Sterile specimens of *R. maritima* and *Potamogeton pectinatus* resemble each other closely but they may be readily distinguished by the character of the stipules.

3. Zannichellia

1. **Z. palustris** L. HORNED PONDWEED. In streams and ponds of fresh or brackish water: San Antonio Creek; near Dillons Beach.

NAJADACEAE. NAJAS FAMILY

1. Najas

1. **N. guadalupensis** (Spreng.) Morong. In Marin County known only from the freshwater pond behind the beach at Stinson Beach.

ZOSTERACEAE. EELGRASS FAMILY

a. Plants monoecious, flowers unisexual; leaves generally more than 5 mm. wide, usually 5- or 7-nerved ....1. *Zostera*
a. Plants dioecious; leaves generally less than 5 mm. wide, usually 1- or 3-nerved ....2. *Phyllospadix*

1. Zostera. EELGRASS

1. **Z. marina** L. var. **latifolia** Morong. On sandy or muddy bottoms of shallow bays and inlets along the ocean: Bolinas Bay; Tomales Bay; Dillons Beach.

2. Phyllospadix. SURFGRASS

a. Leaves flat and thin, the ventral side of sheath and auricles spongy; peduncle 1–6 cm. long ....1. *P. Scouleri*
a. Leaves usually longer and narrower, thicker and elliptic in cross section, the ventral side of sheath and auricles coriaceous; peduncle more than 1 dm. long ....2. *P. Torreyi*

1. **P. Scouleri** Hook. On rocks, in surf of exposed ocean shores: Dillons Beach, acc. Stacey.

2. **P. Torreyi** Wats. In surf on reefs and wave-cut terraces of rocky ocean shores: Muir Beach; Stinson Beach; Tomales Bay.

LILAEACEAE. FLOWERING QUILLWORT FAMILY

1. Lilaea. FLOWERING QUILLWORT

1. **L. scillioides** (Poir.) Haum. Low places in fields and along roads where water stands in the spring, or rarely in shallow ponds: Tiburon; Mill Valley; Lagunitas Meadows, Mount Tamalpais; between Santa Venetia and Chinese Camp; Ignacio; Bolinas; Salmon Creek Canyon; east of Aurora School. Plants growing in a pond near Mill Valley had styles from basal flowers 3 dm. long.—*L. subulata* Humb. & Bonpl.

JUNCAGINACEAE. ARROWGRASS FAMILY

1. Triglochin. ARROWGRASS

a. Fruit nearly globose, consisting of 3 fertile carpels alternating with 3 inconspicuous sterile carpels ....3. *T. striata*
a. Fruit elongate, consisting normally of 6 fertile carpels ....b

b. Plants tall (mostly more than 3 dm.) with stout short-creeping rhizomes; ligules simple ....1. *T. maritima*
b. Plants low (mostly less than 3 dm.) with slender elongate rhizomes; ligules 2-parted to the base ....2. *T. concinna*

1. **T. maritima** L. Along tidal sloughs and about seepages in the upper part of the salt marshes: Sausalito, acc. Stacey; Tamalpais Valley; Escalle; Burdell; Drakes Estero; Inverness. Although usually confined to saline or subsaline habitats in our district, *T. maritima* has been found on the wet springy slope in the serpentine area east of Bootjack on Mount Tamalpais.

2. **T. concinna** Burtt Davy. Forming broad lawnlike patches in the lower salicornia belt in the salt marshes or growing sparsely with the salicornia: Tiburon; Almonte; Escalle; Burdell; Stinson Beach; Drakes Estero; Inverness.— *T. maritima* var. *debilis* of authors in part.

Triglochin maritima and *T. concinna* sometimes grow together but only in the salt marshes below Larkspur has a suspected hybrid between the two been seen. This plant, which was intermediate in size, had the stout rhizome characteristic of *T. maritima* but the ligule of *T. concinna*.

3. **T. striata** R. & P. In Marin County known only from the marshy pond behind the beach at Dillons Beach. This plant combines the slender habit of *T. concinna* with the simple ligule of *T. maritima*, but from both it differs in its distinctive fruit.

ALISMACEAE. WATER PLANTAIN FAMILY

a. Leaf blades sagittate; carpels spirally arranged in many series ...1. *Sagittaria*
a. Leaf blades not sagittate; carpels in 1 series ....b
b. Petals about 0.5 cm. long; carpels rounded, inconspicuously beaked 2. *Alisma*
b. Petals about 1 cm. long; carpels long-beaked ....3. *Damasonium*

1. Sagittaria. ARROWHEAD

1. **S. latifolia** Willd. TULE POTATO. Fresh water marshes, rare in Marin County: Santa Venetia; laguna in Chileno Valley.

2. Alisma

1. **A. Plantago-aquatica** L. var. **triviale** (Pursh) Farwell. WATER PLANTAIN. Occasional in marshy places: Mill Valley; Mount Tamalpais (Hidden Lake, *Leschke*, Lake Lagunitas, *Sutliffe*); Santa Venetia; Hamilton Field; Stinson Beach; Bolinas; Olema Marshes; Aurora School. The American form of this widespread species has also been called subsp. *brevipes* (Greene) Samuelsson.

3. Damasonium

1. **D. californicum** Torr. In Marin County known only from the marsh at the Sonoma County line on the road from Petaluma to Tomales.—*Machaerocarpus californicus* (Torr.) Small.

HYDROCHARITACEAE. FROGBIT FAMILY
1. Anacharis. ELODEA; WATERWEED

1. **A. canadensis** (Michx.) Planch. In Marin County known only from a pond between Mill Valley and Tamalpais High School, probably an escape from cultivation.—*Elodea canadensis* Michx.; *Philotria canadensis* (Michx.) Britt.

GRAMINEAE. Grass Family
Key to the Tribes

a. Spikelets with 1 to several florets; imperfect or sterile florets, if present, above the fertile florets. (A few species are dioecious in the *Festuceae*. In *Arrhena-therum* in the *Aveneae* the lower floret is staminate and the upper perfect.) b

a. Spikelets with 1 perfect terminal floret and with 1 or 2 imperfect or rudimentary florets below. (In the *Andropogoneae* numerous spikelets in an inflorescence are rudimentary or only staminate and in *Andropogon* the palea of the perfect floret is lacking.) ....f

b. Spikelets pedicellate (pedicels very short or lacking in *Brachypodium, Cynosurus,* and *Dactylis* in the *Festuceae* and in *Phleum* in the *Agrostideae*) ...c

b. Spikelets sessile (or sometimes when there are 2 or more spikelets at a joint of the rachis, one or more may be pedicellate) ....e

c. Spiklets 1-flowered. (See also *Melica Torreyana*.) ....Tribe 4. *Agrostideae*

c. Spikelets 2- to several-flowered ....d

d. Glumes shorter than the first floret (except sometimes in *Melica*)
....Tribe 1. *Festuceae*

d. Glumes equaling the first floret, usually as long as the spikelet or longer
....Tribe 3. *Aveneae*

e. Spikelets not in 1-sided spikes, the spikes solitary, terminal Tribe 2. *Hordeae*

e. Spikelets in 1-sided spikes, the spikes several to many in a branched inflorescence ....Tribe 5. *Chlorideae*

f. Spikelets paired, the first sessile with a perfect floret, the second pedicellate and either rudimentary or staminate; lemmas hyaline
....Tribe 8. *Andropogoneae*

f. Spikelets not paired, or if more than one at a place, then each spikelet with a perfect floret; fertile lemma indurate ....g

g. Glumes nearly equal, or if unequal (as in *Anthoxanthum*), the upper glume much longer than the sterile lemmas ....Tribe 6. *Phalarideae*

g. Glumes very unequal, the upper glume about equaling the sterile lemma or shorter ....Tribe 7. *Paniceae*

Tribe 1. Festuceae. Fescue Tribe

a. Glumes papery and scarious-margined ....b

a. Glumes not papery ....d

b. Inflorescence racemose, the pedicels short. ....7. *Pleuropogon*

b. Inflorescence paniculate, sometimes narrow c

c. Lemmas broadly rounded and erose-denticulate at the apex. Plants of streams, swamps, and lakes ....6. *Glyceria*

c. Lemmas tapering to a narrowed apex, the apex acute, obtuse, emarginate, or erose. Plants of moist or dry woods and rocks ....13. *Melica*

d. Spikelets in clusters or panicles arranged on one side of the rachis ....e

d. Spikelets not in 1-sided clusters. ....f

e. Spikelets of 1 kind, densely clustered, the clusters in an openly branched panicle ....11. *Dactylis*

e. Spikelets of 2 kinds, fertile and sterile intermixed in a closely branched panicle ....12. *Cynosurus*

f. Lemmas awned (sometimes awnless in *Festuca californica*) ....g

f. Lemmas awnless i

g. Spikelets mostly 2–4 in spikelike racemes, the pedicels of the lateral spikelets less than 1 mm. long ....2. *Brachypodium*

g. Spikelets in broad or narrow panicles, the pedicels longer ....h

h. Lemma awned from the minutely bifid apex; spikelets large ......1. *Bromus*

h. Lemma not bifid; spikelets smaller ....3. *Festuca*

i. Florets spreading horizontally along the rachilla; lemma as broad as long ....9. *Briza*

i. Florets erect-imbricate; lemma longer than wide ....j

j. Flowers unisexual, plants dioecious. Grass of saline flats and salt marshes ....10. *Distichlis*

j. Flowers perfect (unisexual in *Poa Douglasii* and *P. confinis,* species of maritime dunes) ....k

k. Nerves of lemma faint, parallel, not converging near the apex. Tufted perennial of coastal salt marshes ....5. *Puccinellia*

k. Nerves of lemma converging near the apex l

l. Pedicels of spikelets short and thick; low annual with stiff panicles 4. *Scleropoa*

l. Pedicels of spikelets slender and frequently elongate; annual or usually perennial plants with rather soft panicles ....8. *Poa*

TRIBE 2. **Hordeae.** BARLEY TRIBE

a. Spikelets 2 to several at each node of the rachis (or rarely some nodes with only 1 spikelet in *Elymus*), 2- to several-flowered except in *Hordeum* ......b

a. Spikelets 1 at each node of the rachis (or sometimes 2 at a node in robust specimens of *Scribneria*) ....e

b. Rachis readily disarticulating at maturity (except in *Hordeum vulgare*); glumes usually elongate and bristle-like (or narrowly lanceolate in *Sitanion Hansenii*) c

b. Rachis continuous; glumes, if present, not elongate bristles ....d

c. Spikelets 3 at each node, the middle one fertile, the lateral pedicellate, sterile, and generally reduced to awns (or in *Hordeum vulgare* the lateral sessile and sometimes fertile), the middle fertile spikelet 1-flowered 20. *Hordeum*

c. Spikelets usually 2 at each node, fertile, 2- to several-flowered ...18. *Sitanion*

d. Glumes reduced to small awns or lacking; spikelets loosely ascending ....19. *Hystrix*

d. Glumes conspicuous; spikelets approximate, appressed-ascending 17. *Elymus*

e. Spikelets 2- to several-flowered; rachis continuous ....f

e. Spikelets 1-flowered; spikes cylindric with spikelets sunken in cavities of the rachis; rachis disarticulating at maturity i

f. Spikelets placed edgewise to the rachis; glume 1 except in terminal spikelet ....21. *Lolium*

f. Spikelets placed sidewise to the rachis; glumes 2. (*Brachypodium* of the *Festuceae,* with pedicels less than 1 mm. long, may be sought here) ....g

g. Plants perennial ....14. *Agropyron*

g. Plants annual ....h

h. Glumes broad; spikelets thick ....15. *Triticum*

h. Glumes narrow; spikelets compressed ....16. *Secale*

i. Lemma short-awned ....24. *Scribneria*

i. Lemma awnless ....j
j. Glume 1 ....22. *Monerma*
j. Glumes 2 ....23. *Parapholis*

Tribe 3. Aveneae. OAT Tribe

a. Glumes more than 1 cm. long ....b
a. Glumes less than 1 cm. long c
b. Plants annual; spikelets numerous ....29. *Avena*
b. Plants perennial; spikelets few ....32. *Danthonia*
c. Lemmas toothed or bifid at the apex ....d
c. Lemmas entire ....f
d. Lemmas 4-toothed ....27. *Deschampsia*
d. Lemmas bifid ....e
e. Annuals with culms 3 dm. tall or less; spikelets 2.5–4 mm. long .....28. *Aira*
e. Perennials with culms 4 dm. tall or more; spikelets 6–8 mm. long 26. *Trisetum*
f. Spikelets 7–9 mm. long, with a conspicuous exserted awn ..30. *Arrhenatherum*
f. Spikelets 3–4 mm. long, lemma with a short awn or awnless ....g
g. Glumes a little shorter than the uppermost floret; lemmas awnless or rarely
    with a short straight awn ....25. *Koeleria*
g. Glumes exceeding the florets; upper lemma with a short hooklike awn
    ....31. *Holcus*

Tribe 4. Agrostideae. TIMOTHY Tribe

a. Spikelets strongly compressed laterally, the glumes keeled ....b
a. Spikelets not strongly compressed c
b. Glumes not awned; lemmas awned; spikelets disarticulating below the glumes
    ....36. *Alopecurus*
b. Glumes short-awned; lemmas not awned; spikelets disarticulating above the
    glumes ....38. *Phleum*
c. Glumes copiously plumose-hairy. Lemmas bearing a dorsal awn and 2 elongate
    apical bristles ....40. *Lagurus*
c. Glumes not copiously plumose-hairy ....d
d. Glumes long-awned; spikelets disarticulating below the glumes 37. *Polypogon*
d. Glumes usually not awned, sometimes acuminate-tipped or rarely awn-tipped;
    spikelets usually disarticulating above the glumes ....e
e. Plants annual; lemma about ¼ as long as the shorter of the very unequal
    glumes ....39. *Gastridium*
e. Plants mostly perennial (annual in two species of *Agrostis*); lemma at least ½
    as long as the glumes ....f
f. Spikelets small, 4 mm. long or less (excluding awns), if longer, the palea re-
    duced to a minute nerveless scale or lacking ....35. *Agrostis*
f. Spikelets larger, generally 4–20 mm. long (excluding awns), if shorter, the palea
    well developed and nearly as long as the lemma ....g
g. Lemma not awned ....34. *Ammophila*
g. Lemma awned ....h
h. Lemmas not becoming indurate or terete; rachilla prolonged as a hairy bristle
    ....33. *Calamagrostis*
h. Lemmas becoming indurate and terete; rachilla not prolonged .....41. *Stipa*

Tribe 5. **Chlorideae.** Chloris Tribe

a. Plants annual; spikes in a rather loose narrow panicle. Spikelets laterally flattened, almost circular, falling entire ....43. *Beckmannia*

a. Plants perennial with elongate rhizomes or stolons; spikes not paniculate ...b

b. Spikes digitate, radiating from the top of short erect culms; grass escaped from cultivation in waste places ....42. *Cynodon*

b. Spikes closely appressed in a narrow erect spikelike raceme; grass of salt marshes ....44. *Spartina*

Tribe 6. **Phalarideae.** Canarygrass Tribe

a. Panicle openly branched. Spikelets not compressed; glumes nearly equal ....45. *Hierochloe*

a. Panicle dense, spikelike ....b

b. Spikelets not laterally compressed; glumes unequal; sterile lemmas awned ....46. *Anthoxanthum*

b. Spikelets laterally compressed; glumes equal; lemmas not awned ..47. *Phalaris*

Tribe 7. **Paniceae.** Millet Tribe

a. Spikelets subtended by several bristles ....53. *Setaria*

a. Spikelets not subtended by bristles ....b

b. Spikelets pedicellate in a diffuse open panicle ....51. *Panicum*

b. Spikelets sessile or very shortly pedicellate, arranged along one side of the rachis c

c. Inflorescence simple, terminal and axillary, the rachis conspicuously broadened and flattened ....49. *Stenotaphrum*

c. Inflorescence with 2 to many branches (rarely 1 in *Paspalum distichum*), rachis not conspicuously broad and flat ....d

d. Inflorescence paniculate with numerous short congested racemose branches; rachis not winged ....52. *Echinochloa*

d. Inflorescence of 1 to several digitately or racemosely arranged racemes, the rachis narrowly winged ....e

e. Racemes several, digitate or nearly so ....48. *Digitaria*

e. Racemes 1 to several, racemose if more than 2 ....50. *Paspalum*

Tribe 8. **Andropogoneae.** Sorghum Tribe

a. Racemes of numerous joints, slender, the rachis conspicuously villous with long spreading hairs; spikelets small, the pedicellate spikelet sterile and rudimentary ....54. *Andropogon*

a. Racemes of 1–5 joints, the rachis short-pubescent; spikelets turgid, the pedicellate spikelet staminate ....55. *Sorghum*

Tribe 1. **Festuceae.** Fescue Tribe

1. **Bromus.** Bromegrass

a. Plants perennial (or if annual, then the lemmas carinate-compressed) ......b

a. Plants annual ....g

b. Spikelets compressed, glumes and lemmas keeled c

b. Spikelets not compressed, glumes and lemmas rounded on back ....f

c. Leaf blades narrow (mostly 3 mm. wide or less), involute, the blades and sheaths canescent; panicle narrow, the branches ascending ....1. *B. breviaristatus*
c. Leaf blades broader (mostly 3 mm. wide or more), flat, blades and sheaths pilose, scabrous, or glabrous; panicle narrow or open ....d
d. Panicle narrow, the branches very short, strictly erect; plant of maritime dunes ....2. *B. maritimus*
d. Panicle narrow or broad, the branches elongate; plants of grassland and woodland ....e
e. Awns 8 mm. long or less; panicle narrow, the branches strongly ascending ....3. *B. marginatus*
e. Awns 8 mm. long or more; panicle branches spreading-ascending to deflexed ....4. *B. carinatus*
f. First glume 1-nerved ....5. *B. vulgaris*
f. First glume 3-nerved ....6. *B. laevipes*
g. Lemmas acute or rounded, not acuminate ....h
g. Lemmas acuminate, conspicuously awned ....j
h. Branches of panicle spreading or drooping; lemmas with narrow chartaceous-hyaline margin and truncate or erose apex, glabrous ......9. *B. commutatus*
h. Branches of panicle suberect; lemmas with rather broad hyaline margin and more or less deeply bifid apex i
i. Lemmas pubescent ....7. *B. mollis*
i. Lemmas glabrous ....8. *B. racemosus*
j. Branches of panicle erect or suberect ....k
j. Branches of panicle spreading and drooping l
k. Culms pilose with short spreading hairs near the top; panicle dense; glumes rather broadly lanceolate; teeth of lemmas 3–5 mm. long ......10. *B. rubens*
k. Culms generally glabrous or scabrous near the top; panicle rather loosely branched; glumes linear-lanceolate; teeth of lemmas 2–3 mm. long ....11. *B. madritensis*
l. Spikelets glabrous or nearly so; awns more than 3 cm. long .....12. *B. rigidus*
l. Spikelets conspicuously hairy; awns less than 2 cm. long ....13. *B. tectorum*

1. **B. breviaristatus** Buckl. Serpentine slopes at south end of Carson Ridge. The narrow canescent leaf blades and the narrow panicle together distinguish this widespread western American grass from the many forms of *B. carinatus* which are so common in Marin County.

2. **B. maritimus** (Piper) Hitchc. Abundant perennial on the fixed dunes and ocean bluffs: Rodeo Lagoon; Dillons Beach; Point Reyes Peninsula, the type locality. This distinctive relative of *B. carinatus* is found along the California coast from Bodega Point, Sonoma County, to Point Sur, Monterey County.—*B. marginatus maritimus* Piper.

3. **B. marginatus** Nees. Grassy places on wooded hills: Sausalito; Tiburon; Mill Valley; Carson country; Inverness Ridge. Although the length of the awn may seem a purely arbitrary mark by which this grass is separated from *B. carinatus,* this character serves to demark in Marin County a plant with a distinctive inflorescence. Beyond Marin County, however, *B. marginatus* is described with broad open panicles as well as the narrow kind which is typical of the species.

4. **B. carinatus** H. & A. Common and widespread from waste places around habitations to grassy, brushy, or wooded hills: from Sausalito, Tiburon, and Mount

Tamalpais to San Rafael Hills, Point Reyes Peninsula, and Tomales. This is one of the most variable plants in Marin County and, with further study, it may be divided into several varieties. In nearly half of all the Marin County specimens examined, both the leaf sheaths and lemmas are hairy while smaller numbers have one or both structures glabrous or merely scabrous. Probably a much more significant variant is a giant form with tall, thick culms, broad leaves, and large inflorescences. In a specimen of this sort from moist woods near Inverness, the culms were about 6 feet tall and nearly a half inch in diameter and the blades were nearly an inch broad.

5. **B. vulgaris** (Hook.) Shear. Rocky slopes in woods and shaded canyons: Mount Tamalpais (Cascade Canyon, Steep Ravine, Laurel Dell, Rifle Camp); Corte Madera, acc. Stacey; Papermill Creek; Inverness Ridge; Point Reyes, acc. Davy.

6. **B. laevipes** Shear. Rather common and widespread on open or wooded rocky slopes, well developed on high serpentine ridges: Mount Tamalpais (Bootjack, Laurel Dell, Northside Trail, Fish Grade); San Rafael Hills; Fairfax; White Hill. This species, although superficially resembling the preceding, is generally more robust in habit and is coarser in all its parts.

7. **B. mollis** L. SOFT CHESS. Common weed of waste places about towns and in open grasslands: Sausalito and Tiburon to Mount Tamalpais, San Rafael, and Inverness. Introduced from Europe.

8. **B. racemosus** L. Occasional in grassland on the edge of woods: Rock Spring and West Peak, Mount Tamalpais; Shell Beach near Inverness. Native of Europe. As understood here, *B. racemosus* seems scarcely more than a glabrous form of *B. mollis*.

9. **B. commutatus** Schrad. Waste ground and grassy meadows: Almonte; Lagunitas Meadows; near Olema; north of Dillons Beach. Introduced from Europe.

10. **B. rubens** L. FOXTAIL CHESS. Rather widespread but not common in weedy places and on grassy slopes: Angel Island; Sausalito; West Point and head of Cataract Creek, Mount Tamalpais; Carson country; San Rafael; Big Rock Ridge. Introduced from Europe.

11. **B. madritensis** L. MADRID BROME. Occasional in partly shaded or exposed brushy places: Sausalito; Angel Island; Mount Tamalpais (Bootjack, Corte Madera Ridge, Phoenix Lake) ; Fairfax and San Rafael hills. Naturalized from Europe.

12. **B. rigidus** Roth. RIPGUT GRASS. Abundant weed around towns, in grassland, and on wooded or brushy slopes: Sausalito, Angel Island, Tiburon, and Mount Tamalpais (Cascade Canyon, Bootjack) to San Rafael and Point Reyes. At San Rafael, Inverness, and other places, the variant with more widely branching panicle, var. *Gussonii* (Parl.) Coss., has been seen. Both the typical form and the variety have been introduced from Europe.

13. **B. tectorum** L. DOWNY CHESS. Occasional weedy grass in waste places or on brushy slopes: along railroad near Manzanita and at San Rafael; West Peak, Mount Tamalpais; Inverness. Native of Europe.

2. Brachypodium

1. **B. distachyon** (L.) Beauv. Locally common on open grassy hillsides: Tiburon Peninsula; Devils Gulch near Camp Taylor. Introduced from Europe.

3. Festuca. FESCUE

a. Plants perennial ....b
a. Plants annual ....f
b. Collar of sheaths pubescent; lemmas short-awned or merely acuminate
   ....1. *F. californica*
b. Collar of sheaths not pubescent; lemmas conspicuously awned (sometimes
   short-awned in *F. rubra*) c
c. Panicle loose and open; awns generally as long as, or longer than, the lemmas ....d
c. Panicle narrow and contracted; awns generally shorter than lemmas ....e
d. Blades, or some of them, 2–4 mm. wide, scabrous or pubescent on upper surface; lemmas hispidulous ....2. *F. Elmeri*
d. Blades not more than 1 mm. wide, smooth and glabrous; lemmas a little scabrous near apex ....3. *F. occidentalis*
e. Culms loosely tufted and decumbent at base; sheaths reddish brown with conspicuous pale nerves; blades quite smooth ....4. *F. rubra*
e. Culms closely tufted, not decumbent; sheaths not reddish at base; blades
   usually more or less scabrous ....5. *F. idahoensis*
f. Inflorescence narrow, branches ascending or appressed-ascending (rarely spreading in *F. octoflora*) ....g
f. Inflorescence broader, the branches spreading or at maturity reflexed ....j
g. The lower glume 2–4 mm. long, at least half as long as the upper glume ....h
g. Glumes very unequal, the lower about 1–2 mm. long i
h. Spikelets usually more than 5-flowered; lemmas 3–5 mm. long, the awns usually
   5 mm. long or less ....6. *F. octoflora*
h. Spikelets usually 4- or 5-flowered; lemmas 5–7 mm. long, the awns more than
   5 mm. long ....7. *F. dertonensis*
i. Lemma conspicuously long-ciliate near apex ....8. *F. megalura*
i. Lemma not ciliate ....9. *F. myuros*
j. Pedicels appressed to the rachis at maturity ....k
j. Pedicels reflexed or spreading at maturity l
k. Lemmas glabrous ....10. *F. pacifica*
k. Lemmas pubescent ....11. *F. Grayi*
l. Lemmas glabrous ....12. *F. reflexa*
l. Lemmas pubescent ....m
m. Glumes glabrous ....13. *F. microstachys*
m. Glumes pubescent ....14. *F. Eastwoodae*

1. **F. californica** Vasey. Widely distributed and rather common from wet coastal
swales to wooded slopes and chaparral-covered ridges: Angel Island; Sausalito;
Tiburon; Mount Tamalpais; San Rafael Hills; Big Rock Ridge; Point Reyes
Peninsula. A handsome species that is particularly abundant in the serpentine area
near Bootjack on Mount Tamalpais.

2. **F. Elmeri** Scribn. & Merr. Occasional on wooded or brushy slopes in rocky
soil: Sausalito; Mount Tamalpais (Bootjack, Phoenix Lake, Camp Handy); Larkspur, acc. Stacey; Deer Park; San Rafael Hills.

3. **F. occidentalis** Hook. Rare on Mount Tamalpais along the edge of redwood
groves in moist rocky soil: Tenderfoot Trail; Fern Canyon; Lake Lagunitas.

4. **F. rubra** L. Common and widespread from salt marshes and marine bluffs

to moist hillsides and boggy meadows: Angel Island; Tiburon; Rodeo Lagoon; Mount Tamalpais (Potrero Meadow, Rifle Camp, Rock Spring); Escalle Marshes; Fairfax and San Rafael hills; Burdell Marshes; Bolinas; Inverness Ridge; Point Reyes; Dillons Beach. Occasional tufted forms closely resemble *F. idahoensis* but the present species can be readily distinguished by the sheaths of the basal leaves which are reddish brown with pale striate nerves.

5. **F. idahoensis** Elmer. Occasional but rather widespread on open grassy slopes or in the drier parts of meadows: Angel Island, Sausalito, Tiburon, and Mount Tamalpais (Bootjack, Rifle Camp, Lagunitas Meadows) to Lagunitas, San Rafael Hills, and Black Point.

6. **F. octoflora** Walt. Clearings or burns on dry brushy hillsides or in the chaparral, occasional: Mount Tamalpais (West Peak, near Barths Retreat, Berry Trail); Carson Ridge; San Rafael Hills; Inverness Ridge. Sometimes this grass is difficult to distinguish from the following, though in Marin County the sizes of lemmas and awns as given in the key are reliable.

7. **F. dertonensis** (All.) Aschers. & Graebn. Common and variable, in waste places about towns, on grassy slopes and flats, and in the chaparral: Angel Island; Tiburon; Sausalito; Mill Valley; Mount Tamalpais (Corte Madera Ridge, East and West peaks, Berry Trail) ; Fairfax; San Rafael; Inverness Ridge; Point Reyes. Introduced from Europe.—*F. bromoides* of authors, not L.

A depauperate form in which the low slender culms tend to be tufted and in which the simple inflorescences bear only 1 to 5 spikelets is probably *Vulpia dertonensis* var. *gracilis* (Lange) Hegi.

8. **F. megalura** Nutt. Abundant on meadowy flats and brushy slopes from Angel Island, Sausalito, Tiburon, and Mount Tamalpais (Blithedale Canyon, Bootjack, West Peak, Rifle Camp) to San Rafael Hills and Point Reyes Peninsula.

On the south side of Mount Tamalpais plants of both *F. megalura* and *F. dertonensis* have been found with open flowers. Usually fertilization takes place in closed flowers.

9. **F. myuros** L. Occasional in waste places: Angel Island; Sausalito; Mill Valley; Fairfax. In the San Rafael Hills a tall vigorous form has been collected in which there are 7 to 9 florets in each spikelet. Introduced from Europe.

10. **F. pacifica** Piper. Open slopes and woods in gravelly soil around rocks or brush: Angel Island; Tiburon; Mount Tamalpais (Bootjack, Potrero Meadow, Laurel Dell, West Peak); Fairfax Hills; Carson country; Chileno Valley. In Marin County this is the most common and widespread species in the group related to *F. reflexa*. Although the flowers in these annual fescues are usually cleistogamous, many specimens representing several species have been observed with exserted anthers and stigmas.

11. **F. Grayi** (Abrams) Piper. Open gravelly slopes around rocks or brush: Tiburon; Bootjack and Laurel Dell, Mount Tamalpais; Big Carson; San Anselmo Canyon.

12. **F. reflexa** Buckl. Dry open slopes in gravelly or rocky soil: Sausalito; Angel Island; Tiburon; Mount Tamalpais (Bootjack, Northside Trail, Berry Trail); San Rafael Hills; Carson country.

13. **F. microstachys** Nutt. A single plant from Big Carson Canyon is doubtfully referable to this rare species. The plant is listed for Mount Tamalpais by Stacey but no specimen has been seen from there.

14. **F. Eastwoodae** Piper. Two collections of this rare fescue have been seen

from Mount Tamalpais, one from West Peak, the other from the Laurel Dell Trail near Rock Spring.

4. Scleropoa

1. **S. rigida** (L.) Griseb. Sidewalk weed along shaded streets: Sausalito; Mill Valley; San Rafael. Introduced from Europe.

5. Puccinellia

1. **P. grandis** Swallen. In the *Distichlis-Salicornia* association of coastal salt marshes on Point Reyes Peninsula: Tomales Bay; Drakes Estero.—*P. nutkaensis* of California references.

6. Glyceria. MANNAGRASS

a. Branches of panicle drooping or flexuous; spikelets ovate, about 6-flowered
....1. *G. pauciflora*
a. Branches of panicle stiff, erect or spreading; spikelets slender, oblong-linear, about 8–10-flowered ....b
b. Lemmas about 3 mm. long ....2. *G. leptostachya*
b. Lemmas about 5 mm. long ....3. *G. occidentalis*

1. **G. pauciflora** Presl. Occasional in swampy ground, mostly near the coast: Potrero Meadow, Mount Tamalpais; Lily Lake; Olema, acc. Stacey; Chileno Valley; Point Reyes Peninsula (Inverness, Mud Lake, Ledum Swamp).

This widespread western American species has been referred to the recently described genus, *Torreyochloa*, as *T. pauciflora* (Presl) G. L. Church. In *Glyceria* proper the leaf sheaths are connate, the upper glume is 1-nerved, and the styles are stigmatic above the middle, while in *Torreyochloa* the leaf sheaths are not connate, the upper glume is 3-nerved, and the styles are stigmatic nearly or quite to the base.

2. **G. leptostachya** Buckl. Occasional in wet springy places in the hills or along streams: Elk Valley; Mill Valley; between Mill Valley and Muir Woods; Lake Lagunitas and Alpine Lake, Mount Tamalpais; Lagunitas Creek at Tocaloma, *Leschke;* Olema Marshes; Chileno Valley. This species is not known south of Marin County.

3. **G. occidentalis** (Piper) J. C. Nels. In Marin County known only from the marshy borders of a pond between Bolinas and Olema. This is the southernmost station known.

7. Pleuropogon. SEMAPHORE GRASS

a. Plants annual; awn 3–6 mm. long ....1. *P. californicus*
a. Plants perennial; awn 1–3 mm. long ....2. *P. Hooverianus*

1. **P. californicus** (Nees) Benth. Wet fields or depressions where water has stood: Ross, acc. Davy; San Rafael, acc. Stacey; Hamilton Field; Black Point; east of Aurora School; Inverness Ridge. In Marin County, *P. californicus* is especially characteristic of low fields between the hills and salt marshes bordering the bay.

2. **P. Hooverianus** (L. Benson) J. T. Howell. Rare in grassy places on the edge of redwoods: Ross; Lagunitas Meadows; Woodacre, acc. Hoover; midway between Nicasio and San Geronimo, *Leschke.* Hoover's semaphore grass has been reported only from Mendocino and Marin counties.

8. **Poa.** BLUEGRASS

a. Plants annual ....b
a. Plants perennial c
b. Culms generally 2 dm. tall or less; sheaths smooth; lemmas more or less pubescent near the base but not webbed ....1. *P. annua*
b. Culms generally more than 2 dm. tall; sheaths retrorsely roughened; lemmas webbed at the base ....2. *P. Howellii*
c. Rhizomes present ....d
c. Rhizomes lacking ....h
d. Culms distinctly flattened and 2-edged ....3. *P. compressa*
d. Culms terete ....e
e. Plants dioecious; panicle contracted; lemmas scarcely webbed ....f
e. Plants not dioecious; panicle open and usually widely branching; lemmas copiously webbed ....g
f. Plants coarse with dense, spikelike panicles; lemmas 5–6 mm. long
....4. *P. Douglasii*
f. Plants delicate with slender and somewhat looser panicles; lemmas 3–4 mm. long ....5. *P. confinis*
g. Lemmas glabrous except for the web ....6. *P. Kelloggii*
g. Lemmas pubescent above the web ....7. *P. pratensis*
h. Lemmas with a conspicuous web; panicle open ....8. *P. trivialis*
h. Lemmas without a web but more or less pubescent at the base on the keel and nerves; panicle usually contracted (or open in anthesis) i
i. Lemmas distinctly keeled ....9. *P. unilateralis*
i. Lemmas rounded on back and scarcely keeled ....10. *P. scabrella*

1. **P. annua** L. From late winter to early summer, common and widespread in moist soil along roads, in cultivated areas, and about dwellings: Sausalito, Tiburon, and Mount Tamalpais (Phoenix Lake, Rock Spring) to San Rafael, Tomales, and Point Reyes. Introduced from Europe.

2. **P. Howellii** Vasey & Scribn. Occasional on wooded or brushy slopes, frequently vigorous and locally abundant in loose soil of cleared areas and burns: Sausalito; Angel Island; Mount Tamalpais (Northside Trail, Corte Madera Ridge, Artura Trail, Phoenix Lake); San Rafael Hills; Carson country; Inverness Ridge, Salmon Creek.

The Howell bluegrass is usually characteristic of shaded or partially shaded places but rarely it also grows in full sunlight of exposed ridges, as on West Peak of Mount Tamalpais and in chaparral burns in San Anselmo Canyon west of Fairfax. It varies greatly in size and habit: usually the culms are few but occasionally they are densely tufted, the plants then simulating a perennial bunchgrass; the inflorescence generally is slender but sometimes the branches spread widely; plants may be only 2.5 inches tall in depauperate specimens in the chaparral, but in loose rich soil (as in the wooded canyon below Phoenix Lake) the plants may be 3 feet tall.

3. **P. compressa** L. CANADA BLUEGRASS. Rare in Marin County: Mount Tamalpais, East Peak near the tavern, *Leschke;* near Aurora School, *Yates.* Native of Europe.

4. **P. Douglasii** Nees. Generally confined to coastal dunes (Rodeo Lagoon, Stinson Beach, Point Reyes, Dillons Beach), but also occurring at the southern end of

Tiburon Peninsula. The Douglas bluegrass is an endemic of the middle California coast from Mendocino County to Monterey County.

5. **P. confinis** Vasey. Known in Marin County only from sandy flats back of the dunes near the radio station on Point Reyes Peninsula. This is the southernmost known station for a rather rare grass that has a distribution extending to British Columbia.

6. **P. Kelloggii** Vasey. Moist, shaded woods: Big Carson; Corte Madera and Mill Valley, acc. Stacey. In the Santa Cruz Mountains, plants of *Poa Kelloggii* and *Festuca occidentalis* Hook. are characteristic and abundant along the edge of redwood forests.

7. **P. pratensis** L. KENTUCKY BLUEGRASS. Common in waste places about towns and in wet places in the hills: Sausalito; Angel Island; Mill Valley; San Rafael; Fairfax; Elk Valley; Rifle Camp and Potrero Meadow, Mount Tamalpais; Inverness Ridge; Point Reyes. Although certain specimens collected near dwellings undoubtedly represent naturalized European strains of this species that have escaped from cultivation, the plant appears to be indigenous as it occurs on Mount Tamalpais and near the coast.

8. **P. trivialis** L. Rare on moist slopes, Inverness Park, Point Reyes Peninsula. Introduced from Europe.

9. **P. unilateralis** Scribn. Occasional around rocks or in shallow soil on bluffs near the ocean: Sausalito; Stinson Beach, acc. Stacey; Tomales; Point Reyes Peninsula (summit of Inverness Ridge, Point Reyes, McClure Beach).

10. **P. scabrella** (Thurb.) Benth. Open grassy hills or around rocks: Angel Island; Sausalito; Tiburon; Rifle Camp and Northside Trail, Mount Tamalpais; Fairfax Hills; San Rafael Hills.

9. **Briza.** QUAKING GRASS

a. Branches of panicle drooping; spikelets 10 mm. long or more . . 1. *B. maxima*
a. Branches of panicle erect or spreading; spikelets 3–4 mm. long . . 2. *B. minor*

1. **B. maxima** L. RATTLESNAKE GRASS. Becoming widespread and locally common on grassy slopes and flats: Sausalito; Belvedere; Mill Valley; Mountain Theater and Potrero Meadow, Mount Tamalpais; Fairfax; San Rafael Hills; Big Rock Ridge; Novato. Introduced from Europe.

2. **B. minor** L. Common and widespread through the hills of Marin County: Sausalito, Tiburon, and Mount Tamalpais (Rock Spring, Phoenix Lake) to San Rafael, Tomales, and Inverness. Robust individuals growing under especially favorable conditions become 2 feet tall. Introduced from Europe.

10. **Distichlis.** SALTGRASS

1. **D. spicata** (L.) Greene. Common in salt marshes along the bay and ocean and on saline flats above the marshes: Rodeo Lagoon; Tiburon; Manzanita; Escalle; Santa Venetia; Black Point; Stinson Beach; Inverness.

The Marin County plant has been called var. *stolonifera* Beetle and was differentiated from the typical form on the Atlantic coast by the lower habit and congested oval inflorescences, but tall plants with slender inflorescences are common in marshes in the northern part of the county. At Bolinas Lagoon plants may be as much as 2 feet tall.

11. Dactylis

1. **D. glomerata** L. ORCHARD GRASS. Common in waste places around habitations and in grassy meadows: Angel Island; Sausalito; Mill Valley; Mount Tamalpais (West Point, Potrero Meadow, Phoenix Lake) ; Santa Venetia; Inverness. A native of Europe, widely cultivated as a pasture grass.

12. Cynosurus. DOGTAIL

1. **C. echinatus** L. Widespread and sometimes abundant grass in open places and on the edge of thickets: Sausalito; Tiburon; Mount Tamalpais (Rock Spring, Potrero and Lagunitas meadows) ; Ross; Fairfax; San Rafael. Introduced from Europe.

13. Melica. MELICGRASS

a. Glumes much shorter than the narrow spikelets ....b
a. Glumes as long as the broad spikelets or a little shorter ....d
b. Lemmas rounded-obtuse or emarginate, short-awned, the awns 3–4 mm. long; stems not bulbous ....1. *M. Harfordii*
b. Lemmas acuminate, subacute, or bifid, awnless or awn-pointed (awn 2 mm. long or less) ; stem bulbous c
c. Lemmas acuminate, pilose on the nerves, the tips more or less penicillate-ciliate ....2. *M. subulata*
c. Lemmas generally subacute or bifid, sometimes awn-pointed, scabrous on the back ....3. *M. Geyeri*
d. Spikelets about 10 mm. long; lemmas scabrous ....4. *M. californica*
d. Spikelets 4–6 mm. long; lemmas hairy ....5. *M. Torreyana*

1. **M. Harfordii** Boland. Common in woods and on shaded slopes on Mount Tamalpais: Wheeler Trail; Fern Canyon; West Point; Bootjack; Lake Lagunitas; Alpine Dam.

2. **M. subulata** (Griseb.) Scribn. ALASKA ONIONGRASS. Occasional on wooded slopes: Mount Tamalpais (Steep Ravine, Kent Ravine, Laurel Dell Trail) ; Big Carson; Inverness. This species has been reported also from Sausalito and San Rafael, but it is probable that these records may refer to the following which is closely related and more common.

3. **M. Geyeri** Munro. The typical form, in which the lemmas are almost or quite awnless, is rare in shaded woodland or on open coastal hills: near Mountain Theater, Mount Tamalpais; Salmon Creek Canyon; Tomales.

Much more common and widespread in dry woods or brushy places is var. *aristulata* J. T. Howell in which the lemma bears a short awn up to 2 mm. long: Sausalito; Lagunitas and Potrero meadows, Mount Tamalpais; Big Carson; Fairfax; San Rafael Hills, the type locality. No specimen of the variety has been seen beyond Marin County.

4. **M. californica** Scribn. An attractive, rather common grass in open rocky places, occasional on grassy slopes: Sausalito; Angel Island; Tiburon; Bootjack and Rifle Camp, Mount Tamalpais; Fairfax Hills; San Rafael Hills; Big Rock Ridge. Corte Madera is the type locality.—*M. bulbosa* of some treatments, not *M. bulbosa* Geyer.

5. **M. Torreyana** Scribn. Common on rocky slopes in shade or sun: Sausalito,

Tiburon, and Mount Tamalpais (Fern Canyon, Bootjack, Northside Trail) to Fairfax, San Rafael Hills, and Tomales. This species grows as commonly on serpentine as on other formations.

TRIBE 2. Hordeae. BARLEY TRIBE

14. Agropyron. WHEATGRASS

a. Creeping rhizomes present; culms 2 dm. tall or less; blades involute
....1. *A. arenicola*
a. Creeping rhizomes absent; culms more than 2 dm. tall; blades flat
....2. *A. trachycaulum*

1. **A. arenicola** Davy. Fixed dunes: Stinson Beach; Point Reyes Peninsula, the type locality. This California grass is restricted to relatively few coastal stations from Point Sur on the south to Fort Bragg on the north.

The plant is typically low with culms scarcely exceeding the leaves, but growing near it on the Point Reyes dunes is a much taller broader-leaved plant. In this plant the spikelets are solitary as in *Agropyron* but the general habit and appearance suggest that it may have originated by hybridization between *A. arenicola* and a species of *Elymus*. At Dillons Beach a plant was found that resembles *A. arenicola* but in it the spikelets were frequently paired. The fact that this plant was sterile suggests that it too may be of hybrid origin.

2. **A. trachycaulum** (Link) Malte. SLENDER WHEATGRASS. In Marin County known only from Tiburon; widespread across North America.—*A. pauciflorum* (Schwein.) Hitchc.

15. Triticum

1. **T. aestivum** L. WHEAT. Occurring locally along roads and probably not persistent: Mount Tamalpais, acc. Stacey; San Rafael; Ignacio.—*T. vulgare* Vill.

16. Secale

1. **S. cereale** L. RYE. Perhaps not persistent but occasionally spontaneous along roads and in waste places as at Ignacio and San Antonio.

17. Elymus. WILD-RYE

a. Lemmas conspicuously awned, the awn as long to twice as long as the lemma
....b
a. Lemmas mucronate, awn-tipped, or short-awned c
b. Blades and sheaths glabrous, scabrous, or hispidulous ....1. *E. glaucus*
b. Blades and sheaths pubescent ....1a. *E. glaucus* var. *Jepsonii*
c. Plants without rhizomes ....2. *E. virescens*
c. Plants with elongate rhizomes ....d
d. Spikelets more or less pubescent, generally densely so; glumes coriaceous or papery, not rigid. Plants of coastal dunes ....5. *E. mollis*
d. Spikelets more or less scabrous or the glumes sparsely villous; glumes rigid ...e
e. Glumes linear-lanceolate, the margins not thin, the nerves not evident; plants widespread, generally in clay soil ....3. *E. triticoides*
e. Glumes lanceolate, the margins thin and somewhat scarious, the nerves conspicuous; plants of beaches and dunes ....4. *E. vancouverensis*

1. **E. glaucus** Buckl. BLUE WILD-RYE. A common and widely distributed grass on wooded and brushy slopes from Sausalito, Angel Island, Tiburon, and Mount Tamalpais to the San Rafael Hills and Point Reyes Peninsula. Plants are generally green, only occasionally glaucous; culms usually cespitose, rarely spreading-stoloniferous.

1*a*. **E. glaucus** Buckl. var. **Jepsonii** Davy. A robust plant, occasional on grassy slopes near woods in hills north of Mount Tamalpais: Deer Park near Fairfax; San Rafael Hills; Santa Venetia; Hamilton Field; Point Reyes.

In var. *Jepsonii* the blades as well as the sheaths are pubescent, but in Marin County a form with blades glabrous or scabrous is rather common: Tiburon; Fern Canyon and Laurel Dell, Mount Tamalpais; Phoenix Lake; Fairfax Hills. This is perhaps the form described from Olema as *E. hispidulus* Davy.

2. **E. virescens** Piper. Occasional on grassy slopes: Tiburon; Willow Meadow and Berry Trail, Mount Tamalpais; Carson Ridge; Ledum Swamp, Point Reyes Peninsula. At the south end of Carson Ridge and on Mount Tamalpais, *E. virescens* is a common grass on serpentine outcrops.

Elymus pubescens Davy, described from Point Reyes, has been treated as a synonym of *E. virescens,* and the short-awned *E. glaucus* var. *breviaristatus* Davy, also described from Point Reyes, probably belongs here too.

3. **E. triticoides** Buckl. Occasional on clay flats and near woods, also weedy and gregarious along roads: Rodeo Lagoon; Angel Island; Tiburon; Mill Valley; San Rafael; San Anselmo Canyon; Bolinas and Abbotts Lagoon, Point Reyes Peninsula.

4. **E. vancouverensis** Vasey. Sandy beaches and dunes: Stinson Beach; Point Reyes Peninsula along Tomales Bay and at head of Drakes Bay; Dillons Beach.

5. **E. mollis** Trin. AMERICAN DUNE-GRASS. Maritime dunes: Point Reyes; Stinson Beach.

18. **Sitanion.** SQUIRRELTAIL

a. Spike about as broad as long; glumes bristle-like .... 1. *S. jubatum*
a. Spike slender, much longer than broad; glumes narrowly lanceolate, subulate-
   awned .... 2. *S. Hansenii*

1. **S. jubatum** J. G. Smith. Widely distributed and rather common on grassy slopes in thin rocky soil: Sausalito, Angel Island, and Tiburon to Carson Ridge, San Rafael, and Black Point; on Mount Tamalpais a characteristic element in grassland at Potrero Meadow, Rock Spring, and Bootjack.

2. **S. Hansenii** (Scribn.) J. G. Smith. Occasional on open grassy slopes and on the edge of brush, always occurring with *S. jubatum* and *Elymus glaucus:* Angel Island; Tiburon; near Muir Woods; Mount Tamalpais (Bootjack, Rock Spring, Potrero Meadow) ; Fairfax Hills; Devils Gulch near Camp Taylor. Since all the plants of *S. Hansenii* that have been examined have been sterile, they undoubtedly represent hybrid derivatives of the two species with which they are always found and between which they are intermediate in morphological characters.

19. **Hystrix**

1. **Hystrix californica** (Boland.) Kuntze. Moist wooded slopes and flats, usually in coniferous forests or their border woodlands, rarely on brushy coastal slopes: Sausalito, acc. Davy; Mill Valley, acc. Stacey; Muir Woods; Kent Trail, Fish Grade, and Lake Lagunitas, Mount Tamalpais; Fairfax Hills; Big Carson; Lagunitas;

north of Tocaloma; Olema, acc. Davy; Inverness; near head of Drakes Bay; Salmon
Creek. Marin County is probably the type locality.

This handsome species is commonly found in redwood, pine, or Douglas fir
woodlands, but along the road to Point Reyes near Drakes Estero the tall graceful
culms conspicuously overtop the dense brush that covers the windy treeless slopes.
Although *Hystrix* is rather common in Marin County, it is one of the rarer California grasses and its presently known distribution extends as a narrow belt near
the coast from southern Sonoma County to Santa Cruz County. It generally occurs
as solitary or paired individuals in a given locality, but occasionally, as on a flat
at the head of Muir Woods, a luxuriant colony will dominate several hundred
square feet.

20. Hordeum. BARLEY

a. Glumes glabrous or scabrous . . . . .b
a. Glumes conspicuously ciliate . . . . .f
b. Spikes slender, usually about 1 cm. broad c
b. Spikes broader, usually more than 1 cm. broad. Plants annual . . . . .e
c. Plants annual . . . . .3. *H. depressum*
c. Plants perennial . . . . .d
d. Leaves usually glabrous or scabrous; pedicels of lateral spikelets usually curved,
   the concave side inward; anthers 1.5 mm. long or less . . 1. *H. brachyantherum*
d. Leaves usually more or less pubescent; pedicels of lateral spikelets erect and
   straight; anthers 1.5–3 mm. long . . . . .2. *H. californicum*
e. Rachis disarticulating at maturity; plants low, with spreading stems
   . . . . .4. *H. Hystrix*
e. Rachis not disarticulating at maturity; plants tall (more than 5 dm.) , with
   erect stems . . . . .7. *H. vulgare*
f. Spikes usually more than 1 cm. broad, the spikelets coarse and rather loosely
   arranged; rachis scabrous or scabrous-ciliate; anthers of central floret exserted
   at anthesis, 0.8–1.5 mm. long, auriculate at base . . . . .5. *H. leporinum*
f. Spikes 1 cm. broad or less, the spikelets finer and more appressed; rachis ciliate; central floret cleistogamous, the anthers 0.5 mm. long or less, at base
   rounded and not auriculate . . . . .6. *H. Stebbinsii*

1. **H. brachyantherum** Nevski. MEADOW BARLEY. Widespread in meadows and
along forest borders: Sausalito, Angel Island, and Almonte to San Rafael and
Point Reyes Peninsula; on Mount Tamalpais occurring in Potrero and Lagunitas
meadows.—*H. nodosum* of California references in part.

2. **H. californicum** Covas & Stebbins. Occasional on open grassy slopes or in
open woodland: Tiburon; San Anselmo Canyon; Devils Gulch near Camp Taylor;
Drakes Estero, Point Reyes Peninsula.—*H. nodosum* of California references in
part.

3. **H. depressum** (Scribn. & Smith) Rydb. Occasional in waste ground, and, according to Covas and Stebbins, perhaps introduced in Marin County: Almonte;
Tiburon; Fairfax.

4. **H. Hystrix** Roth. MEDITERRANEAN BARLEY. In clay soil of grasslands, particularly common in northern Marin County: Sausalito; Tiburon; Almonte; Potrero
Meadow, Mount Tamalpais; Little Carson; Olema, acc. Davy; San Antonio; Chileno Valley. Introduced from Europe.—*H. Gussoneanum* Parl.

5. **H. leporinum** Link. FARMER'S FOXTAIL. Common and widespread on grassy flats and slopes and in weedy places about towns: Sausalito, Angel Island, Tiburon, and Mount Tamalpais (Potrero Meadow) to San Rafael, Black Point, and Point Reyes Peninsula (Inverness, Bolinas). Introduced from Europe.—*H. murinum* of California references, in part.

6. **H. Stebbinsii** Covas. Known in Marin County only from Angel Island, Tiburon, Hamilton Field, and Ignacio, but probably more widespread and mistaken for the preceding. This foxtail barley, which has been included in *H. murinum* in American references, is a native of the Old World where it was called *H. murinum* L. var. *pedicellatum* Pau & Font Quer. Field evidence would seem to indicate that it is in flower at least two weeks later than *H. leporinum*.

7. **H. vulgare** L. BARLEY. Occasional along roads but not widely persistent: Sausalito; Angel Island; Tiburon; San Rafael; Lagunitas; Inverness; Dillons Beach. Introduced from the Old World.

Hordeum jubatum L., foxtail barley, has been listed by Stacey from Mount Tamalpais and San Rafael but no specimens have been seen to substantiate his records.

21. Lolium

a. Glumes at least as long as the spikelet .... 3. *L. temulentum*
a. Glumes shorter than the spikelets .... b
b. Lemmas usually awned; plants annual .... 1. *L. multiflorum*
b. Lemmas awnless or nearly so; plants perennial .... 2. *L. perenne*

1. **L. multiflorum** Lamk. ITALIAN RYEGRASS. Common in grassy places, especially around towns: Sausalito; Tiburon; Mill Valley; San Rafael; Bolinas; Inverness. Introduced from Europe. An anomalous form with branching inflorescence has been found on Angel Island and near Bolinas. On Angel Island and in the salt marsh at Stinson Beach an awnless form, var. *muticum* DC., was found growing with the typical awned form.

2. **L. perenne** L. PERENNIAL RYEGRASS. Occasional, usually in towns: Tiburon; Mill Valley; San Anselmo; San Rafael; Ignacio; Inverness and Bolinas, Point Reyes Peninsula. Introduced from Europe.

3. **L. temulentum** L. DARNEL. Widespread and occasional around towns or on grassy slopes in the hills: Rattlesnake and Bootjack, Mount Tamalpais; Mill Valley; San Rafael; Fairfax Hills; Inverness. Introduced from Europe.

Lolium temulentum L. var. *leptochaeton* A. Br., a form in which the lemmas are awnless, has been collected at Kentfield by Miss Eastwood.

22. Monerma

1. **M. cylindrica** (Willd.) Coss. & Dur. Occasional in subsaline situations bordering salt marshes: Rodeo Lagoon; Angel Island; Tiburon; Sausalito; Dillons Beach. Also locally established along the coast road near Muir Beach and on Mount Tamalpais at Lake Lagunitas and in the serpentine area above Bootjack. Introduced from the Old World.—*Lepturus cylindricus* (Willd.) Trin.

23. Parapholis

1. **P. incurva** (L.) C. E. Hubbard. SICKLE GRASS. In subsaline soil bordering salt marshes: Rodeo Lagoon; Tiburon; Manzanita; San Rafael; near Ignacio; San

Antonio; near Olema; Stinson Beach; Inverness; Drakes Bay; Point Reyes Lighthouse; Dillons Beach. Native of Europe.—*Pholiurus incurvus* (L.) Schinz &
Thell.

24. Scribneria

1. **S. Bolanderi** (Thurb.) Hack. Rare in thin soil of rocky ridges, generally
where water seeps or stands in the spring: Mount Tamalpais (near Bootjack, between Potrero Meadow and Mountain Theater, on serpentine above Camp
Handy) ; Carson Ridge; Big Carson, *Eastwood*.

Usually in *Scribneria* there is only one spikelet at each node of the rachis but
in robust individuals there are frequently two spikelets at the middle nodes.
When two spikelets are present, one is subsessile while the other is raised on a
stout pedicel that is about half the length of the lower spikelet. In some forms
of *S. Bolanderi* the awn on the lemma is conspicuously elongate, but in the
Marin County plants the awn is shorter, sometimes scarcely surpassing the glumes.

TRIBE 3. **Aveneae.** OAT TRIBE

25. Koeleria

1. **K. cristata** (L.) Pers. Widespread on grassy slopes and in dry meadows but
not very abundant: Sausalito, Angel Island, Tiburon, and Mount Tamalpais
(Bootjack, Rifle Camp) to San Rafael Hills, Fairfax Hills, and Point Reyes Peninsula.

26. Trisetum

1. **T. canescens** Buckl. Common and widespread on the edge of thickets and in
open or dense woodland: Sausalito, Tiburon, and Mount Tamalpais (Corte Madera Ridge, Rock Spring, Potrero and Lagunitas meadows) to San Rafael Hills,
Dillons Beach, and Point Reyes Peninsula (Inverness, Ledum Swamp, Point
Reyes) .

27. Deschampsia. HAIRGRASS

a. Plants annual; leaves sparse, not tufted at base of culms. Glumes 6 mm. long
   . . . . .1. *D. danthonioides*
a. Plants perennial; leaves numerous and mostly tufted at base of culms . . . . . . .b
b. Leaves filiform; inflorescence elongate, very slender, the branches appressed-
   ascending; glumes 4 mm. long . . . . .2. *D. elongata*
b. Leaves not filiform; inflorescence broader, the branches ascending or somewhat
   spreading; glumes 4–6 mm. long . . . . .3. *D. holciformis*

1. **D. danthonioides** (Trin.) Munro. In clay soil, generally in low places where
water has seeped or stood: Tiburon; Potrero and Lagunitas meadows, Mount
Tamalpais; Carson Ridge; Chileno Valley; Hamilton Field; Black Point.

2. **D. elongata** (Hook.) Munro. Rather common on moist flats in brushy or
wooded hills: Sausalito; Rock Spring and Kent Trail, Mount Tamalpais; Fairfax
Hills; Carson country; San Rafael Hills; Bolinas and Inverness, Point Reyes Peninsula.

3. **D. holciformis** Presl. Usually in marshy places in hills near the coast, occasionally on drier slopes: Rodeo Lagoon; Potrero and Lagunitas meadows,
Mount Tamalpais; Mount Vision and Ledum Swamp, Point Reyes Peninsula.

This hairgrass is one of the several handsome grasses with robust tufted habit common in wet coastal swales. Others are *Festuca californica* Vasey and *Calamagrostis nutkaensis* (Presl) Steud.

28. Aira

a. Spikelets 2–2.5 mm. long, usually with one exserted awn ......1. *A. capillaris*
a. Spikelets about 3 mm. long, with 2 exserted awns ....b
b. Inflorescence diffuse, the branches slender and elongate ....2. *A. caryophyllea*
b. Inflorescence compact, the branches short and appressed-ascending 3. *A. praecox*

1. **A. capillaris** Host. Occasional on grassy hills or on the edge of brush: Mount Tamalpais, acc. Stacey; Kentfield; San Anselmo; San Rafael Hills; Hamilton Field; Chileno Valley; Tocaloma, acc. Davy. Introduced from Europe.
2. **A. caryophyllea** L. Common throughout the grass- or brush-covered hill country of Marin County, rare in the shade of woods: Sausalito, Tiburon, and Mount Tamalpais to San Rafael Hills, Tomales, and Point Reyes Peninsula. Introduced from Europe.
Between this species and the preceding there are occasional puzzling intermediates in which the spikelets are small and long-pedicellate but 2-awned.
3. **A. praecox** L. Occasional in sandy soil on maritime bluffs and mesas on Point Reyes Peninsula: near the lighthouse and behind the dunes near the radio station.

29. Avena. OAT

a. Lemmas awn-tipped ....1. *A. barbata*
a. Lemmas not awn-tipped ....b
b. Lemmas usually conspicuously hairy, the awn twisted and bent ....2. *A. fatua*
b. Lemmas glabrous, the awn straight or lacking ....3. *A. sativa*

1. **A. barbata** Brot. In Marin County the most abundant species of oat in grassland, on brushy slopes, and in waste places around towns: Sausalito, Tiburon, and Mount Tamalpais to San Rafael, Black Point, and Point Reyes Peninsula. Introduced from Europe.
2. **A. fatua** L. Frequently with *A. barbata* but not so common: Sausalito; Tiburon; Bootjack; Mount Tamalpais; San Rafael; Fairfax; Inverness. Introduced from Europe.
3. **A. sativa** L. Occasional as an escape from cultivation: Rodeo Lagoon; Sausalito; Mill Valley; Berry Trail, Mount Tamalpais; Carson country; Ignacio; Point Reyes Peninsula. Native of the Old World.

30. Arrhenatherum

1. **A. elatius** (L.) J. & C. Presl. Introduced from Europe and escaped from cultivation: Mill Valley, acc. Stacey; San Geronimo.

31. Holcus

1. **H. lanatus** L. VELVET GRASS. Very common near the coast on grassy slopes and along the edge of thickets, less common towards the interior: Sausalito and Mount Tamalpais (Bootjack) to Stinson Beach, Dillons Beach, and Point Reyes Peninsula. The attractive form with violet-tinted inflorescences, var. *coloratus* Rchb.,

is occasional. At Bolinas and Ledum Swamp on Point Reyes Peninsula plants
have been found in which the spikelets are viviparous.

32. Danthonia

1. **D. californica** Boland. Widespread and sometimes abundant in grassland
of open hills and meadows: Sausalito; Angel Island; Tiburon; Mount Tamalpais
(Throckmorton, Azalea Flat, Bootjack, Rifle Camp); Fairfax Hills; San Rafael
Hills. The form found on serpentine near Barths Retreat, Mount Tamalpais, has
smaller spikelets that are usually solitary.

Much less common than the glabrous-leaved type is var. *americana* (Scribn.)
Hitchc. with pilose leaves: Sausalito; Angel Island; Tiburon; Bootjack and
Phoenix Lake, Mount Tamalpais.

Tribe 4. Agrostideae. Timothy Tribe
33. Calamagrostis. Reedgrass

a. Hairs of the callus and rachilla about as long as the lemma. Glumes 3–4 mm.
   long ....5. *C. crassiglumis*
a. Hairs of the callus and rachilla ⅓–½ the length of the lemma ....b
b. Plants with elongate rhizomes; collar of sheaths hairy; glumes 3–5 mm. long
   ....1. *C. rubescens*
b. Plants tufted, with or without short rhizomes; collar of sheaths glabrous or
   scabrous; glumes 5–7 mm. long c
c. Awn of lemma shorter than the glumes; callus hairs about 2 mm. long; robust
   plants usually 1 m. or more tall growing in moist places near the coast
   ....2. *C. nutkaensis*
c. Awn of lemma equaling or longer than the glumes; callus hairs about 1 mm.
   long; tufted plants of dry rocky banks and hillsides ....d
d. Awn about equaling the glumes ....3. *C. koelerioides*
d. Awn exceeding the glumes by 1–2 mm. ....4. *C. purpurascens*

1. **C. rubescens** Buckl. Occasional on shady or partially shady slopes in chaparral or woodland: Mill Valley; Mount Tamalpais (Tenderfoot Trail, Azalea Flat,
Willow Meadow); near Liberty Spring. Not infrequently in a given season broad
patches of this grass will produce no flowering culm.

The length of the spikelet in plants from the chaparral of Mount Tamalpais
is appreciably smaller than that usually recorded for the species.

2. **C. nutkaensis** (Presl) Steud. Not uncommon in freshwater marshes near the
coast and occasional around springs farther inland: Sausalito; Azalea Flat, Mount
Tamalpais; Ledum Swamp and swales among dune hills, Point Reyes Peninsula.

3. **C. koelerioides** Vasey. Rocky slopes on the edge of brush: north of East Peak,
Mount Tamalpais; near Liberty Spring. The plants from the latter station may
represent a robust form of *C. rubescens* with sheath collars merely scabrous or
puberulent.

4. **C. purpurascens** R. Br. var. **ophitidis** J. T. Howell. An attractive bunchgrass
that in Marin County is usually common on rocky slopes of serpentine areas:
Tiburon; Mount Tamalpais (Bootjack, Artura Trail, Collier Spring); Carson
Ridge. The type locality is on the trail between Rock Spring and Laurel Dell.

This grass, which is known from the serpentine areas of Lake and Marin counties, has not been generally distinguished from the species which is widespread through the northern part of North America. Two other montane plants which are widely distributed in western North America and which reach Marin County along with the *Calamagrostis* are *Cheilanthes gracillima* D. C. Eat. and *Erigeron inornatus* Gray.

5. **C. crassiglumis** Thurb. In Marin County known only from swales near the radio station on Point Reyes Peninsula, the southernmost station for this rare coastal reedgrass.

34. Ammophila

1. **A. arenaria** (L.) Link. BEACHGRASS (plate 16). Widely naturalized on coastal dunes: Point Reyes Peninsula; Stinson Beach; Dillons Beach. Originally introduced from Europe as a sand binder in Golden Gate Park, San Francisco, and now planted extensively along the Pacific coast in dune areas.

35. Agrostis. BENTGRASS

a. Palea present, conspicuous, generally half as long as the lemma or longer ...b

a. Palea lacking, or if present at most ⅓ the length of the lemma ....f

b. Lemma hairy on the back; rachilla prolonged behind the palea; rhizomes lacking ....1. *A. retrofracta*

b. Lemma not hairy on the back; rachilla not prolonged; rhizomes or stolons generally developed c

c. Floret about half as long as the glumes; palea nearly as long as the lemma ....2. *A. semiverticillata*

c. Floret more than half the length of the glumes; palea much shorter than the lemma ....d

d. Ligule inconspicuous, usually about 1 mm. long ....5. *A. tenuis*

d. Ligule conspicuous, usually 2 mm. long or more ....e

e. Branches of panicle appressed-ascending ....3. *A. palustris*

e. Branches of panicle spreading ....4. *A. alba*

f. Rhizomes present, long-creeping ....g

f. Rhizomes lacking (or rarely present and very short) i

g. Inflorescence closely branched; coastal dunes ....8. *A. pallens*

g. Inflorescence loosely branched; coastal slopes and interior hills ....h

h. Spikelets 3.5–4.5 mm. long; lemma 3 mm. long; hairs at base of lemma more than 1 mm. long ....6. *A. Hallii*

h. Spikelets 2.5–3.5 mm. long; lemma 2–2.5 mm. long; hairs at base of lemma 1 mm. long or less ....7. *A. diegoensis*

i. Branches of panicle loosely ascending, very slender, not bearing spikelets near the base ....15. *A. longiligula*

i. Branches of panicle appressed-ascending, at least some branches bearing spikelets from near the base ....j

j. Leaves densely tufted, the blades involute, filiform; inflorescence very narrow, the branches strictly appressed ....9. *A. Blasdalei*

j. Leaves not densely tufted, the blades flat; inflorescence broader, loose to dense, scarcely linear-elongate ....k

k. Lemmas awnless or the awns when present scarcely exceeding the glumes ...l

k. Lemmas awned, the awns conspicuously exceeding the glumes ....m

1. Culms usually erect and tall; panicles loose to dense, usually well exserted from
    the uppermost leaf sheaths; glumes scabrous on the keel, smooth or scabrous
    on the sides; palea usually less than 0.5 mm. long ....10. *A. exarata*
1. Culms frequently spreading-ascending, low and stout; panicles dense, spikelike,
    at least some partly enclosed or closely subtended by the somewhat inflated
    uppermost leaf sheaths; glumes scabrous on keel and sides; palea about 0.5
    mm. long ....12. *A. californica*
m. Anthers about 0.5 mm. long ....n
m. Anthers about 1 mm. long to nearly 2 mm. long ....o
n. Plants perennial, the culms usually stout and tall ..10*a*. *A. exarata* var. *pacifica*
n. Plants annual, the culms slender, not very tall ....13. *A. microphylla*
o. Plants perennial; palea 0.5 mm. long or less ....11. *A. ampla*
o. Plants annual; palea about 1 mm. long ....14. *A. aristiglumis*

    1. **A. retrofracta** Willd. Sidewalk weed, Fairfax. Introduced from the southwest
Pacific area.
    2. **A. semiverticillata** (Forsk.) C. Christ. Widespread weedy grass of rather com-
mon occurrence, generally in low places along streets or roads where water collects
in the winter and spring: Sausalito, Angel Island, and Tiburon to San Rafael and
Inverness. Native of Europe.—*A. verticillata* Vill.
    3. **A. palustris** Huds. Low moist soil near the coast, in Marin County only
known from the east side of Bolinas Lagoon at McKennans Landing. Along the
coast, this is the southernmost known station.
    4. **A. alba** L. REDTOP. Escaping from cultivation along roads, in meadows, and
in waste places around habitations: Sausalito, Tiburon, and Mount Tamalpais
(West Point, Lagunitas Meadows) to Santa Venetia, Inverness, and Dillons Beach.
Introduced from Europe.
    5. **A. tenuis** Sibth. Occasional in Marin County, probably escaping from cul-
tivation: shaded canyon, Cascade Drive, Mill Valley; moist meadowy patch near
dunes and in swales, Point Reyes Peninsula. Introduced from Europe.
    6. **A. Hallii** Vasey. Widely distributed but not common, occurring on wooded
or brushy slopes usually in partial shade: Rodeo Lagoon; Sausalito; Mount Tamalpais (Pipeline Trail, West Point, above Bootjack, Potrero Meadow) ; San Anselmo Canyon; San Rafael Hills; Gallinas Valley, *Leschke;* Inverness Ridge; Bolinas; Dillons Beach.
    7. **A. diegoensis** Vasey. Common and widespread on partially shaded flats and
banks in woods, chaparral, and meadows: Rodeo Lagoon; Angel Island; Tiburon;
Mount Tamalpais; San Rafael Hills; Black Point; Carson country; Inverness,
Point Reyes Peninsula. In Marin County this species and the preceding are nearly
confluent and at times are difficult to distinguish, although *A. diegoensis* is generally more delicate in foliage and inflorescence. At maturity the inflorescence is
quite slender with the branches appressed-ascending, so that the difference between this species and the maritime *A. pallens* is not readily apparent.
    8. **A. pallens** Trin. Maritime dunes on Point Reyes Peninsula. This plant, prob-
ably representing a dune form of the preceding, is known south of Marin County
only in San Francisco.
    9. **A. Blasdalei** Hitchc. Occasional in shallow sandy or gravelly soil or on dune
hills: Point Reyes Peninsula (Point Reyes, Drakes Estero, dunes near the radio

station, north of Abbotts Lagoon) ; Dillons Beach. Not known south of Marin County.

10. **A. exarata** Trin. Moist or wet places in the hills: Sausalito; Tiburon; Muir Woods; Barths Retreat, Mount Tamalpais; San Anselmo Canyon; Point Reyes Peninsula (Bolinas, Inverness, Ledum Swamp) .

10*a*. **A. exarata** Trin. var. **pacifica** Vasey. Wet or moist soil along streams or in summer-dried streambeds, rather common: Mount Tamalpais (Laurel Dell, Potrero Meadow, Willow Meadow, Lagunitas Meadows, Phoenix Lake) ; Deer Park, Santa Venetia; Chileno Valley; Point Reyes Station.

11. **A. ampla** Hitchc. Rare along streams or on coastal flats: Lagunitas Meadows, Mount Tamalpais; Bolinas.

12. **A. californica** Trin. Occasional on rocky slopes and clay flats of coastal mesas and dune hills: Rodeo Lagoon; Point Reyes Peninsula (Bolinas, Drakes Estero, Point Reyes, McClure Beach) ; Dillons Beach.

13. **A. microphylla** Steud. Rare, in thin soil overlying serpentine and in sandy places near the coast: Tiburon; Carson Ridge; Chileno Valley; Point Reyes Peninsula (dunes near the radio station, rocks north of Abbotts Lagoon) .

14. **A. aristiglumis** Swallen. Known only from the type locality on Point Reyes Peninsula, a slope of gravelly soil near Drakes Estero west of Mount Vision.

15. **A. longiligula** Hitchc. var. **australis** J. T. Howell. Marshy places on Point Reyes Peninsula at Ledum Swamp, the type locality, and Shell Beach. The Marin County plant, var. *australis,* differs from the typical form in the very short awns which do not exceed the glumes and in the notably long ligules which may be more than 1 cm. in length. The typical form with slender exserted awns is found along the Oregon and California coast as far south as Mendocino County, while the species as a whole reaches its southern distributional limit in Marin County.

36. Alopecurus

a. Plant perennial; spikelet 2.5 mm. long, the awn exserted 2–3 mm.
. . . . .1. *A. geniculatus*
a. Plant annual; spikelet about 4 mm. long, the awn exserted 5–6 mm.
. . . . .2. *A. Howellii*

1. **A. geniculatus** L. WATER FOXTAIL. Widespread through the northern hemisphere but in Marin County rare in marshy places on Point Reyes Peninsula: Bolinas; Bear Valley, acc. Davy; near the radio station.

2. **A. Howellii** Vasey. Rare in northern Marin County along the borders of vernal ponds: laguna in Chileno Valley; east of Aurora School.

37. Polypogon

a. Lemma not awned; glumes with long hairy lobes . . . . .3. *P. maritimus*
a. Lemma awned; glumes entire or with very short lobes at apex . . . . .b
b. Glumes scabrous, the apex entire or bidentulate, the margins not scarious; plants perennial . . . . .1. *P. interruptus*
b. Glumes scabrous and hairy, the apex with short lobes, the margins scarious; plants annual . . . . .2. *P. monspeliensis*

1. **P. interruptus** H.B.K. Common along roadside ditches and in low places in fields and meadows: Angel Island, Sausalito, and Tiburon to Mount Tamalpais,

San Rafael, and Point Reyes Peninsula. Introduced from Europe.—*P. lutosus* and
P. littoralis of American references.

2. **P. monspeliensis** (L.) Desf. RABBITFOOT GRASS. Common in seepages and in
low places that are wet during the rainy season: Sausalito, Tiburon, and Mount
Tamalpais to San Rafael Hills, Point Reyes Peninsula, and Dillons Beach. Introduced from Europe. Both this and the preceding species vary greatly in habit depending on environmental conditions.

3. **P. maritimus** Willd. In Marin County known only on the north side of
Mount Tamalpais at Willow Meadow on a low gravelly flat where water collects
in the spring. Introduced from the Old World.

38. **Phleum.** TIMOTHY

a. Inflorescence linear, several times longer than broad . . . . .1. *P. pratense*
a. Inflorescence oblongish, less than 3 times longer than broad . . . .2. *P. alpinum*

1. **P. pratense** L. Occasionally escaped from cultivation, as at Santa Venetia.
Native of the Old World.

2. **P. alpinum** L. Dune hills and coastal swales: near the radio station, Point
Reyes Peninsula; Dillons Beach. These are the southernmost known stations along
the coast for this high altitude or high latitude timothy. According to some European workers the variant with circumpolar distribution should be called *P.
commutatum* Gaud.

39. **Gastridium**

1. **G. ventricosum** (Gouan) Schinz & Thell. Common in clay soil on summer-dry
slopes and flats: Angel Island; Tiburon; Bootjack and Rock Spring, Mount
Tamalpais; Fairfax Hills; San Rafael Hills; Point Reyes Peninsula.

40. **Lagurus.** HARE'S TAIL GRASS

1. **L. ovatus** L. Cultivated and occasionally naturalized: Sausalito, *Eastwood*.
Native of Europe.

41. **Stipa.** NEEDLEGRASS

a. Spikelet more than 1 cm. long, generally 1.5–2 cm. long (excluding awn)
. . . . .1. *S. pulchra*
a. Spikelet 1 cm. long or less . . . . .2. *S. lepida*

1. **S. pulchra** Hitchc. Common and widespread bunchgrass of slopes and flats:
Sausalito, Angel Island, Tiburon, and Mount Tamalpais (Mountain Theater, Rifle
Camp, Phoenix Lake) to San Rafael Hills, Black Point, and Point Reyes Peninsula.

2. **S. lepida** Hitchc. Common bunchgrass throughout the hilly grasslands of
Marin County: Rodeo Lagoon; Sausalito; Angel Island; Tiburon; Azalea Flat and
Bootjack, Mount Tamalpais; Fairfax Hills; Carson country; San Rafael Hills;
Salmon Creek. In rocky exposed places in shallow soil, the plants become much
reduced in size and the inflorescence few-flowered and slender. This form, which
has been called var. *Andersonii* (Vasey) Hitchc., has been found on the ridge west
of Mill Valley and in the Fairfax Hills. In Marin County the variety intergrades
completely with the typical aspect of the species.

TRIBE 5. **Chlorideae.** CHLORIS TRIBE

42. Cynodon

1. **C. Dactylon** (L.) Pers. BERMUDA GRASS. Commonly occurring about towns as an escape from lawns: Sausalito; Tiburon; Mill Valley; San Anselmo; San Rafael; Inverness. On Mount Tamalpais Bermuda grass has been noted at West Point and Mountain Theater. Introduced from the Old World.

43. Beckmannia

1. **B. Syzigachne** (Steud.) Fern. AMERICAN SLOUGHGRASS. Rare in Marin County on muddy strands, marshy flats, and streambanks: Lake Lagunitas, Mount Tamalpais; Olema Creek; Hamilton Field; Novato; Hicks Valley. These are the southernmost stations in the Coast Ranges.—*B. erucaeformis* of American references.

44. Spartina

1. **S. foliosa** Trin. PACIFIC CORDGRASS. Common in salt marshes bordering San Francisco Bay and occasional on the edge of lagoons along the ocean: Sausalito; Tiburon; San Rafael; Black Point; Burdell; Stinson Beach; Drakes Bay.

Pacific cordgrass is generally the first plant to appear on tidal flats where it frequently establishes broad pure stands. Later it is succeeded by *Salicornia* and a more diversified salt marsh association as higher ground is built up around it. In this later association *Spartina* still occurs as a narrow fringe along tidal sloughs and also occasionally as a localized colony in low areas.—*S. leiantha* Benth.

TRIBE 6. **Phalarideae.** CANARYGRASS TRIBE

45. Hierochloe

1. **H. occidentalis** Buckl. CALIFORNIA VANILLA GRASS. Common in coastal woods on moist shady slopes and flats: Sausalito; Muir Woods; Mount Tamalpais (Cascade Canyon, Cataract Gulch, Fish Grade); San Rafael Hills; Inverness.

Although this attractive and fragrant grass is especially characteristic of the deep shade of redwood groves, it occurs at Sausalito and Inverness where no redwoods are found. "Redwoods of Marin County" is the type locality of *H. macrophylla* Thurb. on which the name *Torresia macrophylla* (Thurb.) Hitchc. is based.

46. Anthoxanthum. VERNALGRASS

a. Plants perennial; longer awn little exceeding the upper glume. .1. *A. odoratum*
a. Plants annual; longer awn much exceeding the upper glume. . .2. *A. aristatum*

1. **A. odoratum** L. In Marin County known only from a hillside between Mill Valley and Muir Woods. Native of the Old World.

2. **A. aristatum** Boiss. Locally abundant in a low field between Ignacio and Hamilton Field. Introduced from Europe.

47. Phalaris. CANARYGRASS

a. All except 1 spikelet of each branch of panicle sterile; sterile lemmas none
. . . . .1. *P. paradoxa*
a. All spikelets alike and fertile; sterile lemmas 1 or 2 . . . . .b

b. Glumes more or less keeled but not winged c
b. Glumes narrowly to broadly winged ....d
c. Plants perennial, forming large robust clumps; lateral nerves of glumes somewhat raised but not keel-like ....2. *P. californica*
c. Plants annual; lateral nerves of glumes prominent and keel-like
.... 3. *P. Lemmonii*
d. Plants perennial, large and robust. Sterile lemma usually solitary
.... 4. *P. tuberosa*
d. Plants annual ....e
e. Sterile lemma 1. Glumes with keel rather broadly winged ....5. *P. minor*
e. Sterile lemmas 2 ....f
f. Glumes broadly winged ....6. *P. canariensis*
f. Glumes narrowly winged ....g
g. Panicle elongate-cylindric; spikelets densely crowded, 3.5–4 mm. long
.... 7. *P. angusta*
g. Panicle ovate to oblong; spikelets not densely crowded, 5–6 mm. long
.... 8. *P. caroliniana*

1. **P. paradoxa** L. Occasional in valleys and on grassy slopes near roads: Summit Avenue above Mill Valley; Ridgecrest Road near Alpine Lake; Burdell; Chileno Valley. Native of Europe.

2. **P. californica** H. & A. Widespread and rather common in moist or wet soil of marshes, springs, and streambeds: Sausalito, Angel Island, Tiburon, and Mount Tamalpais (Bootjack, Potrero Meadow, Rifle Camp) to Lucas Valley and Point Reyes Peninsula. In robust plants the culms may attain a height of 6 feet. Frequently the spikelets are attractively tinged with lavender or purple.

3. **P. Lemmonii** Vasey. Especially characteristic of low wet places in fields bordering salt marshes, sometimes locally abundant: Greenbrae; San Rafael; Hamilton Field; Burdell, *Leschke;* Black Point. Also found on the fire road near the head of Black Canyon, San Rafael Hills, where it was undoubtedly introduced.

4. **P. tuberosa** L. var. **stenoptera** (Hack.) Hitchc. Large clumps with numerous stout culms, occasional along roads: Tiburon; near Stinson Beach; Novato; San Geronimo and Point Reyes Station, *Leschke.* Introduced into California for forage.

5. **P. minor** Retz. Widely distributed and rather common on grassy slopes and flats and in waste places: Sausalito; Angel Island; Tiburon; Mill Valley; above Cascade Canyon, Mount Tamalpais; Greenbrae; San Rafael; Fairfax. Native of the Old World.

6. **P. canariensis** L. CANARYGRASS. Adventive on the edge of thickets in Fairfax near Deer Park. Introduced from the Old World.

7. **P. angusta** Nees. Rare in the northern part of Marin County: Novato Creek; laguna in Chileno Valley.

8. **P. caroliniana** Walt. In Marin County known only from Bootjack on Mount Tamalpais, *L. S. Rose,* where it has probably been introduced.

TRIBE 7. **Paniceae.** MILLET TRIBE

48. **Digitaria**

1. **D. sanguinalis** (L.) Scop. CRABGRASS. Occasional weedy grass about habitations and in waste ground: Mill Valley; Ross; Alpine, Lagunitas, and Phoenix lakes; San Rafael. Native of Europe.

49. Stenotaphrum

1. **S. secundatum** (Walt.) Kuntze. St. Augustine Grass. Escaping from cultivation in Fairfax; locally naturalized on clay slopes and flats west of Mill Valley on the road to Muir Woods. Native of the southeastern United States.

50. Paspalum

a. Plants with long stolons; racemes 1–3, usually 2 . . . . .1. *P. distichum*
a. Plants without stolons; racemes several, usually 3–5 . . . . .2. *P. dilatatum*

1. **P. distichum** L. Knotgrass. Occasional in wet places, generally along roads: Tiburon; Stinson Beach; San Rafael; Ignacio; Point Reyes Station; Inverness.

2. **P. dilatatum** Poir. In Marin County known only in low ground at McKennans Landing on Bolinas Lagoon near Stinson Beach. Introduced from South America as a forage grass.

51. Panicum

a. Spikelets about 2 mm. long . . . . .1. *P. pacificum*
a. Spikelets 4.5–5 mm. long . . . . .2. *P. miliaceum*

1. **P. pacificum** Hitchc. & Chase. In moist meadows or on slopes where water seeps: Mount Tamalpais (Bootjack, Azalea Flat, Potrero Meadow); Point Reyes Peninsula.

2. **P. miliaceum** L. Broomcorn Millet. This widely cultivated species has been listed from Mount Tamalpais by Stacey but no Marin County specimen has been seen.

52. Echinochloa

1. **E. Crus-galli** (L.) Beauv. Occasional in wet ground in waste places: Mill Valley; Pipeline Trail, Mount Tamalpais; Fairfax Hills; Novato.

Escaping from cultivation and less common than the species is var. *frumentacea* (Roxb.) Wight. The variety, which may be distinguished from the species by the awnless lemmas, has been found on San Antonio Creek and near Inverness.

53. Setaria

1. **S. geniculata** (Lamk.) Beauv. Locally established on the strand of Phoenix Lake where undoubtedly it has been introduced.—*Chaetochloa geniculata* (Lamk.) Millsp. & Chase; *S. gracilis* H.B.K.

Tribe 8. Andropogoneae. Sorghum Tribe

54. Andropogon. Beardgrass

1. **A. glomeratus** (Walt.) B.S.P. Moist or wet springy slopes in serpentine areas: Laurel Dell Trail and Bootjack, Mount Tamalpais; Liberty Spring. Indigenous in southern California but undoubtedly introduced in Marin County and elsewhere in northern California. The grass has become a conspicuous feature at the west end of Mount Tamalpais but the earliest collection seen from there is one made in 1938.

55. Sorghum

a. Plants perennial; culms slender; panicle openly branched . . . . .1. *S. halepense*
a. Plants annual; culms thick; panicle compactly branched . . . . .2. *S. vulgare*

1. **S. halepense** (L.) Pers. JOHNSON GRASS. In Marin County seen only at Novato and on San Antonio Creek; a common weed or forage grass in the interior valleys of California. Native of the Old World.

2. **S. vulgare** Pers. Occasional but probably not persisting. Specimens from along the railroad in Sausalito and San Rafael are referable to the variety known as Egyptian corn that is cultivated for seed.

CYPERACEAE. SEDGE FAMILY

a. Flowers unisexual, the pistillate enclosed in a saclike bract (perigynium)
.... 5. *Carex*
a. Flowers mostly perfect, if unisexual, the pistillate flowers not enclosed by a saclike bract ....b
b. Spikelets laterally flattened, the scales in 2 rows ....1. *Cyperus*
b. Spikelets not laterally flattened, the scales spirally arranged c
c. Spikelet solitary, terminating a smooth, apparently leafless culm; involucral bract none ....2. *Eleocharis*
c. Spikelets usually more than 1 and variously clustered, or if only 1, then subtended by an involucral bract ....d
d. Base of style not much enlarged ....3. *Scirpus*
d. Base of style much enlarged and persisting on the achene ...4. *Rhynchospora*

1. Cyperus

a. Plants annual; spikelets brownish-purple, about 1 mm. wide ..1. *C. difformis*
a. Plants perennial; spikelets yellowish-green or whitish blotched with brown, 2–4 mm. wide ....b
b. Spikelets 2–4 mm. wide; inflorescence subtended by 3–6 leaves
....2. *C. Eragrostis*
b. Spikelets about 2 mm. wide; inflorescence subtended by 10 or more leaves
....3. *C. alternifolius*

1. **C. difformis** L. In Marin County known only from Mount Tamalpais in a seepage above Muir Woods at the lower end of the Sierra Trail. This species, which is a native of the Old World, is a common weed in California rice fields.

2. **C. Eragrostis** Lamk. Common in wet places, around springs, and along streams, from Tiburon, Sausalito, and Mount Tamalpais (West Point, Phoenix Lake) to San Rafael, Ignacio, and Point Reyes Peninsula.—*C. vegetus* Willd.; *C. virens* of authors, not Michx., as to Marin County plant.

3. **C. alternifolius** L. UMBRELLA PLANT. Occasionally escaping from cultivation and becoming established in wet places as at Inverness. Introduced from Madagascar.

2. Eleocharis. SPIKE-RUSH

a. Style branches 3; achenes 3-angled ....b
a. Style branches 2; achenes biconvex ....d
b. Culms less than 2 dm. tall, very slender; persistent base of style distinct in color and texture from the achene ....3. *E. acicularis*
b. Culms usually 2 dm. tall or more, stouter; persistent base of style confluent in texture with the achene c
c. Sterile culms not rooting at the tip; roots reddish ....1. *E. pauciflora*

c. Sterile culms rooting at the tip; roots pale and spongy.........2. *E. rostellata*
d. Plants perennial; persistent base of style much narrower than the achene
....4. *E. macrostachya*
d. Plants annual; persistent base of style as broad as the achene
....5. *E. Engelmannii*

1. **E. pauciflora** (Lightf.) Link. One of the rare occurrences of this widely distributed boreal species in the California Coast Ranges is in swales near the radio station on Point Reyes Peninsula. The plants are relatively robust in habit (about 1 dm. tall) and have been called var. *Suksdorfiana* (Beauverd) Svenson by Svenson although they lack the underground bulblets characteristic of that variety.

2. **E. rostellata** (Torr.) Torr. WALKING SEDGE. Locally common in marshes and wet meadows: Tiburon; Mount Tamalpais (Azalea Flat, near Rock Spring, Lake Lagunitas); south end of Carson Ridge; Ledum Swamp, Point Reyes Peninsula.

The sterile culms frequently arch over and, touching the ground, root at the tip. The sensation produced by walking through a patch of this sedge when the culms have so rooted is unmistakable.

3. **E. acicularis** (L.) R. & S. Beds of ponds or low ground inundated during the rainy season: Hidden Lake, Mount Tamalpais; east of Aurora School; Point Reyes, acc. Stacey. Our form of this widespread species is var. *occidentalis* Svenson.

4. **E. macrostachya** Britt. Widespread and rather common in wet meadows or shallow water: Rodeo Lagoon; Tiburon; Phoenix Lake and Lake Lagunitas, Mount Tamalpais; San Rafael; west of Novato; Olema Marshes; Mount Vision, Point Reyes Peninsula; Dillons Beach.—*E. palustris* of authors, not *Scirpus palustris* L.

This species presents several aspects in Marin County and according to some students more than one species might be recognized. The texture of the stem (broad and soft or round and firm), the color of the scales (purplish or pale buff), and the size of the tubercle on the achene (as long as wide or longer) are the variables that have been noted but they have not been successfully aligned with named segregates of this variable complex.

5. **E. Engelmannii** Steud. Marshy ground bordering Hidden Lake on the north slope of Mount Tamalpais; apparently not otherwise known from the Coast Ranges. The Tamalpais plant corresponds to the form found in the middle and northern Sierra Nevada that has been called *E. Engelmannii* var. *monticola* (Fern.) Svenson (*E. monticola* Fern.).

3. **Scirpus.** BULRUSH; TULE

a. Plants annual; culms slender, mostly less than 2 dm. tall ....b
a. Plants perennial; culms stout, mostly over 3 dm. tall c
b. Bract a little shorter or a little longer than the spikelets; scales not keeled
....1. *S. cernuus*
b. Bract much longer than the spikelets; scales prominently keeled
....2. *S. koilolepis*
c. Bract subtending the inflorescence not leaflike, simulating a continuation of the culm ....d
c. Bracts subtending the inflorescence several and leaflike i

d. Culms sharply 3-angled ....e
d. Culms obtusely triangular or round ....f
e. Culms firm, not spongy, the sides flat; bract much longer than the spikelets
....3. *S. americanus*
e. Culms soft and spongy, the sides deeply concave-channeled; bract little longer
than the spikelets ....4. *S. Olneyi*
f. Culms 3-sided; rays of inflorescence smooth or nearly so, frequently extended
and forming a loosely branched panicle; perianth bristles reddish, plumose-
hairy ....5. *S. californicus*
f. Culms round; rays of inflorescence scabrous with short upwardly pointing
trichomes, the panicle loosely to compactly branched; perianth bristles reddish, brown, or white, retrorsely barbed ....g
g. Achenes usually much shorter than the subtending scales, about 2 mm. long,
the scales up to 4 mm. long. Spikelets narrowly ovate or oblong ...6. *S. acutus*
g. Achenes equaling or little shorter than the scales, about 2.5–3 mm. long, the
scales 2.75–3 mm. long ....h
h. Spikelets narrowly ovate or oblong; style 2- or 3-cleft ......7. *S. rubiginosus*
h. Spikelets plumply ovate; style 2-cleft ....8. *S. validus*
i. Spikelets about 0.5 cm. long ....11. *S. microcarpus*
i. Spikelets more than 1 cm. long ....j
j. Style mostly 2-cleft; scales not reddish-puncticulate ....9. *S. paludosus*
j. Style mostly 3-cleft; scales reddish-puncticulate ....10. *S. robustus*

1. **S. cernuus** Vahl. Wet soil in marshy meadows or along roads and trails, common in hills near the ocean: Rodeo Lagoon; Tiburon; Elk Valley; Stinson Beach; Laurel Dell Trail and Willow Meadow, Mount Tamalpais; Kentfield Marshes; Point Reyes Peninsula (Bolinas, Inverness, Ledum Swamp); coastal bluffs at Dillons Beach.

As the species occurs along the Pacific coast, it differs in habit from the typical form in Europe and has been called var. *californicus* (Torr.) Beetle.

2. **S. koilolepis** (Steud.) Gleason. Occasional in wet soil of meadows and road-sides: Lagunitas Meadows, Mount Tamalpais; east of Aurora School; Ledum Swamp, Point Reyes Peninsula.—*S. carinatus* (H. & A.) Gray.

3. **S. americanus** Pers. THREE-SQUARE. Locally common on the higher tidal flats or in brackish water: Rodeo Lagoon; Tiburon; Stinson Beach; Tomales; Dillons Beach; Drakes Estero and McClure Beach, Point Reyes Peninsula.

The California plant, usually with 3 or more leaves, is var. *polyphyllus* (Boeckl.) Beetle, a widely distributed form occurring in North and South America, New Zealand, and Tasmania.

4. **S. Olneyi** Gray. Rather rare in brackish water along tidal sloughs and salt marshes: Burdell; Olema Marshes; Inverness.

Scirpus Olneyi is closely related to the South American *S. chilensis* Nees & Mey. but in the northern plant the styles are 2-cleft and in the southern 3-cleft.

5. **S. californicus** (C. A. Mey.) Steud. CALIFORNIA TULE. Occasional on tidal flats or along the edge of lakes: Rodeo Lagoon; Phoenix Lake; Olema Marshes; Drakes Estero; McClure Beach road. In the Coast Ranges, the California tule is not found north of Marin County.

6. **S. acutus** Muhl. COMMON TULE. In Marin County occurring locally in salt

marshes along San Pablo Bay (as near Santa Venetia), but very abundant in the upper reaches of the bay to the north and east.

7. **S. rubiginosus** Beetle. Tidal flat at the south end of Bolinas Lagoon near Stinson Beach, the type locality.

Marked by characters intermediate between *S. acutus* and *S. validus,* this plant may have originated as a hybrid between those two species. This theory finds further support in the almost complete sterility of plants at the type station and in the hybrid-like vigor exhibited by culms over 10 ft. tall. To *S. rubiginosus* may also belong the tules in the laguna in Chileno Valley and in low ground east of the Aurora School.

8. **S. validus** Vahl. Ponds and brackish marshes: Hidden Lake, Mount Tamalpais; south end of Tomales Bay. Perhaps also at Rodeo Lagoon. According to Beetle the achene of *S. validus* is lenticular, that of *S. rubiginosus* trigonous.

9. **S. paludosus** A. Nels. Common along tidal sloughs and in brackish water: Rodeo Lagoon and Tiburon to Point Reyes Peninsula (Shell Beach, Drakes Estero).—*S. campestris* Britt.; *S. pacificus* Britt.

10. **S. robustus** Pursh. Locally common in salt marshes along San Francisco and San Pablo bays: San Rafael; Santa Venetia; Burdell; Black Point.

11. **S. microcarpus** Presl. Common about springs, along streams, and in fresh water marshes, especially among hills near the coast: Tiburon; Mount Tamalpais (Bootjack, Potrero Meadow, Phoenix Lake); Stinson Beach; Inverness and Ledum Swamp, Point Reyes Peninsula.

4. Rhynchospora

1. **R. californica** Gale. Wet soil of coastal marsh near Ledum Swamp, Point Reyes Peninsula. This very rare species is known only from Marin and Sonoma counties.

5. Carex. SEDGE

a. Spikelets composed of staminate and pistillate flowers (or in *C. praegracilis* and *C. simulata* the plants often dioecious), the spikelets sessile, closely aggregated or discrete, or the spikelets solitary . . . . .b

a. Spikelets either staminate or pistillate, or if flowers are mixed in some spikelets, then the spikelets discrete or distant and more or less pedunculate; plants not dioecious . . . . .s

b. Spikelets androgynous, *i.e.,* the staminate flowers above the pistillate in the spikelet, or sometimes the plants dioecious c

b. Spikelets gynaecandrous, *i.e.,* the pistillate flowers above the staminate in the spikelet; plants not dioecious i

c. Spikelets solitary; perigynia beakless . . . . .18. *C. leptalea*

c. Spikelets several to numerous; perigynia beaked . . . . .d

d. Rootstocks conspicuous; inflorescence spicate, lower spikelets somewhat discrete, the upper aggregated . . . . .e

d. Rootstocks short, the culms cespitose; inflorescence more or less branched, compact or loose, the branches densely aggregated or distant . . . . .g

e. Plants not dioecious; beak of perigynium conspicuously bidentate with subulate teeth . . . . .3. *C. tumulicola*

e. Plants tending to be dioecious; beak of perigynium bidentulate to nearly entire . . . . .f

f. Perigynia unequally biconvex, the beak ⅕–⅓ length of perigynium; rootstocks slender, not blackish ....1. *C. simulata*

f. Perigynia plano-convex, the beak ⅓–½ length of perigynium; rootstocks stout,
blackish ....2. *C. praegracilis*

g. Inflorescence loosely branched, at least near the base, the branches more or
less elongate and evident; perigynia blackish-brown, gradually narrowed into
a broadly triangular bidentulate beak ....6. *C. Cusickii*

g. Inflorescence compactly branched, the branches capitate, sessile or nearly so;
perigynia reddish brown tinged, contracted into a slender bidentate beak ..h

h. Ventral side of perigynia with conspicuous nerves ....4. *C. densa*

h. Ventral side of perigynia smooth or with inconspicuous nerves ...5. *C. vicaria*

i. Perigynia without thin winglike margins ....j

i. Perigynia with thin winglike margins l

j. Perigynia spreading or reflexed ....7. *C. phyllomanica*

j. Perigynia appressed-ascending ....k

k. Beak of perigynium bidentulate ....8. *C. leptopoda*

k. Beak of perigynium strongly bidentate ....9. *C. Bolanderi*

l. Bracts subtending the inflorescence conspicuous and generally longer than the
uppermost spikelet. Perigynia flattened, the achene distending the ventral
face as well as the dorsal face ....17. *C. athrostachya*

l. Bracts subtending the inflorescence not conspicuous and elongate ....m

m. Sheaths of leaves green and strongly nerved ventrally, hyaline only at the
mouth ....16. *C. feta*

m. Sheaths of leaves hyaline ventrally, not green and nerved ....n

n. Perigynia small, 3.5 mm. long or less ....o

n. Perigynia longer, 3.5–5 mm. long ....p

o. Perigynia about 3 mm. long, ventral face nerved ....10. *C. montereyensis*

o. Perigynia generally 3.25–3.5 mm. long, ventral face smooth ....11. *C. subfusca*

p. Ventral face of perigynia smooth or nerved only near the base ....q

p. Ventral face of perigynia conspicuously nerved ....r

q. Spikelets congested in a dense oblong or ovate head; perigynia numerous,
10–20 in a spikelet ....12. *C. subbracteata*

q. Spikelets approximate or separate in a loose narrow head; perigynia few, 2–12
in a spikelet ....13. *C. gracilior*

r. Scales not completely covering the perigynia; spikelets congested in a dense
ovate head ....14. *C. Harfordii*

r. Scales almost or quite covering the perigynia; spikelets loosely approximate in
a slender oblong head ....15. *C. Tracyi*

s. Stigmas 3 ....t

s. Stigmas 2. Perigynia glabrous, the beak sometimes hispidulous ....E

t. Perigynia from densely hairy to sparsely scabrous-ciliate, or if glabrous, then
the beak more or less hispidulous. Beak mostly 1 mm. long or less. (Perigynia
sometimes entirely glabrous in *C. luzulina* in which the perigynia are
smooth and not inflated and the beak is less than 1 mm. long.) ....u

t. Perigynia glabrous, smooth or minutely papillate; beak not hispidulous ....B

u. Lowest pistillate spikelets arising among basal leaves v

u. Lowest pistillate spikelets not arising among basal leaves ....w

v. Perigynia with conspicuous nerves as well as 2 lateral ribs ......19. *C. globosa*

v. Perigynia with 2 lateral ribs but without conspicuous nerves . .20. *C. brevicaulis*
w. Bracts (at least the lowest) long-sheathing x
w. Bracts short-sheathing or sheathless ....A
x. Rootstocks and base of plants conspicuously fibrous-coated; pistillate scales and perigynia scarcely hairy ....25. *C. luzulina*
x. Rootstocks and base of plants scaly but not conspicuously fibrous; pistillate scales hairy or perigynia hairy ....y
y. Basal leaves conspicuously hairy; pistillate spikelets oblong, mostly more than 5 mm. thick ....22. *C. gynodynama*
y. Basal leaves sparsely hairy or glabrous; pistillate spikelets linear, less than 5 mm. thick ....z
z. Pistillate spikelets rather lax and spreading; upper part of perigynia scabrous-ciliate ....23. *C. debiliformis*
z. Pistillate spikelets strictly erect; upper part of perigynia sparsely pubescent and ciliate ....24. *C. mendocinensis*
A. Perigynia densely hairy ....26. *C. lanuginosa*
A. Perigynia glabrous except the scabrous-hispidulous beak ...28. *C. serratodens*
B. Perigynia scarcely inflated, densely papillate; beak less than 0.5 mm. long ....29. *C. Buxbaumii*
B. Perigynia more or less inflated at maturity, smooth; beak 1–3 mm. long C
C. Beak of perigynium strongly hyaline at the oblique apex, not conspicuously bidentate ....27. *C. amplifolia*
C. Beak of perigynium not hyaline, bidentate with stiff subulate teeth ......D
D. Perigynia nervose-ribbed, the beak 2–3 mm. long ....34. *C. exsiccata*
D. Perigynia nerved, the beak 1–2 mm. long ....35. *C. rostrata*
E. Lowest bract of the inflorescence sheathing; lowest spikelet frequently widely separated from the rest; perigynia almost or quite beakless. Perigynia papillate ....21. *C. Hassei*
E. Lowest bract of the inflorescence not sheathing; lowest spikelet not widely separated from the rest; perigynia generally with a conspicuous beak ....F
F. Perigynia membranaceous, attenuate into the short, entire beak . .31. *C. nudata*
F. Perigynia thick-coriaceous, abruptly narrowed into the beak ....G
G. Beak of perigynium stout, bidentate, the teeth scabrous-ciliate; achenes not constricted at the middle ....30. *C. barbarae*
G. Beak of perigynium subentire, glabrous or a little hairy; achenes constricted at the middle ....H
H. Perigynia smooth and shining; plants of springy places and fresh water marshes ....32. *C. obnupta*
H. Perigynia papillate and dull; plants of tidal flats ....33. *C. Lyngbyei*

1. **C. simulata** Mkze. Very rare in the Coast Ranges; in Marin County known only from a fresh water marsh just north of Stinson Beach, where it occurs with *Berula erecta* (Huds.) Cov., and from Abbotts Lagoon on Point Reyes Peninsula.

2. **C. praegracilis** W. Boott. Occasional in wet places on slopes and flats or between the hills and salt marshes: Tiburon; Greenbrae Hills; east of Aurora School.

3. **C. tumulicola** Mkze. Widespread and rather common on grassy slopes in open oak-madroño woodland, sometimes forming extended patches: Angel Island; Tiburon; Rodeo Lagoon; Elk Valley; Mount Tamalpais (Mountain Theater,

Laurel Dell, Hidden Lake Fire Trail); Big Carson; San Rafael; Big Rock Ridge; Point Reyes Peninsula (Mud Lake, Ledum Swamp, Shell Beach).

4. **C. densa** Bailey. The most common and widespread species of springy slopes and low wet fields and meadows: Tiburon and Mount Tamalpais (Steep Ravine, Potrero Meadow, Lake Lagunitas) to San Rafael, Tomales, and Bolinas.

As accepted here *C. densa* is a complex and variable entity and perhaps includes forms better referred to *C. Dudleyi* Mkze. and *C. breviligulata* Mkze. *Carex Dudleyi* is usually well marked by the elongate awns on all the pistillate scales but, although some of the Marin County plants referred here to *C. densa* have awned scales, the awns of the uppermost pistillate scales are scarcely long enough to allow a definite recognition of this species. *Carex breviligulata* has been separated from *C. densa* by the much shorter ligule (in *C. densa* "as long as wide"), but the character is doubtfully diagnostic and in any case very difficult to apply. All of these plants, together with the following, need critical study.

5. **C. vicaria** Bailey. In Marin County known only from a moist wooded slope at Inverness Park, the southernmost station yet reported for a species ranging northward to Washington.

6. **C. Cusickii** Mkze. Restricted to wet meadows on Mount Tamalpais (Rock Spring, Potrero Meadow) and swales on Point Reyes Peninsula near the radio station.

7. **C. phyllomanica** W. Boott. Rare in coastal marshes: Ledum Swamp and Abbotts Lagoon, Point Reyes Peninsula.

8. **C. leptopoda** Mkze. Occasional in wet soil usually shaded by trees or brush: Nicasio road, *Leschke;* Olema, acc. Mackenzie; Shell Beach and Ledum Swamp, Point Reyes Peninsula.

9. **C. Bolanderi** Olney. Near streams and springs: Sausalito; Mount Tamalpais (Blithedale Canyon, Fern Canyon, Collier Spring, Potrero Meadow); Inverness Ridge, Point Reyes Peninsula. The Bolander sedge is variable in size and low slender shade forms may simulate *C. leptopoda* in appearance. Such individuals, however, with their prominent, coarse teeth on the beak of the perigynium, are readily referable to *C. Bolanderi*.

10. **C. montereyensis** Mkze. Rather rare on the edge of brush and pine woods: Sausalito; Point Reyes Peninsula (Bolinas, Inverness Ridge, Ledum Swamp). This species, restricted to the California coast, has not been reported north of Marin County.

11. **C. subfusca** W. Boott. Rare about springs: Potrero Meadow, Mount Tamalpais; Warm Spring, Big Carson Canyon.

12. **C. subbracteata** Mkze. Common and widespread through the hills of Marin County, generally in moist soil of shaded brushy slopes, also on open flats near the ocean: Rodeo Lagoon; Tiburon and Mill Valley, acc. Stacey; Mount Tamalpais (Rock Spring, Potrero Meadow, Lagunitas Meadows); San Rafael; Big Carson; Tocaloma, *Leschke;* Dillons Beach; Point Reyes Peninsula (Shell Beach, Ledum Swamp, fixed dunes near the radio station). In a large series of specimens, the differences between *C. subbracteata* and *C. pachystachya* Cham. become negligible and it would seem that all Marin County material might be referred to that wide-ranging and variable species.

13. **C. gracilior** Mkze. Occasional in open fields or on the edge of brush, in

soil moist in the spring: Angel Island; Tiburon; Mill Valley; Greenbrae Hills; San Rafael Hills; Ignacio; Woodacre; Olema, acc. Stacey; Inverness Park; east of Aurora School. In the Greenbrae Hills overlooking the salt marshes, a peculiar sterile individual with androgynous spikelets was collected which may have been a hybrid between *C. gracilior* and *C. praegracilis* W. Boott.

14. **C. Harfordii** Mkze. Widespread but not common, from fixed dunes and low ground bordering salt marshes to brushy and wooded slopes: Sausalito; Mill Valley; Point Reyes Peninsula (fixed dunes near radio station, Ledum Swamp, Inverness Ridge). It is not easy to distinguish *C. Harfordii* from those forms of *C. subbracteata* in which the nerves on the ventral face of the perigynium are sometimes evident. Only specimens with several nerves strongly developed are here regarded as *C. Harfordii*.

15. **C. Tracyi** Mkze. Occasional in wet ground: Potrero Meadow, Mount Tamalpais; Mount Vision, Point Reyes Peninsula; north of Dillons Beach. These are the southernmost stations known for this species.

16. **C. feta** Bailey. Low wet parts of meadows on Mount Tamalpais: Potrero Meadow; Lagunitas Meadows.

17. **C. athrostachya** Olney. Marshy places in low valleys: the laguna in Chileno Valley; east of Aurora School. These are the southernmost stations in the Coast Ranges.

18. **C. leptalea** Wahl. In wet soil of coastal marsh, Ledum Swamp, Point Reyes Peninsula, the only locality known in Marin County and the southernmost station in California.

19. **C. globosa** Boott. Common and widespread on moist shaded rocks near redwood forests or their border woodland, rarely on exposed rocks or in open meadows: Sausalito; Mount Tamalpais (Blithedale Canyon, Cataract Gulch, Potrero and Lagunitas meadows); Bolinas Ridge; San Rafael Hills.

A distinctive form of *C. globosa,* simulating in appearance *C. brevicaulis,* is not uncommon in the serpentine areas of Mount Tamalpais and Carson Ridge. Not only is the plant more condensed with shorter leaves and peduncles, but the terminal inflorescence is either entirely staminate or produces only 1 (or rarely 2) pistillate flowers.

20. **C. brevicaulis** Mkze. Not uncommon on slopes near the coast in shallow soil or around rocks: Angel Island; Sausalito; Elk Valley; Steep Ravine, Mount Tamalpais; Bolinas Ridge; Point Reyes Peninsula (Inverness, Mud Lake, Ledum Swamp, Point Reyes); Dillons Beach. Some of the plants from the pine woods on Point Reyes Peninsula are more robust and in habit resemble *C. globosa,* but usually the plants are low and inconspicuous, forming turflike patches.

21. **C. Hassei** Bailey. In Marin County known only from Ledum Swamp on Point Reyes Peninsula.

22. **C. gynodynama** Olney. Moist soil of low meadows or on wet brushy slopes: Mount Tamalpais (Rock Spring, Potrero Meadow, Willow Meadow, Lagunitas Meadows); Ledum Swamp and Shell Beach, Point Reyes Peninsula; Dillons Beach.

23. **C. debiliformis** Mkze. Not uncommon in wet soil around springs on Mount Tamalpais and in the Carson country, generally in serpentine: Azalea Flat; Bootjack; Potrero Meadow; Willow Meadow; Camp Handy; Little Carson. Marin County marks the southern distributional limit.

24. **C. mendocinensis** Olney. This rare plant, known from only three or four

stations in the North Coast Ranges, is probably a hybrid between *C. gynodynama* Olney and *C. debiliformis* Mkze. On Mount Tamalpais, where it was found growing with those two species in Willow Meadow, *C. mendocinensis* was sterile while the suspected parents were fertile and normal.

25. **C. luzulina** Olney. Rare in marshes on Mount Tamalpais (Azalea Flat) and on Point Reyes Peninsula (Ledum Swamp and near the radio station). *Carex luzulina* has not been collected south of Marin County.

26. **C. lanuginosa** Michx. Rare in marshy places on Point Reyes Peninsula (Ledum Swamp and near the radio station). Although *C. lanuginosa* is common and widespread in the Sierra Nevada and in southern California, it is rare in the North Coast Ranges and has apparently not been found along the coast south of the Golden Gate.

27. **C. amplifolia** Boott. Wet swampy meadows or stream banks: Rodeo Lagoon; Lagunitas Creek below Lake Lagunitas.

28. **C. serratodens** W. Boott. Common on moist slopes and about springs, generally in serpentine areas: Tiburon; Mount Tamalpais (Azalea Flat, Rock Spring, Potrero Meadow); Carson country; Lucas Valley.—*C. bifida* Boott.

29. **C. Buxbaumii** Wahl. Very rare and local in cold swales near the radio station on Point Reyes Peninsula, the southernmost station in the Coast Ranges.

30. **C. barbarae** Dewey. Common and widespread on open or brushy slopes that are usually dry by late spring: Angel Island; Tiburon; Mill Valley; Bootjack and Dipsea Trail, Mount Tamalpais; Stinson Beach; San Rafael; Salmon Creek; Bolinas.

31. **C. nudata** W. Boott. Common on rocky streambeds: Mount Tamalpais (Bootjack, Laurel Dell Trail, Alpine Dam); Ross; San Anselmo Canyon; Lagunitas Creek. Marin County is probably the type locality.

An anomalous plant that suggests a hybrid grows along the Matt Davis Trail near Bootjack. One of the possible parents is undoubtedly *C. nudata* which grows in the adjacent stream and the other parent may be either *C. barbarae* Dewey or *C. obnupta* Bailey, both of which occur in the vicinity of Bootjack.

32. **C. obnupta** Bailey. Common in coastal swales and marshes: Sausalito; Muir Beach; Azalea Flat and Potrero Meadow, Mount Tamalpais; Tomales; Dillons Beach; east of Aurora School; Point Reyes Peninsula (Inverness, Ledum Swamp, Drakes Estero).

Although *C. obnupta* and *C. barbarae* Dewey are usually quite distinctive in habit and in characters of fruit and inflorescence, it is not always easy to place some sterile specimens that are more or less intermediate. It may be that such intermediates are of hybrid origin.

33. **C. Lyngbyei** Hornem. In the *Salicornia* belt of the salt marsh along Tomales Bay near Inverness. This is the southernmost known California station.

34. **C. exsiccata** Bailey. Occurring locally in marshy ground: Potrero and Willow meadows, Mount Tamalpais; pond north of Bolinas; Olema Marshes; laguna in Chileno Valley.

35. **C. rostrata** Stokes. In Marin County this wide-ranging species is restricted to Point Reyes Peninsula where it occurs in Ledum Swamp and in swales near the radio station. Except for a doubtful station designated by Bolander as "San Francisco," these are the southernmost stations known in the Coast Ranges.

Carex rostrata, C. lanuginosa Michx., and *C. Hassei* Bailey are common in the mountains of California outside the Coast Ranges and the three are known in Marin County only on Point Reyes Peninsula. *Carex Buxbaumii* Wahl., *C. Lyngbyei* Hornem., and *C. leptalea* Wahl. are likewise restricted to Point Reyes Peninsula but, although these species are widespread in North America, they are among the rarest California sedges.

ARACEAE. Calla Family

1. Zantedeschia

1. **Z. aethiopica** (L.) Spreng. Common Calla. Occasionally escaped from cultivation and naturalized in coastal marshes and in springy places in the hills: Tiburon; Phoenix Lake and Wheeler Trail, Mount Tamalpais; Stinson Beach; Bolinas; Marshalls. Native of South Africa.

LEMNACEAE. Duckweed Family

a. Blades with several roots ....1. *Spirodela*
a. Blades with one root ....2. *Lemna*

1. Spirodela

1. **S. polyrhiza** (L.) Schleid. Surface of quiet ponds: Lily Lake, *Eastwood, Sutliffe;* near Point Reyes Station, *Stacey.*

2. Lemna. Duckweed

a. Blades almost hemispheric, flat above and strongly swollen-convex below
....1. *L. gibba*
a. Blades flat or low-convex above and below ....b
b. Blades pale green, oblongish, cellular-translucent, asymmetric at base
....4. *L. valdiviana*
b. Blades bright green, elliptic to roundish, rather dense and not translucent, nearly symmetric at base c
c. Blades smooth and shining, 3–5 mm. long ....2. *L. minor*
c. Blades with several median papillae, not shiny, 1–3 mm. long..3. *L. minima*

1. **L. gibba** L. Lake Ranch, Point Reyes Peninsula, acc. H. L. Mason. This is the only station known for this widespread duckweed in Marin County, but it may be expected elsewhere since it has been collected on the road between Tomales and Petaluma only one mile east of the Marin County line.

2. **L. minor** L. Pools and streams: Rodeo Lagoon; Ross; Olema Marshes.

3. **L. minima** Philippi. Shallow water of streams and marshes: Sausalito; bog on east side of Bolinas Lagoon and in stream above Stinson Beach; Olema Marshes; Abbotts Lagoon, Point Reyes Peninsula.

4. **L. valdiviana** Philippi. Shallow streams and marshy ground: Sausalito; Azalea Flat and Nora Trail, Mount Tamalpais.—*L. cyclostasa* of American references.

The blades of the Mount Tamalpais plant have the shape of the species but they are thin and translucent as in var. *abbreviata* Hegelm.

COMMELINACEAE. Spiderwort Family

1. Tradescantia. Spiderwort

1. **T. fluminensis** Vell. Wandering Jew. Escaped from cultivation and abundantly naturalized on moist slopes under redwoods: Cascade and Blithedale canyons, Mill Valley. Native of South America.

JUNCACEAE. Rush Family

a. Capsule many-seeded; leaves not shreddy-filamentose, rarely grasslike
....1. *Juncus*
a. Capsule 3-seeded; leaves grasslike, shreddy-filamentose near the base of the blades especially when young ....2. *Luzula*

1. Juncus. Rush

a. Flowers not in heads, subtended by 2 scarious bracts, the clusters loose or congested ....b
a. Flowers in heads, subtended by a single bract, the heads solitary or numerous, consisting of few to many flowers ....j
b. Inflorescence appearing lateral, subtended by an elongate subcylindric bract simulating a continuation of the stem c
b. Inflorescence appearing terminal, subtended by a flattened or channeled leaf-like bract ....h
c. Rhizomes little, if at all, prolonged, the plants densely cespitose; anthers about as long as the filaments or shorter ....d
c. Rhizomes long-creeping; anthers much longer than the filaments ....f
d. Stamens 6 ....2. *J. patens*
d. Stamens 3 ....e
e. Perianth brown, 2–2.5 mm. long, the segments not rigid
....1a. *J. effusus* var. *brunneus*
e. Perianth straw-color, 2.5–3 mm. long, the segments rigid
....1b. *J. effusus* var. *pacificus*
f. Some basal leaf sheaths blade-bearing. Perianth about 5 mm. long
....3. *J. mexicanus*
f. Basal leaf sheaths not blade-bearing ....g
g. Perianth about 5 mm. long ....4. *J. balticus*
g. Perianth about 6 mm. long ....5. *J. Leseurii*
h. Plants perennial, generally 2 dm. tall or more ....6. *J. occidentalis*
h. Plants annual, usually less than 2 dm. tall i
i. Flowers about 4 mm. long; capsule usually less than 3 mm. long; seeds less than 0.5 mm. long, somewhat shining ....7. *J. sphaerocarpus*
i. Flowers about 5 mm. long; capsule usually more than 3 mm. long; seeds 0.5 mm. long or more, dull ....8. *J. bufonius*
j. Plants annual with very slender culms less than 1 dm. tall; leaves basal, delicate and almost threadlike ....9. *J. Kelloggii*
j. Plants perennial, mostly more than 1 dm. tall, the culms usually stout; leaves not threadlike, generally 1 or more cauline ....k
k. Leaves grasslike, the blades dorsi-ventrally flattened, without transverse septa.1

k. Leaves not grasslike, the blades subcylindric or laterally flattened, with more or less conspicuous septa ....m
l. Capsule shorter than the perianth ....10. *J. falcatus*
l. Capsule longer than the perianth ....11. *J. Covillei*
m. Stamens 3; leaves subcylindric with complete septa ....12. *J. Bolanderi*
m. Stamens 6; leaves more or less laterally flattened with incomplete septa....n
n. Perianth mostly 3 mm. long; anthers shorter than, or equaling, the filaments
....14. *J. xiphiodes*
n. Perianth mostly 4–6 mm. long; anthers longer than the filaments ....o
o. Flowers many in 1–3 heads ....13. *J. phaeocephalus*
o. Flowers few in numerous heads ....13*a. J. phaeocephalus* var. *paniculatus*

1*a*. **J. effusus** L. var. **brunneus** Engelm. Swamps and swales generally near the ocean, forming large tussocks: Tiburon; Sausalito; Frank Valley; Point Reyes Peninsula (Bolinas, Inverness, Ledum Swamp).

1*b*. **J. effusus** L. var. **pacificus** Fern. & Wieg. Beds of intermittent streams and springy slopes of hills, mostly away from the coast: Sausalito, Tiburon, and Mount Tamalpais (Azalea Flat, Potrero Meadow, Rock Spring) to the Carson country and San Rafael. On Mount Tamalpais the plants frequently vary towards var. *brunneus.*

2. **J. patens** E. Mey. Common on grassy or brushy slopes and flats that are wet in the spring: Tiburon, Sausalito, and Ross to San Rafael and Point Reyes Peninsula. The pale blue-green or gray-green culms of this rush are a characteristic feature of the drier valleys around Mount Tamalpais but the plant is rare or lacking on the higher parts of the mountain.

3. **J. mexicanus** Willd. Meadows and slopes in clay soil, occasional: Tiburon; Rodeo Lagoon; Azalea Flat and Laurel Dell, Mount Tamalpais; Escalle; Carson country; Olema Marshes.

4. **J. balticus** Willd. In Marin County this species is not readily distinguished from *J. Leseurii,* although certain collections from the north side of Mount Tamalpais (Laurel Dell, Barths Retreat, Lagunitas Meadows, Phoenix Lake) are referable here. The meadowy flats and clay soil where these plants grow are quite different from the salt marsh and sand dune habitats of *J. Leseurii.* The specimens from Lagunitas Meadows have been determined as *J. balticus* var. *montanus* Engelm. by F. J. Hermann.

5. **J. Leseurii** Boland. Common along the upper reaches of salt marshes and in moist places on coastal dunes: Rodeo Lagoon; Tiburon; Tamalpais Valley; Escalle; Olema Marshes; Bolinas and Drakes Estero, Point Reyes Peninsula.

The spelling of the specific name is that given in the original description. Variants are *Lesueurii* and *Lescurii.*

6. **J. occidentalis** (Cov.) Wieg. Particularly characteristic of meadows and grassy slopes that are wet in the spring but dry in the summer: Angel Island, Tiburon, and Mount Tamalpais (Rock Spring, Potrero Meadow) to Ross, San Rafael, and Point Reyes Peninsula (Bolinas, Ledum Swamp).—*J. tenuis* Willd. var. *congestus* Engelm.

7. **J. sphaerocarpus** Nees. Rare on Mount Tamalpais in gravelly soil: near Laurel Dell; Kent Trail. The only collection of *J. sphaerocarpus* reported for the South Coast Ranges is from Antioch, Contra Costa County.

8. **J. bufonius** L. TOAD RUSH. Common in low wet places in shade or sun, variable: Sausalito, Tiburon, and Mount Tamalpais (Azalea Flat, Rock Spring, Potrero Meadow) to San Rafael Hills, Tomales, and Point Reyes Peninsula.

9. **J. Kelloggii** Engelm. Rare on Mount Tamalpais in low wet places in meadows: Willow Meadow; Lagunitas Meadows.

10. **J. falcatus** E. Mey. The only station known for Marin County is in wet sandy soil back of the dunes near the radio station on Point Reyes Peninsula.

11. **J. Covillei** Piper. Wet meadows and gravelly beds of summer-dry water courses at the west end of Mount Tamalpais: near Bootjack; Rock Spring; Laurel Dell; Potrero Meadow; Alpine Dam. Marin County is the southern distributional limit of Coville's rush.—*J. falcatus* E. Mey. var. *paniculatus* Engelm.; *J. obtusatus* of authors in part.

12. **J. Bolanderi** Engelm. Rather common in wet or swampy ground near the coast, occasional about springs on Mount Tamalpais: Mount Tamalpais (Azalea Flat, Rock Spring, Potrero Meadow); Inverness and Ledum Swamp, Point Reyes Peninsula; Dillons Beach. Except for a record from Orange County in southern California, the Bolander rush reaches its southern limit in Marin County.

13. **J. phaeocephalus** Engelm. Common in coastal marshes and in low places among dune hills: Point Reyes Peninsula (Bolinas, Inverness, Ledum Swamp, near the radio station); Dillons Beach. Also locally abundant on the strand of Lake Lagunitas as the waters recede in the summer.

13a. **J. phaeocephalus** Engelm. var. **paniculatus** Engelm. Common and widespread in moist ground of hills and meadows away from the coast: Tiburon and Mount Tamalpais (Rock Spring, Potrero Meadow, Hidden Lake) to San Rafael, Fairfax, Aurora School, and Inverness Ridge.

14. **J. xiphioides** E. Mey. Common in marshy or springy places in the hills or along the coast: Mount Tamalpais (Azalea Flat, Bootjack, Potrero Meadow, Phoenix Lake); San Rafael; Carson country; Inverness. In the Tamalpais area, the inflorescence of *J. xiphioides* varies nearly as much as the inflorescence of *J. phaeocephalus* but in *J. xiphioides* the extremes of variation have not been named.

2. **Luzula.** WOOD-RUSH

1. **L. multiflora** (Retz.) Lejeune. Grassland and borders of brush and woods, common: Tiburon, Sausalito, and Mount Tamalpais (Rock Spring, Potrero Meadow, Throckmorton Ridge, Blithedale Canyon) to San Rafael Hills, Tomales. and Point Reyes Peninsula.—*L. campestris* of most American references.

LILIACEAE. LILY FAMILY

a. Inner and outer perianth segments essentially alike ....b
a. Inner and outer perianth segments dissimilar in size or color ....n
b. Foliage leaves basal or mostly so, cauline leaves if present (as in *Clintonia*, *Xerophyllum*, and *Zigadenus*) reduced in size or few c
b. Foliage leaves cauline ....j
c. Flowers in racemes or panicles ....d
c. Flowers in umbels, or almost in heads when short-pedicellate ....g
d. Styles 3, distinct nearly to the base; perianth segments creamy white or white ....e

d. Style 1, stigma 3-lobed or 3-parted; perianth segments blue or whitish with purplish veins ....f

e. Basal leaves evergreen, very numerous, rough, elongate-linear; perianth segments without a conspicuous gland ....1. *Xerophyllum*

e. Basal leaves not surviving the summer, smooth, linear; perianth segments with a conspicuous gland near base ....2. *Zigadenus*

f. Flowers in a raceme; perianth segments blue ....3. *Camassia*

f. Flowers in a panicle; perianth segments white with a purplish midvein ....4. *Chlorogalum*

g. Umbels terminal and lateral, on a stem bearing 1 or several foliaceous bracts ....13. *Clintonia*

g. Umbel solitary, terminating a leafless stem ....h

h. Perianth segments united into a distinct tube ....7. *Brodiaea*

h. Perianth segments divided nearly or quite to the base i

i. Plants with the odor and taste of onions ....5. *Allium*

i. Plants without the odor or taste of onions ....6. *Muilla*

j. Leaves alternate, not whorled or aggregated, narrowly to broadly ovate; fruit a berry ....k

j. Leaves more or less aggregated, the stems generally bearing 1 or more distinct whorls, linear-lanceolate to narrowly ovate; fruit a capsule. ....m

k. Leaves few, usually 2 or 3; perianth segments 4 ....15. *Maianthemum*

k. Leaves numerous; perianth segments 6 l

l. Stem unbranched below the inflorescence; flowers less than 1 cm. long ....14. *Smilacina*

l. Stem branched; flowers 1–2 cm. long ....16. *Disporum*

m. Perianth segments generally 4 cm. long or more; stigma 3-lobed ...8. *Lilium*

m. Perianth segments less than 4 cm. long; stigma deeply parted ...9. *Fritillaria*

n. Basal foliage leaves none; flower 1, subtended by 3 broad cauline leaves ....12. *Trillium*

n. Basal foliage leaves 1–3; flowers generally more than one ....o

o. Leaves linear; inner perianth segments broader than the outer; flowers subtended by leaflike bracts ....10. *Calochortus*

o. Leaves elliptic-oblong, mottled; inner perianth segments much narrower than the outer; flowers solitary on naked stalks ....11. *Scoliopus*

1. Xerophyllum

1. **X. tenax** (Pursh) Nutt. SQUAW GRASS; FIRE LILY. Rocky slopes and ridges in the chaparral: Mount Tamalpais (Corte Madera Ridge, Throckmorton Ridge, Kent Trail, Berry Trail); near Lily Lake; San Geronimo Ridge; San Rafael Hills. Although plants of squaw grass commonly bloom each season from northern California northward, on Mount Tamalpais they rarely bloom except after a fire. So numerous were the blooms after the fire of 1929 that the south side of Mount Tamalpais above Cascade Canyon was colored by the creamy-white flower clusters. A fine display also followed the fire of 1945. At that time the inflorescences were unusually vigorous, one specimen attaining a height of 8¼ feet.

2. Zigadenus

a. Summer-blooming plants of wet places; perianth segments less than 1 cm. long ....1. *Z. fontanus*

a. Spring-blooming plants of grassland, chaparral, or woods; perianth segments
usually more than 1 cm. long ....b
b. Inflorescence branched ....2. *Z. Fremontii*
b. Inflorescence simple c
c. Plants tall, stout, of subsaline flats; flowers rather numerous
....2a. *Z. Fremontii* var. *salsus*
c. Plants low, dwarfed, of open grassy hills; flowers few, less than 10
....2b. *Z. Fremontii* var. *minor*

1. **Z. fontanus** Eastw. MARSH ZIGADENE. Frequently abundant in low boggy
meadows or along streamlets, generally near serpentine: Tiburon; Mount Tamalpais (Bootjack, the type locality, Laurel Dell Trail, Potrero and Lagunitas meadows); Carson country. Beyond Marin County the marsh zigadene is rare and is
known from only a few scattered stations from San Benito County north to
Sonoma and Lake counties.

2. **Z. Fremontii** (Torr.) Torr. STAR LILY. Common and widespread on wooded
or brushy slopes: Angel Island, Sausalito, Tiburon, and Mount Tamalpais (Corte
Madera and Throckmorton ridges, Nora Trail) to the Fairfax Hills and San
Rafael. In the Sausalito Hills some of the plants are low with simple inflorescence
as in var. *minor*.

The Fremont zigadene is one of the earliest plants to bloom in the spring
and its creamy-white, starlike flowers are always beautiful whether they are
found on an open summit or in a brushy draw. The vigor and abundance of
blossoming plants on chaparral burns frequently produce startling effects amid
the charred shrubbery.

2a. **Z. Fremontii** (Torr.) Torr. var. **salsus** Jeps. Rare along the edge of salt
marshes, as at San Antonio near the mouth of San Antonio Creek, where it
grows with *Distichlis* and *Salicornia*.

2b. **Z. Fremontii** (Torr.) Torr. var. **minor** (H. & A.) Jeps. Rare in open grassy
places on windswept maritime hills: east side of Tomales Bay north of Point Reyes
Station, *Cantelow*; Dillons Beach.

3. Camassia. CAMASS

1. **C. Quamash** (Pursh) Greene var. **linearis** (Gould) J. T. Howell. Low boggy
meadows and coastal swales: Point Reyes Peninsula (Inverness, acc. Stacey, Ledum
Swamp, near the radio station); near Dillons Beach.—*Quamasia Quamash* (Pursh)
Cov. in part.

These stations in Marin County represent the southernmost distributional limit
for the genus in the Coast Ranges.

4. Chlorogalum

1. **C. pomeridianum** (DC.) Kunth. SOAP PLANT. Common and characteristic
along grassy or brushy woodland borders in hills away from the coast: Blithedale
and Fern canyons, Mount Tamalpais; Ross; San Rafael; Black Point.

In typical form, the soap plant is about 3 feet tall or more with slender
elongate branches, but in Marin County, a low form a foot or so tall is even more
frequent, occurring on coastal slopes and on rocky ridges in the chaparral. This
dwarf soap plant, which differs not only in its lower stature but also in the

divaricate branches which frequently arise near the ground, is var. *divaricatum* (Lindl.) Hoover and has been seen in the following places: Sausalito; Angel Island; Tiburon; Azalea Flat and Throckmorton Ridge, Mount Tamalpais; Carson Ridge; Mud Lake and Point Reyes, Point Reyes Peninsula.

In both forms of this interesting plant, the medium-sized spider-like flowers are closed from morning until late afternoon when they open to attract vespertine insects.

5. **Allium.** ONION

a. Scape terete, not flattened or angled ....b
a. Scape flattened or 3-angled ....e
b. Perianth less than 1 cm. long, whitish or pale pink c
b. Perianth 1–1.5 cm. long, rosy-pink to purplish ....d
c. Ovary with 3 low crests; markings on outer bulb coats rectangular; perianth segments often with rose midvein ....1. *A. lacunosum*
c. Ovary with 6 crests; markings on bulb coats undulate; perianth segments without rose midvein ....2. *A. amplectens*
d. Perianth segments obtuse or subacute; anthers about 1 mm. long; ovary crested ....3. *A. dichlamydeum*
d. Perianth segments acute or subacuminate; anthers 1.5–2 mm. long; ovary not crested ....4. *A. unifolium*
e. Scape flattened ....5. *A. falcifolium*
e. Scape sharply 3-angled ....6. *A. triquetrum*

1. **A. lacunosum** Wats. Shallow soil on serpentine slopes, Tiburon Peninsula. This is the only known occurrence of the species in the North Coast Ranges.

2. **A. amplectens** Torr. On rocky slopes or in shallow soil overlying rocks: Mount Tamalpais (Fern and Baltimore canyons, Northside Trail, between Mountain Theater and Potrero Meadow); Carson country; Kentfield, *M. E. Parsons;* Big Rock Ridge, *Robbins.*

3. **A. dichlamydeum** Greene. Open rocky slopes, frequently maritime: Angel Island; Sausalito; Dillons Beach; Point Reyes Lighthouse.

4. **A. unifolium** Kell. Moist stream banks or low meadows that are marshy in the spring, rather widespread but not common: Tiburon; Mount Tamalpais (Rifle Camp, Willow and Lagunitas meadows); San Anselmo Canyon; Dillons Beach; Point Reyes Peninsula.

5. **A. falcifolium** H. & A. Gravelly slopes and shallow soil pockets on serpentine: Potrero Meadow and Northside Trail, Mount Tamalpais; Carson Ridge. These Marin County stations mark the southern distributional limit of the species, unless one regards the closely related *A. Breweri* Wats. of the South Coast Ranges as conspecific.

6. **A. triquetrum** L. An attractive white-flowered onion with a propensity for escaping from cultivation and becoming naturalized on moist shady slopes and flats: Sausalito; Mill Valley; Ross; San Anselmo; San Rafael.

6. **Muilla**

1. **M. maritima** (Torr.) Wats. In Marin County this is known only from Point Reyes Peninsula where it was first collected by Bigelow in 1854. With its rather inconspicuous but delightfully fragrant flowers it has been recently recol-

lected on sandy flats behind the dunes near the radio station. Although the plant
has been found in the Sacramento Valley and is widespread in the South Coast
Ranges, it is not known beyond Marin County in the North Coast Ranges.

7. Brodiaea

a. Pedicels shorter than the perianth, the umbels subcapitate or subracemose..b
a. Pedicels elongate, usually longer than the perianth, the umbels loose and
open c
b. Stamens with anthers 6 ....1. *B. pulchella*
b. Stamens with anthers 3 ....2. *B. congesta*
c. Stamens with anthers 6 ....d
c. Stamens with anthers 3, alternating with conspicuous staminodia ....f
d. Ovary not stipitate; perianth open-campanulate ....3. *B. hyacinthina*
d. Ovary stipitate; perianth funnelform-campanulate ....e
e. Perianth lilac to violet-purple; ovary purplish ....4. *B. laxa*
e. Perianth whitish with purplish midvein; ovary yellow .....5. *B. peduncularis*
f. Staminodia acute, shorter than the anthers ....6. *B. elegans*
f. Staminodia acute, obtuse, or 3-dentate, equaling or longer than the anthers
....7. *B. terrestris*

1. **B. pulchella** (Salisb.) Greene. COMMON BRODIAEA; GRASS NUT. Grassy or rocky
slopes of open or wooded hills, common and widespread: Sausalito, Tiburon, and
Mount Tamalpais to San Rafael, Big Rock Ridge, and Tomales.—*B. capitata*
Benth.; *Dichelostemma pulchellum* (Salisb.) Hel.

2. **B. congesta** Smith. Canyonsides and hills in brush and open woodland: north
side of Mount Tamalpais (Fish Grade, Lagunitas Meadows, Rifle Camp, Rock
Spring); San Rafael Hills; San Anselmo Canyon; Point Reyes Peninsula near
Mud Lake.—*B. pulchella* and *Dichelostemma pulchellum* of authors.

3. **B. hyacinthina** (Lindl.) Baker. WHITE BRODIAEA. Occasional in low places
that are wet in the spring: Sausalito, acc. Stacey; Mount Tamalpais (Bootjack,
Rock Spring, Potrero and Lagunitas meadows); Fairfax, Greenbrae, and San
Rafael hills; Corte Madera and Olema, acc. Hoover; north of Tomales.—*Hes-
peroscordum hyacinthinum* Lindl.

4. **B. laxa** (Benth.) Wats. TRITELEIA. Common on dry slopes or flats in grassland,
brush, or open woodland: Sausalito, Tiburon, and Mount Tamalpais (Summit
Avenue, Throckmorton Ridge, Bootjack, Potrero Meadow) to San Rafael, Dillons
Beach, and Point Reyes Peninsula (Inverness Ridge, dunes near the radio station,
Point Reyes).—*Triteleia laxa* Benth.

The flowers of this attractive plant vary from lavender to deep violet-purple.
In Marin County most of the flowers are smaller than usual, a form described
from Tiburon Peninsula as *Triteleia angustiflora* Hel. and from the Marin coast
as *B. laxa* var. *nimia* Jeps.

5. **B. peduncularis** (Lindl.) Wats. MARSH TRITELEIA. In wet ground around seepages and in meadows, frequently near serpentine: Rodeo Lagoon, acc. Eastwood;
Tiburon; Bootjack and Potrero Meadow, Mount Tamalpais; Ledum Swamp and
radio station, Point Reyes Peninsula.—*Triteleia peduncularis* Lindl.

The plant at Bootjack is a form with the inner perianth segments trifid and
has been named *Triteleia peduncularis* var. *trifida* Eastw.

6. **B. elegans** Hoover. HARVEST BRODIAEA. Dry grassland of open or brushy slopes: Tiburon; Mount Tamalpais (Azalea Flat, Bootjack, Rock Spring, Potrero Meadow); San Rafael Hills; San Anselmo Canyon; Mud Lake and Inverness Ridge, Point Reyes Peninsula.—*B. coronaria* and *Hookera coronaria* of authors in part.

7. **B. terrestris** Kell. DWARF BRODIAEA. Meadows and grassy slopes: Lagunitas Meadows, Mount Tamalpais; west of Novato; Inverness Ridge and near the radio station, Point Reyes Peninsula.—*Hookera terrestris* (Kell.) Britten; *Brodiaea coronaria* (Salisb.) Engler var. *macropoda* (Torr.) Hoover.

Brodiaea Ida-maia (Wood) Greene, the showy firecracker plant, with bright red perianth tube, is generally reported as far south as Marin County but no definite station is known south of Mendocino and Lake counties.

8. Lilium

1. **L. pardalinum** Kell. CALIFORNIA TIGER LILY. Wet meadows and springs: Sausalito Hills above Rodeo Lagoon; Bootjack and Potrero Meadow, Mount Tamalpais; Lagunitas Canyon; Carson country; Ledum Swamp, Point Reyes Peninsula. The tiger lily is rather common around springs and seepages in the serpentine areas at the west end of Mount Tamalpais where it is generally found among protecting thickets of azalea.

Two other lilies have been reported from Marin County, *L. maritimum* Kell. and *L. rubescens* Wats., but neither specimens nor definite collection data have been seen to substantiate the reports. The type locality of *L. maritimum* was given as "in the vicinity of San Francisco" to which Abrams adds "probably Marin County." Abrams also gives "Marin County" as the type locality of *L. rubescens* probably because Watson states its distribution as "California (Coast Range, Marin to Humboldt counties)." From the character of the original description of *L. rubescens,* however, it seems proper to consider the type of that species the same as that of *L. Washingtonianum* Kell. var. *purpureum* Masters which is cited as a synonym and which was described as a "native of Humboldt County in California." Both species are known from recent collections from Sonoma County and they may have occurred in Marin County, but positive evidence is at present lacking.

9. Fritillaria

a. Flowers cream-white ....1. *F. liliacea*
a. Flowers brownish-purple, more or less mottled with green ....2. *F. lanceolata*

1. **F. liliacea** Lindl. Occurring locally on grassy hills near the coast: between Nicasio and Tomales; Tomales Bay; Olema, acc. Stacey.

2. **F. lanceolata** Pursh. MISSION BELLS. Widespread and rather common on brushy or wooded slopes: Sausalito, Tiburon, and Mount Tamalpais (Corte Madera, Cataract Gulch, Phoenix Lake) to the Carson country and San Rafael.— *F. mutica* Lindl.; *F. lanceolata* var. *gracilis* Wats.

The type specimen of *F. lanceolata* var. *gracilis* Wats. was collected by Hartweg at Corte Madera. On Point Reyes Peninsula on exposed rocky slopes near the lighthouse grows the low form with dark brown-purple flowers that was described from the "Marin coast" as var. *tristulis* A. L. Grant.

10. Calochortus

a. Flowers white to lilac or pale violet ....b
a. Flowers yellow ....d
b. Petals hairy on the inner side from the gland to the apex ....1. *C. Tolmiei*
b. Petals hairy only on the lower half of the inner side near the gland c
c. Flowers whitish or pale lavender; plants of moist hills and rocky slopes
.... 2. *C. umbellatus*
c. Flowers lilac or pale violet; plants of low wet meadows ....3. *C. uniflorus*
d. Flowers open-campanulate, the flowers and fruits erect ....4. *C. luteus*
d. Flowers subglobose, the flowers and fruits nodding ....5. *C. amabilis*

1. **C. Tolmiei** H. & A. HAIRY STAR TULIP; PUSSY-EARS. Moist grassy slopes along the coast: Ledum Swamp and Drakes Bay, Point Reyes Peninsula; Dillons Beach. —*C. caeruleus* (Kell.) Wats. var. *Maweanus* (Leicht.) Jeps.; *C. Maweanus* Leicht.

2. **C. umbellatus** Wood. Moist wooded and brushy hills or gravelly open slopes, frequently on or near serpentine: Tiburon, *Orr;* Mount Tamalpais (Muir Woods, Pipeline Trail, Rattlesnake Camp, Bootjack, Berry Trail); Carson Ridge.

3. **C. uniflorus** H. & A. Low grassy flats that are wet in spring: Lagunitas Meadows, Mount Tamalpais.

4. **C. luteus** Dougl. YELLOW MARIPOSA. Open grassy summer-dried hills: Tiburon; Bootjack and Lake Lagunitas, Mount Tamalpais; San Rafael Hills.

5. **C. amabilis** Purdy. GOLDEN FAIRY LANTERN. Wooded or brushy hills: Corte Madera.

The Golden Fairy Lantern, one of California's most beautiful woodland flowers, is restricted to the Coast Ranges north of San Francisco Bay, although it is closely related to the Mount Diablo Fairy Lantern, *C. pulchellus* Dougl.

11. Scoliopus

1. **S. Bigelovii** Torr. FETID ADDER'S TONGUE. Moist shaded slopes, generally under redwoods: Sausalito; Muir Woods; Mount Tamalpais (Blithedale Canyon, Cataract Gulch, Fish Grade); Bolinas Ridge; San Geronimo Ridge; San Rafael Hills; Tomales, acc. Leschke.

One of the first plants to bloom after the beginning of the rainy season, the Fetid Adder's Tongue thrusts its queer ill-scented flowers from the pair of closely rolled leaves as soon as they are above ground. By the time the attractive brown-spotted leaves are developed, the first fruits are already well formed at the ends of elongate sprawling twisting pedicels. This remarkable plant was dis covered by Bigelow at "Tamul Pass" in 1854.

12. Trillium

a. Flower on a slender pedicel ....1. *T. ovatum*
a. Flower sessile ....2. *T. chloropetalum*

1. **T. ovatum** Pursh. TRILLIUM; WAKE-ROBIN. Rather common on moist shaded slopes of canyons under redwoods: Muir Woods; Mount Tamalpais (Cascade Canyon, Steep Ravine, Cataract Gulch, Phoenix Lake); Kentfield; Camp Taylor, acc. Leschke.

2. **T. chloropetalum** (Torr.) Howell. Occasional on moist wooded slopes or on

the edge of brush: Dipsea Trail, Mount Tamalpais, *Leschke;* Stinson Beach, acc. Stacey; San Anselmo Canyon; Carson country, acc. Leschke; Black Canyon, San Rafael Hills, acc. H. M. Pollard; near Inverness, acc. Gregory Lyon; Tomales, *Leschke.* The type collection was made by Bigelow in the redwoods of Marin County along his route from Corte Madera to Point Reyes.—*T. sessile* L. var. *giganteum* H. & A.

13. Clintonia

1. **C. Andrewsiana** Torr. Moist shaded slopes and flats, generally under redwoods: Sausalito, acc. Lovegrove; Muir Woods; Mill Valley; Mount Tamalpais (Cascade Canyon, Steep Ravine, Cataract Gulch); Bolinas Ridge; Lagunitas Canyon; Corte Madera; Lagunitas, acc. Stacey.

The deep rose flowers of this handsome plant are followed by remarkable blue berries that appear more porcelaneous than juicy. Marin County is probably the type locality of the clintonia, since that is where Bigelow collected the first cited specimen and probably also where Andrews made his collection.

14. Smilacina

a. Flowers in a raceme ....1. *S. sessilifolia*
a. Flowers in a panicle ....2. *S. racemosa*

1. **S. sessilifolia** Nutt. SLIM SOLOMON. Common and widespread on moist wooded or brushy hills: Sausalito, Tiburon, and Mount Tamalpais (Cascade Canyon, Bolinas Ridge) to San Rafael Hills and Tomales.—*Vagnera sessilifolia* (Nutt.) Greene.

2. **S. racemosa** (L.) Desf. var. **amplexicaulis** (Nutt.) Wats. FAT SOLOMON. Moist slopes, more or less shaded by trees or brush, rather common: Sausalito; Tiburon; Mount Tamalpais (Blithedale and Cascade canyons, Cataract Gulch); Stinson Beach, acc. Stacey; San Anselmo Canyon; San Rafael Hills; Tomales.—*S. amplexicaulis* Nutt.; *Vagnera amplexicaulis* (Nutt.) Greene.

15. Maianthemum

1. **M. dilatatum** (Wood) Nels. & Macbr. Rare on moist slopes or in coastal swales: Sausalito; Bolinas Ridge, acc. Eastwood; Ledum Swamp and near radio station, Point Reyes Peninsula.—*M. bifolium* (L.) F. W. Schm. var. *kamtschaticum* (Gmel.) Jeps., in part; *Unifolium dilatatum* (Wood) Howell.

Marin County is the southern distributional limit for this species.

16. Disporum

a. Perianth pale greenish, less than 1.5 cm. long ....1. *D. Hookeri*
a. Perianth whitish, generally more than 1.5 cm. long ....2. *D. Smithii*

1. **D. Hookeri** (Torr.) Britt. FAIRY BELLS. Widespread and rather common on moist wooded or brushy hills: Sausalito, Tiburon, and Mount Tamalpais to San Rafael Hills and Point Reyes Peninsula (near Mud Lake).

2. **D. Smithii** (Hook.) Piper. Deep shade of wooded canyons: Muir Woods; Blithedale Canyon and Steep Ravine, Mount Tamalpais; Lagunitas Canyon.

IRIDACEAE. Iris Family

a. Perianth segments similar .... 1. *Sisyrinchium*
a. Perianth segments in 2 dissimilar series .... 2. *Iris*

1. Sisyrinchium

a. Perianth segments purplish blue .... 1. *S. bellum*
a. Perianth segments yellow .... 2. *S. californicum*

1. **S. bellum** Wats. Blue-eyed Grass. Open hills and meadows, mostly in grassland, widespread, sometimes locally abundant and forming colorful patches: Angel Island; Tiburon; Sausalito; Azalea Flat and Potrero Meadow, Mount Tamalpais; Ross; San Rafael; Big Rock Ridge; Tomales; Point Reyes Peninsula (Bolinas, Inverness Ridge, Ledum Swamp).

2. **S. californicum** (Ker.) Dryand. Yellow-eyed Grass. Marshes, generally near the coast: Sausalito, acc. Lovegrove; Elk Valley; Azalea Flat, Mount Tamalpais; Stinson Beach, acc. Stacey; Ledum Swamp, Point Reyes Peninsula; Dillons Beach. —*Hydastylus californicus* (Ker.) Salisb.

The plant collected by Bigelow on Point Reyes Peninsula was described as *S. lineatum* Torr.

2. Iris

a. Perianth tube less than 1 cm. long .... 1. *I. longipetala*
a. Perianth tube more than 1 cm. long .... b
b. Stems usually less than 2 dm. tall; leaves with upper and lower surfaces alike;
    perianth tube usually 4–6 cm. long .... 3. *I. macrosiphon*
b. Stems more than 2 dm. tall; leaves arching, bifacial, shiny above, dull below;
    perianth tube usually about 2 cm. long c
c. Flowers deep blue .... 2. *I. Douglasiana*
c. Flowers variable in color, from ivory-white to pale yellow and light blue
    .... 2a. *I. Douglasiana* var. *major*

1. **I. longipetala** Herbert. Common on open hills and in low wet fields: Angel Island; Mill Valley, acc. Stacey; Nicasio; near Aurora School; Tomales. In the Marin County plant, the flowers are not as large as in plants from the Coast Ranges to the southward but they are delicately beautiful and at Nicasio make a fine display. Although this species is common in northern Marin County, no record of it from Sonoma County has been seen.

2. **I. Douglasiana** Herbert. Douglas Iris. Open hills and downs near the coast: Ledum Swamp and dune hills near the radio station, Point Reyes Peninsula; Tomales.

2a. **I. Douglasiana** Herbert var. **major** Torr. Marin Iris. Widespread and rather common through the wooded hill country, in sun or partial shade: Tiburon, Sausalito, and Mount Tamalpais (Corte Madera Ridge, Azalea Flat, Rattlesnake, Cataract Gulch) to the Carson country, San Rafael Hills, and Big Rock Ridge.

The delicacy of form and the fine shadings of color combine to make the Marin iris one of the chief floral delights of the vernal woodland. This variety, which was originally collected by Bigelow at Corte Madera in 1854, is generally more slender than the Douglas iris which occurs only in the immediate vicinity of the coast.

3. **I. macrosiphon** Torr. GROUND IRIS. Forming restricted or broad clumps on open grassy hills and along the edge of woods and brush: Tiburon; Sausalito; Mill Valley; Phoenix Lake, Mount Tamalpais; Corte Madera, the type locality; San Rafael; Big Rock Ridge; Black Point; Tomales.

The small beautiful flowers are fragrant but they bloom so close to the earth one must stoop low to perceive the scent.

ORCHIDACEAE. ORCHID FAMILY

a. Plants without green foliage leaves ....7. *Corallorhiza*
a. Plants with green foliage leaves ....b
b. Flowers relatively large, 1 or several, racemose when more than 1 c
b. Flowers smaller, generally numerous, spicate ....e
c. Flower solitary; foliage leaf 1, basal ....1. *Calypso*
c. Flowers generally several; foliage leaves several, cauline ....d
d. Lip inflated, saccate ....2. *Cypripedium*
d. Lip not inflated or saccate ....3. *Epipactis*
e. Flowers spurred ....4. *Habenaria*
e. Flowers not spurred ....f
f. Foliage leaves basal and cauline; flowers in a spirally twisted spike 5. *Spiranthes*
f. Foliage leaves basal; spike not spirally twisted ....6. *Goodyera*

1. Calypso

1. **C. bulbosa** (L.) Oakes. Moss-covered logs and rocks in moist shady woods: Bolinas Ridge, acc. Davy; above Lily Lake, acc. Sutliffe; north side of Mount Tamalpais, acc. Eastwood. These are the southernmost California stations for this favorite orchid.—*Cytherea bulbosa* (L.) House.

2. Cypripedium

1. **C. californicum** Gray. CALIFORNIA LADYSLIPPER. Moist canyon on south side of Mount Tamalpais above Muir Woods, *Eastwood, Leschke*. This rare plant is not known south of Marin County.

3. Epipactis

1. **E. gigantea** Dougl. STREAM ORCHIS. Not uncommon among rocks of stream-beds and springy slopes: Mill Valley, acc. Stacey; Mount Tamalpais (Azalea Flat, northeast of the Mountain Theater, Cataract Gulch, Camp Handy); Lagunitas Canyon; Big Carson Canyon; swales near the radio station, Point Reyes Peninsula, *Leschke.—Serapias gigantea* (Dougl.) A. A. Eaton.

4. Habenaria. REIN-ORCHIS

a. Flowers white; plants of marshy places ....1. *H. leucostachys*
a. Flowers green or greenish white; plants of grassy or wooded slopes dry in summer ....b
b. Perianth parts white and greenish white; flowers fragrant ......2. *H. Greenei*
b. Perianth parts green; flowers scarcely fragrant c

c. Spur equaling the ovary or shorter ....5. *H. unalascensis*
c. Spur longer than the ovary ....d
d. Stem stout, with numerous bractlike leaves; lip about as broad as long, auriculate at base ....3. *H. Michaeli*
d. Stem slender, with fewer bractlike leaves; lip longer than broad, not auriculate at base ....4. *H. elegans*

1. **H. leucostachys** (Lindl.) Wats. Rare in marshy meadows: Potrero Meadow, Mount Tamalpais; Ledum Swamp, Point Reyes Peninsula; Dillons Beach.—*Limnorchis leucostachys* (Lindl.) Rydb.

2. **H. Greenei** Jeps. Rare on brushy or grassy slopes near the coast: Sausalito, acc. Eastwood; Bolinas; Inverness, *Sutliffe;* Point Reyes, acc. Eastwood; McClure Beach Road.—*H. maritima* Greene, not Raf.; *Piperia maritima* (Greene) Rydb.

A white fragrant rein-orchis that seems to be a small-flowered form of *H. Greenei* grows at Azalea Flat on Mount Tamalpais.

3. **H. Michaeli** Greene. Rare on brushy hills away from the coast: Black Point; Pipeline Trail, *Leschke.*—*Piperia Michaeli* (Greene) Rydb.

The stout stem and the fleshy texture of the perianth would seem to relate this to *H. Greenei* rather than to *H. elegans* to which it is reduced by some botanists.

4. **H. elegans** (Lindl.) Boland. Rather common on wooded or brushy slopes: Mount Tamalpais (Blithedale Canyon, Baltimore Canyon, Azalea Flat, near Rock Spring, Lake Lagunitas); Carson Ridge; San Rafael Hills.—*Piperia elegans* (Lindl.) Rydb.

5. **H. unalascensis** (Spreng.) Wats. In partial shade on wooded slopes: Mount Tamalpais (cypress grove near West Peak, *Leschke,* Phoenix Lake, *L. S. Rose*); San Anselmo Canyon above Fairfax.—*Piperia unalascensis* (Spreng.) Rydb.

The green-flowered rein-orchises that are here reported for Marin County are closely related and by some workers have been reduced to varietal status under *H. unalascensis.* Numerous small differences in both physiologic and morphologic characters adequately demark the entities as they occur in Marin County.

5. **Spiranthes.** LADY'S TRESSES

a. Flowers greenish white; lip without nipple-like swellings at the base
....1. *S. Romanzoffiana*
a. Flowers pale creamy yellowish; lip with two nipple-like swellings at the base
....2. *S. porrifolia*

1. **S. Romanzoffiana** C. & S. Occasional in wet meadows: Mount Tamalpais, San Rafael, and Fairfax, acc. Stacey; Ledum Swamp, Point Reyes Peninsula.—*Ibidium Romanzoffianum* (C. & S.) House.

2. **S. porrifolia** Lindl. Rare in wet meadows on Mount Tamalpais: Rock Spring; Potrero Meadow; Lagunitas Meadows.—*Ibidium porrifolium* (Lindl.) Rydb.

6. **Goodyera**

1. **G. oblongifolia** Raf. RATTLESNAKE PLANTAIN. Rare in shady woodland: Mount Tamalpais (Laurel Dell, Lake Lagunitas, acc. Sutliffe); Lagunitas Creek, acc. Stacey; Bear Valley, Point Reyes Peninsula. The rattlesnake plantain has not been found south of Marin County in the Coast Ranges.—*G. decipiens* (Hook.) F. T. Hubbard; *Peramium decipiens* (Hook.) Piper.

7. Corallorhiza. CORALROOT

a. Lip buff, striped with purple ....1. *C. striata*
a. Lip white, spotted with reddish purple ....2. *C. maculata*

1. **C. striata** Lindl. Occasional in leaf mold of deep woods: Mount Tamalpais (Zigzag Fire Trail, Laurel Dell, Lagunitas Meadows); San Anselmo Canyon; Big Carson, acc. Leschke.—*C. Bigelovii* Wats., the type locality, Corte Madera.

2. **C. maculata** Raf. Rare in deep woods: Mount Tamalpais (Zigzag Fire Trail and on north side of mountain); Big Carson Canyon; near Inverness.

SALICACEAE. WILLOW FAMILY

a. Buds covered by several imbricated scales; floral bracts laciniate ..1. *Populus*
a. Buds covered by a single scale; floral bracts entire or minutely toothed
....2. *Salix*

1. Populus. POPLAR; COTTONWOOD

1. **P. alba** L. WHITE POPLAR. Reproducing by root suckers and becoming locally established along roads and in waste places: Sausalito; Escalle; Ross; San Rafael; San Anselmo; Aurora School; San Antonio Creek; near Inverness. Introduced from Europe.

2. Salix. WILLOW

a. Leaves narrow and elongate, linear-lanceolate or linear-oblong, the upper and lower surfaces of leaves about the same in color or pubescence ....b
a. Leaves broader, lanceolate or oblanceolate to broadly ovate, elliptic, or obovate, the upper and lower surfaces of leaves quite unlike in color or pubescence (except sometimes in *S. Scouleriana*) c
b. Leaves and capsule hairy ....1. *S. Hindsiana*
b. Leaves and capsule glabrous ....2. *S. melanopsis*
c. Leaves densely satiny-pubescent beneath; stamen 1 ....3. *S. Coulteri*
c. Leaves not densely satiny-pubescent beneath; stamens 2–10 ....d
d. Leaves broadest at the middle or above; stamens 2; shrubs and low trees with smooth bark, except in old individuals ....e
d. Leaves broadest near the base; stamens 3–10; trees with rough bark ....f
e. Leaves narrowly to broadly oblong or oblanceolate or rarely obovate; filaments united at base; style present; ovary glabrous ....4. *S. lasiolepis*
e. Leaves broadly oblanceolate to obovate; filaments free; style none; ovary hairy ....5. *S. Scouleriana*
f. Petioles warty-glandular near base of blade ....6. *S. lasiandra*
f. Petioles not glandular ....7. *S. laevigata*

1. **S. Hindsiana** Benth. Flood beds of streams and low places along roads: Ross; Ignacio; San Antonio Creek; Gallinas Valley, *Leschke;* Lagunitas Creek.—*S. sessilifolia* Nutt. var. *Hindsiana* (Benth.) Anders.

2. **S. melanopsis** Nutt. Along railroad north of San Rafael; probably not native in Marin County but introduced with stream gravel on the railroad embankment. Widespread at low and middle elevations in California.

3. **S. Coulteri** Anders. COULTER WILLOW. Occasional along streams in the hills and in marshy places near the coast: Rodeo Lagoon; Blithedale Canyon and Cataract Gulch, Mount Tamalpais; Lagunitas Creek; Dillons Beach; Bear Valley and Ledum Swamp, Point Reyes Peninsula.—*S. sitchensis* Sanson var. *Coulteri* (Anders.) Jeps.

4. **S. lasiolepis** Benth. ARROYO WILLOW. Stream banks, gullies, and springy places, widespread and usually common: Angel Island, Sausalito, Tiburon, and Mount Tamalpais (Blithedale Canyon, Phoenix Lake, Lagunitas Meadows) to San Rafael, Ignacio, Tomales, and Point Reyes Peninsula (Bolinas).

On dunes and in swales near the ocean grows var. *Bigelovii* (Torr.) Bebb with thicker leaves more densely pubescent on the lower surface and with style more elongate: Muir Beach; Stinson Beach; Dillons Beach; Ledum Swamp, Point Reyes Peninsula.

5. **S. Scouleriana** Barratt. SCOULER WILLOW. Rare on the south side of Mount Tamalpais in wet soil of brushy or wooded slopes: Blithedale Canyon; head of Cascade Canyon; above Muir Woods.

6. **S. lasiandra** Benth. YELLOW WILLOW. Rather common along streams at lower elevations: Sausalito, Tiburon, and Mill Valley (Blithedale Canyon) to Stinson Beach, Ignacio, Tomales, and Point Reyes Peninsula (Inverness).

7. **S. laevigata** Bebb. RED WILLOW. On the banks of streams, occasional: Stinson Beach; San Anselmo; Lagunitas Meadows, Mount Tamalpais; Lagunitas Creek; Black Canyon, San Rafael Hills; San Antonio Creek.

At several places in Marin County the yolk willow, *S. vitellina* L., has been planted and along the lower part of San Anselmo Creek it may be naturalized. Although *S. vitellina* somewhat resembles *S. laevigata* in habit and foliage, the bright yellow twigs of the former in contrast to the reddish twigs of the latter immediately distinguish it.

BETULACEAE. BIRCH FAMILY

1. Alnus. ALDER

a. Leaf margin plane, not revolute ....1. *A. rhombifolia*
a. Leaf margin narrowly revolute ....2. *A. rubra*

1. **A. rhombifolia** Nutt. WHITE ALDER (plate 24). Stream banks, frequently adjacent to redwood groves or pockets but away from the immediate coast: Ross Valley; San Anselmo; Lagunitas Meadows, Mount Tamalpais; Little Carson; Lagunitas Creek.

Much of the charm of small rocky creeks in the Coast Ranges is due to the beauty of the alders. To see the green-gold of their blossoming crowns in January is one of the floral treats of the year.

2. **A. rubra** Bong. RED ALDER; OREGON ALDER. Along streams in deep redwood canyons or in marshy places near the coast: Muir Beach; Mill Valley; Steep Ravine; Stinson Beach; Inverness.—*A. oregona* Nutt.

At the mouth of Rio Hondo north of Bolinas the red alder is reduced to shrub size by the almost constant westerly wind.

CORYLACEAE. Hazel Family

1. Corylus. Hazel

1. **C. californica** (A. DC.) Rose. California Hazel. Widespread and common in brush and woodland, generally in moist or partially shaded canyons: Sausalito, Tiburon, and Mount Tamalpais (Blithedale Canyon, Cataract Gulch, Phoenix Lake) to San Rafael Hills and Point Reyes Peninsula (Mud Lake, Inverness).—*C. rostrata* Ait. var. *californica* A. DC.

FAGACEAE. Oak Family

a. Branchlets and lower side of leaves with a close golden scurf; fruit enveloped by a spiny burlike involucre. Catkins erect ....1. *Castanopsis*

a. Branchlets and lower side of leaves without a golden scurf (except in *Quercus chrysolepis*); fruit an acorn in a scaly, cuplike involucre ....b

b. Staminate catkins stiffly erect; leaves evergreen, green above and pale green below, mostly 5–15 cm. long ....2. *Lithocarpus*

b. Staminate catkins slender, pendent or spreading; leaves evergreen or deciduous, if evergreen and pale beneath, then 5 cm. long or less ......3. *Quercus*

1. Castanopsis

1. **C. chrysophylla** (Dougl.) A. DC. Chinquapin. A large shrub or small tree in the chaparral on Mount Tamalpais (Zigzag Fire Trail, Corte Madera Ridge), San Geronimo Ridge, and Inverness Ridge.

Most of the chinquapins in Marin County are referable to var. *minor* (Benth.) A. DC., the low shrubby form of the species. On exposed slopes on Inverness Ridge, it forms an extensive scrub only 1 to 2 feet tall. On the north side of Mount Tamalpais on the trail between Mountain Theater and Potrero Meadow there is a small grove of well-developed chinquapin trees, perhaps the southernmost station for the typical form of the species. The largest specimens here attain a height of about 60 feet and have a trunk diameter of about a foot. The bark of large old trees is rough and dark brownish-gray, while the bark of young trees is silvery-gray and unfissured. A striking and picturesque effect is produced when the pale bark of young trees is first broken by dark longitudinal fissures. In var. *minor*, the silvery-gray bark is not fissured except at the very base in old individuals.

2. Lithocarpus

1. **L. densiflorus** (H. & A.) Rehd. Tanbark Oak. Large trees commonly associated with the madroño and California laurel along the borders of the redwood forest: Mount Tamalpais (common in well-watered canyons on north and south sides of the mountain), Bolinas Ridge, and Inverness Ridge.—*Pasania densiflora* (H. & A.) Oerst.

On exposed rocky ridges the tanbark oak becomes dwarfed and shrubby but it seldom forms part of the chaparral. Although its leaves are rather uniform in size, they may become unusually large in protected redwood canyons or may be much reduced on the edge of chaparral. In the dense forest at the head of Muir Woods, leaves 20 cm. long have been measured, while above, on the open slopes of Mount Tamalpais they may be only 3 cm. long.

Quercus echinacea Torr., a synonym, was based largely on specimens collected by Bigelow at "Tokeloma Creek."

3. Quercus. OAK

a. Leaves somewhat shiny on the lower side and but little paler than above.—Black Oaks ....b

a. Leaves dull on the lower side and much paler than above.—White Oaks .....e

b. Leaves evergreen, not deciduous before spring c

b. Leaves deciduous (tardily so in *Q. morehus*) ....d

c. Leaves convex above, frequently with hairs in angle of midrib and lateral veins on lower side; acorns maturing in one season ....1. *Q. agrifolia*

c. Leaves plane or undulate, entirely glabrous on lower side; acorns maturing in two seasons ....2. *Q. Wislizeni*

d. Lobes of leaves with several bristle-tipped teeth ....3. *Q. Kelloggii*

d. Lobes of leaves with one terminal bristle-like spine ....4. *Q. morehus*

e. Leaves evergreen, teeth spiny-tipped ....f

e. Leaves deciduous, teeth or lobes not spiny ....h

f. Young leaves with golden pubescence beneath; acorns maturing in two seasons; trees, becoming shrubby in the chaparral ....5. *Q. chrysolepis*

f. Young leaves without golden pubescence; acorns maturing in one season; shrubs ....g

g. Leaves plane or undulate; shrubs characteristic of shale or sandstone ridges ....6. *Q. dumosa*

g. Leaves convex above, the margins revolute; shrubs almost confined to serpentine slopes ....7. *Q. durata*

h. Leaves toothed or shallowly lobed, blue-glaucous above ......10. *Q. Douglasii*

h. Leaves deeply lobed, deep green above i

i. Acorns plump, cups shallow; leaves prominently pubescent beneath; buds large, densely pubescent ....8. *Q. Garryana*

i. Acorns slender, cups deep; leaves not prominently pubescent beneath; buds smaller and not densely pubescent ....9. *Q. lobata*

1. **Q. agrifolia** Née. COAST LIVE OAK; ENCINA (plates 5, 18). Round-topped trees of lower hills and high ridges, common along evanescent streams with the California laurel: Angel Island, Sausalito, Tiburon, and Mount Tamalpais to San Rafael Hills, Black Point, and Inverness Ridge. On exposed ridges near the coast, the encina is as wind-pruned as the laurel with which it grows, and, on the open slopes of the Sausalito Hills and Tiburon Peninsula, together with several other oaks, it forms prostrate mats.

A large-leaved, long-petioled individual that is perhaps a hybrid between *Q. agrifolia* and *Q. Kelloggii* has been found with those two species in the Greenbrae Hills.

1*a*. **Q. agrifolia** Née var. **frutescens** Engelm. The shrub form of the coast live oak, 10 feet or less tall, is not common but is occasionally found in the transition belt between grassland and chaparral: Sausalito Hills; Bootjack, Mount Tamalpais; San Geronimo Ridge; Big Rock Ridge.

2. **Q. Wislizeni** A. DC. var. **frutescens** Engelm. CHAPARRAL OAK. The most widespread and abundant oak in the chaparral of Marin County: Angel Island; Sausalito Hills; Mount Tamalpais (very common); San Geronimo Ridge; San

Rafael Hills. On an exposed summit above Sausalito, the chaparral oak is reduced to a brushy mat.

Whereas the shrubby varieties of *Q. agrifolia* and *Q. chrysolepis* may represent only edaphic forms of those species, *Q. Wislizeni* var. *frutescens* is undoubtedly an entity with a genetic basis and perhaps deserves the specific name *Q. parvula* Greene.

3. **Q. Kelloggii** Newb. CALIFORNIA BLACK OAK; KELLOGG OAK. Large trees in moist valleys about Mount Tamalpais, smaller or dwarfed trees on rocky hills: Sausalito; Tiburon; Mill Valley; Ross; San Rafael Hills; Big Rock Ridge; Black Point. The Kellogg oak is a wind-swept depressed dwarf on the Sausalito Hills.

4. **Q. morehus** Kell. ORACLE OAK. Occasional low trees or clustered shrubs on the edge of the chaparral: Sausalito (subprostrate shrubs); Tiburon; Mount Tamalpais (Cascade Canyon, Pipeline Trail, Corte Madera Ridge, above Lake Lagunitas); San Geronimo Ridge; San Rafael Hills.

The oracle oak is generally considered a hybrid between *Q. Wislizeni* and *Q. Kelloggii*. The unusual specific name, *morehus,* is derived from Moreh, the place where the biblical patriarch, Abraham, once dwelt.

Both in the Greenbrae Hills and San Rafael Hills, an oak is found that superficially resembles the oracle oak but seems to be a hybrid between *Q. agrifolia* and *Q. Kelloggii*. In it the leaves tend to be more concave-convex and tufts of stellate hairs in the angles of the veins on the lower leaf surface are not infrequent.

5. **Q. chrysolepis** Liebm. CANYON LIVE OAK; MAUL OAK; GOLDCUP OAK. Common in canyons and on rocky ridges, becoming a large and picturesque tree: Mount Tamalpais (Cascade Canyon, Azalea Flat, Mountain Theater); San Geronimo Ridge; Inverness Ridge.

5a. **Q. chrysolepis** Liebm. var. **nana** Jeps. This is the shrubby form of the maul oak in the chaparral, as along the summit ridge of Mount Tamalpais. On the wind-swept ridges above Sausalito it forms bushy mats.

6. **Q. dumosa** Nutt. CALIFORNIA SCRUB OAK. Brushy slopes on sedimentary rocks, rare in Marin County: Ridgecrest Road above Meadows Club; near Camp Lilienthal in the Fairfax Hills; Fish Grade between Phoenix and Lagunitas lakes; Black Canyon in the San Rafael Hills. Growing with *Q. dumosa* in the San Rafael Hills is a shrub with deeply and spinescently lobed leaves that suggests a hybrid between *Q. dumosa* and *Q. lobata*.

7. **Q. durata** Jeps. LEATHER OAK. A characteristic shrub of the serpentine areas in Marin County: Tiburon; Mount Tamalpais (near West Point, West Peak, Berry Trail); Alpine Lake; Carson Ridge. On the summit of Tiburon Peninsula, the leather oak, along with two other oaks, forms those peculiar shrubby mats that are more reminiscent of timberline than sea level. The leathery, rather than hard and brittle, texture of the leaves suggested the common name.

8. **Q. Garryana** Dougl. GARRY OAK. Small trees on moist or northern slopes, rare: Dipsea Trail at head of Kent Ravine, Mount Tamalpais; Ridgecrest Road near Camp Lilienthal, Fairfax Hills.

In Marin County, near the southern limit of its distribution, *Q. Garryana* has apparently hybridized with the two shrubby white oaks that occur in the same region. The larger number of these hybrid-like individuals appears to represent a cross between *Q. Garryana* and *Q. durata*. It occurs as a shrub or small tree, usually singly in chaparral near serpentine: Hidden Lake, Mount Tamalpais; Ridgecrest Road near Liberty Spring; San Anselmo Canyon. The prostrate white oak on the wind-swept summit above Tiburon that resembles Q. *Garryana* is perhaps an extreme form of this suspected hybrid. The second suspected hybrid is between Q. *Garryana* and Q. *dumosa* and occurs as an arborescent shrub on Fish Grade between Phoenix and Lagunitas lakes.

9. **Q. lobata** Née. VALLEY OAK; CALIFORNIA WHITE OAK; ROBLE. A large handsome tree of valley lands, smaller and sometimes scrubby on dry rocky slopes: Ross; San Anselmo Canyon; San Geronimo; San Rafael; Big Rock Ridge; Black Point.

10. **Q. Douglasii** H. & A. BLUE OAK; DOUGLAS OAK. In Marin County found only on the dry hills near Black Point and on the north slope of the San Rafael Hills. Near Camp Lilienthal in the Fairfax Hills an individual has been found that resembles Q. *Douglasii* in the blue cast of the foliage but perhaps represents a hybrid between that species and either Q. *lobata* or Q. *Garryana*.

MYRICACEAE. BAYBERRY FAMILY

1. Myrica

1. **M. californica** Cham. CALIFORNIA WAX-MYRTLE. Wet slopes about springs or on moist coastal flats: Sausalito; Tiburon; Stinson Beach; Mount Tamalpais (head of Cascade Canyon, Azalea Flat, Cataract Gulch, Willow Meadow); Salmon Creek; Inverness and Ledum Swamp, Point Reyes Peninsula.

In exposed situations the wax-myrtle forms little more than a bushy arborescent shrub but in protected places it becomes a medium-sized tree, much branched and densely leafy, with a gray warty unfissured bark. Plants injured by fire sprout from the heavy root-crown. The California wax-myrtle was described by Chamisso who first collected it at San Francisco in 1816 on the Kotzebue Expedition.

URTICACEAE. NETTLE FAMILY

a. Sepals of pistillate flower distinct, 2 large and 2 small . . . . .1. *Urtica*
a. Sepals of pistillate flower united nearly to the top and closely investing the
    achene . . . . .2. *Hesperocnide*

1. Urtica. NETTLE

a. Plants annual, the stems usually less than 0.5 m. tall . . . . .3. *U. urens*
a. Plants perennial, the stems generally 1–3 m. tall . . . . .b
b. Stems subglabrous; leaves broadly ovate .1. *U. californica*
b. Stems whitish-pubescent; leaves lanceolate to narrowly ovate .2. *U. holosericea*

1. **U. californica** Greene. Common near the coast in brushy thickets in wet soil: Sausalito; Muir Beach; Stinson Beach; Point Reyes Peninsula (Mud Lake, Mount Vision, Inverness); Tomales. This nettle is known only along the California coast from San Mateo County north to Sonoma County. Robust individuals may have the lower leaves as much as 10 inches long and 8 inches wide on petioles up to 6 inches long.

2. **U. holosericea** Nutt. Occasional in wet places in the hills and along the coast: Rodeo Lagoon; Angel Island; Stinson Beach; Little Carson; Ignacio; Lagunitas

Creek; Chileno Valley; Point Reyes Peninsula (near Bolinas, Inverness, Ledum Swamp).—*U. gracilis* Ait. var. *holosericea* (Nutt.) Jeps.

3. **U. urens** L. Widespread weed in gardens and waste ground, occasional in Marin County: Ignacio; Chileno Valley; Olema; Dillons Beach; Point Reyes Peninsula (road to the lighthouse and Drakes Estero).

2. Hesperocnide

1. **H. tenella** Torr. Occasional in shade of brush or trees: Sausalito, acc. Stacey; Steep Ravine, Mount Tamalpais; San Rafael, *Moore;* Nicasio; Mud Lake and Inverness Ridge, Point Reyes Peninsula; Tomales.

ULMACEAE. ELM FAMILY

1. Ulmus. ELM

1. **U. procera** Salisb. ENGLISH ELM. About habitations and in waste ground, becoming a tree weed by numerous root suckers: Belvedere; Mill Valley; Escalle; Ross; San Rafael; Burdell.

This species, a native of Europe, is sometimes called cork-elm because of the conspicuous corky ridges produced on branchlets, but that common name properly belongs to *U. Thomasi* Sarg., an American species.

LORANTHACEAE. MISTLETOE FAMILY

a. Leaves conspicuous, more than 1 cm. long ....1. *Phoradendron*
a. Leaves inconspicuous, reduced to connate scales ....2. *Arceuthobium*

1. Phoradendron. MISTLETOE

a. Stems and leaves glabrous, the leaves less than 2 cm. long; parasitic on cypress
....1. *P. densum*
a. Stems and leaves densely but finely pubescent, the leaves more than 2 cm. long;
    mostly parasitic on oaks ....2. *P. villosum*

1. **P. densum** Torr. CYPRESS MISTLETOE. Not uncommon on *Cupressus Sargentii:* Mount Tamalpais (near Bootjack, Barths Retreat); Carson Ridge; San Geronimo Ridge).—*P. Bolleanum* (Seem.) Eichler var. *densum* (Torr.) Fosberg.

2. **P. villosum** Nutt. OAK MISTLETOE. Occasional on both deciduous and evergreen trees, generally on oaks: Mount Tamalpais (Bootjack, Mountain Theater, Hidden Lake Fire Trail, Phoenix Lake, Lake Lagunitas); Fairfax Hills; Big Rock Ridge.—*P. flavescens* (Pursh) Nutt. var. *villosum* (Nutt.) Engelm.

In Marin County this mistletoe has been observed on the following trees: *Quercus agrifolia, Q. Kelloggii, Q. lobata, Q. Wislizeni, Alnus rhombifolia,* and *Umbellularia californica.* The name mistletoe, which comes from the Anglo-Saxon, rightly belongs to the English plant, *Viscum album* L. The genus *Viscum* is found only in the Old World while *Phoradendron* is restricted to the New World. The name *Phoradendron* is derived from the Greek meaning tree-thief.

2. Arceuthobium. PINE MISTLETOE

1. **A. campylopodum** Engelm. Occasional on *Pinus muricata* at Inverness. Point Reyes Peninsula.—*Razoumofskya campylopoda* (Engelm.) Kuntze.

ARISTOLOCHIACEAE. BIRTHWORT FAMILY

a. Perennial herb with stolons; flowers regular ....1. *Asarum*
a. Woody vine; flowers irregular ....2. *Aristolochia*

1. Asarum. WILD GINGER

1. **A. caudatum** Lindl. Occasional on shaded slopes and flats in deep moist soil: Mount Tamalpais, acc. Stacey; Muir Woods; Kentfield, *Eastwood;* Inverness Park, Point Reyes Peninsula; Tomales.

2. Aristolochia

1. **A. californica** Torr. CALIFORNIA PIPE-VINE. Wooded or brushy hills and valleys, widespread but not common: Angel Island; Sausalito Hills, acc. Sutliffe; Tiburon; Phoenix Lake and Bootjack, Mount Tamalpais; Fairfax Hills; San Rafael; Lagunitas Creek.

Specimens collected by Leschke in Gallinas Valley have leaves 7 inches long and 6 inches wide. The species was described from plants found by Bigelow at Corte Madera.

POLYGONACEAE. BUCKWHEAT FAMILY

a. Leaves alternate, with membranaceous stipules sheathing the stem ....b
a. Leaves without stipules, at least the upper opposite or whorled c
b. Perianth segments 5, similar, not modified in fruit ....1. *Polygonum*
b. Perianth segments 6, the 3 inner enlarging in fruit ....2. *Rumex*
c. Leaves lobed, emarginate, or denticulate; flowers axillary, subtended by a single
    bract that becomes much enlarged in fruit ....5. *Pterostegia*
c. Leaves entire; flowers in cylindric or turbinate involucres ....d
d. Involucres several-flowered, the involucral teeth not spine-tipped 3. *Eriogonum*
d. Involucres mostly 1-flowered, the teeth spine-tipped ....4. *Chorizanthe*

1. Polygonum. KNOTWEED; SMARTWEED

a. Stems twining; leafblades broadly ovate and cordate ....1. *P. Convolvulus*
a. Stems not twining; leafblades not broadly ovate and cordate ....b
b. Leaf blades not jointed to the stipular sheath; flowers more or less congested in
    spikelike inflorescences at the ends of branches c
b. Leaf blades jointed to the stipular sheath; flowers in small clusters in the axils
    of blade-bearing leaves (or the uppermost leaves reduced to bracts in *P.*
    patulum) ....h
c. Lowest leaves in a basal cluster; inflorescence solitary. Flowers white
    ....2. *P. bistortoides*
c. Lowest leaves not clustered at the ground; inflorescences usually several, or if
    solitary, the flowers rose ....d
d. Inflorescences 1 or 2; leaves often conspicuously strigose; flowers deep rose
    ....3. *P. coccineum*
d. Inflorescences several to many; leaves glabrous or sparsely appressed-hairy;
    flowers pink, white, or greenish white ....e
e. Stipular and floral sheaths not bristly-ciliate; branches of inflorescence glandu-
    lar-dotted ....7. *P. lapathifolium*

e. Stipular and floral sheaths bristly-ciliate; branches of inflorescence not glandular-dotted ....f

f. Perianth conspicuously glandular-dotted ....4. *P. punctatum*

f. Perianth not glandular-dotted ....g

g. Plants annual; inflorescence oblong, densely flowered ....5. *P. Persicaria*

g. Plants perennial; inflorescence linear, the flowers not crowded ....6. *P. hydropiperoides*

h. Plants perennial, the stems woody-thickened; midrib of leaves prominent and cordlike on lower side. Margin of leaves strongly revolute ..8. *P. Paronychia*

h. Plants annual; midrib of leaves not cordlike i

i. Stems erect, about 5 cm. long; perianth segments pink, showy, petaloid ....9. *P. spergulariaeforme*

i. Stems erect or prostrate, generally 10 cm. or more long; perianth segments white or rose with green midvein and base ....j

j. Achenes shining and nearly smooth ....k

j. Achenes granular- or striate-roughened l

k. Leaves of inflorescence not reduced ....10. *P. Fowleri*

k. Leaves of inflorescence much reduced and subbracteate ....11. *P. patulum*

l. Stems erect ....12b. *P. aviculare* var. *erectum*

l. Stems procumbent or prostrate ....m

m. Leaves not fleshy ....12. *P. aviculare*

m. Leaves somewhat fleshy-thickened ....12a. *P. aviculare* var. *littorale*

1. **P. Convolvulus** L. Reported from Mill Valley by Stacey. Native of Eurasia.

2. **P. bistortoides** Pursh. In wet soil of coastal marsh: Ledum Swamp, Point Reyes Peninsula, the southernmost station in the Coast Ranges.

3. **P. coccineum** Muhl. Occasional in low places along roads or on the edge of marshes: Stinson Beach; Manor, *Eastwood;* Ignacio; Chileno Valley; Drakes Estero. The form of this variable species that occurs in Marin County is the robust erect plant found in relatively dry ground, forma *terrestre* (Willd.) Stanford.—*P. Muhlenbergii* Wats.

4. **P. punctatum** Ell. The most common and widespread smartweed in wet soil of coastal marshes and springy slopes: Rodeo Lagoon, Elk Valley, and Mill Valley to Phoenix Lake, Bolinas, and Olema Marshes.—*P. acre* H.B.K.

5. **P. Persicaria** L. Occasional in wet soil bordering springs and marshes or along streams: Mill Valley; Black Point; Stinson Beach; Bolinas; Olema Marshes; Inverness. Native of Europe.

6. **P. hydropiperoides** Michx. In low wet ground or in shallow water: between Bolinas and Olema; Olema Marshes; the laguna in Chileno Valley.

7. **P. lapathifolium** L. Rare in Marin County in wet ground along roads bordering marshes: between Almonte and Mill Valley; Stinson Beach; Black Point; Chileno Valley; Point Reyes Station, acc. Stacey.

A smartwood, growing with *P. Persicaria* and *P. lapathifolium* at Black Point, may be a hybrid between those two species.

8. **P. Paronychia** C. & S. Sandy flats on coastal mesas and among dunes: Point Reyes Peninsula; Dillons Beach.

9. **P. spergulariaeforme** Meisn. Dry soil of interior hills and mountains, in Marin County known only from the coastal mesa near Dillons Beach where it occurs with *P. Paronychia*.

10. **P. Fowleri** Robins. Occasional in salt marshes bordering the bay and ocean: Escalle; Burdell; Inverness. These erect leafy plants are not typical of the species as it occurs on the Atlantic coast but they seem nearest to it and have been so determined by J. F. Brenckle.

11. **P. patulum** Biebst. Occasional in salt marshes bordering San Francisco Bay, in Marin County known only from San Antonio. Naturalized from the Old World.

12. **P. aviculare** L. DOORYARD KNOTWEED. Common in hard soil about towns and along roads: Sausalito, Angel Island, and Mill Valley to San Rafael, Bolinas, and Inverness. Introduced from Europe.

Two relatives of this variable weed are worthy of nominal recognition, here given as varieties of the species but sometimes accorded specific rank by other botanists:

12*a*. **P. aviculare** L. var. **littorale** (Link) Koch. On sandy beaches or borders of salt marshes: Almonte; Muir Beach; Inverness.

12*b*. **P. aviculare** L. var. **erectum** Roth. Occasional in shady canyons or in low fields bordering salt marshes: Almonte; Mill Valley; Greenbrae; Fairfax; Bolinas.

2. **Rumex.** DOCK

a. Plants dioecious; leaves generally hastately lobed. Inner sepals entire
....1. *R. Acetosella*
a. Plants monoecious; leaves not hastately lobed ....b
b. Inner sepals entire, crenulate, or denticulate in fruit c
b. Inner sepals with prominent slender lobes or teeth i
c. Stems procumbent or erect, with few to several conspicuous leafy branches
below the terminal inflorescence ....d
c. Stems erect without leafy branches below the branching terminal inflorescence
....g
d. Fruiting calyx about 4 mm. long; callus grain 1, about as large as or even a
little larger than the sepal ....2*c*. *R. salicifolius* var. *crassus*
d. Fruiting calyx about 3 mm. long; callus grain smaller than the sepal or
lacking ....e
e. Inner sepals without callus grains ....2*b*. *R. salicifolius* f. *ecallosus*
e. Inner sepals with 1, 2, or 3 callus grains ....f
f. Callus grain 1 (or rarely 2) ....2. *R. salicifolius*
f. Callus grains generally 3 ....2*a*. *R. salicifolius* f. *transitorius*
g. Inner sepals without callus grains ....3. *R. occidentalis*
g. Inner sepals generally with callus grains ....h
h. Leaves crisped or undulate; inner sepals 4–6 mm. long, much broader than
the callus grains ....4. *R. crispus*
h. Leaves not crisped or undulate; inner sepals 2.5–3 mm. long, almost covered by
the callus grains ....5. *R. conglomeratus*
i. Branches of inflorescence widely spreading; pedicels shorter than the fruiting
calyx, jointed near the middle ....7. *R. pulcher*
i. Branches of inflorescence generally ascending; pedicels equaling the fruiting
calyx or longer, jointed near the base ....j
j. Inner sepals 4–6 mm. long, the teeth shorter than the width of the sepal
....6. *R. obtusifolius*

j. Inner sepals 2–3 mm. long, the teeth usually longer than the width of the sepal ....8. *R. fueginus*

1. **R. Acetosella** L. SHEEP SORREL. Common and widespread in hills and fields from Sausalito, Tiburon, and Mount Tamalpais to San Rafael Hills, Point Reyes Peninsula, and Tomales. Introduced from the Old World.

2. **R. salicifolius** Weinm. Widespread and rather common from low fields and subsaline flats bordering salt marshes to gravelly streambeds and moist places in the hills. Several variants occurring in Marin County have been accorded specific recognition but here they are treated as varieties and forms. The proposals were based on variations in the fruiting calyx which have proved too inconstant to have specific value. Frequently plants of two or more types may be found growing together, entirely alike except for the technical difference in the fruits.

The typical variant, with stems sometimes erect, sometimes procumbent, has been observed in the following localities: Rodeo Lagoon; Sausalito; Almonte; Mill Valley; Laurel Dell and Potrero Meadow, Mount Tamalpais; Corte Madera; Ross; San Rafael; Ignacio; west of Novato.

2*a*. **R. salicifolius** Weinm. f. **transitorius** (Rech. f.) J. T. Howell. Rather rare on clay flats: Rodeo Lagoon; Tiburon; Frank Valley; Lagunitas Meadows, Mount Tamalpais; Bolinas; between Salmon Creek and Chileno Valley.—*R. transitorius* Rech. f.

2*b*. **R. salicifolius** Weinm. f. **ecallosus** J. T. Howell. Occasional in the hills and on the borders of salt marshes: Rodeo Lagoon; Sausalito, the type locality; Tiburon; Mill Valley; Bolinas; Corte Madera; Ross; San Rafael Hills; Tocaloma, *Leschke;* Chileno Valley.

2*c*. **R. salicifolius** Weinm. var. **crassus** (Rech. f.) J. T. Howell. Low ground near the coast or on maritime dunes and bluffs: Rodeo Lagoon; Stinson Beach; Bolinas Lagoon; McClure Beach; Dillons Beach.—*R. crassus* Rech. f.

3. **R. occidentalis** Wats. var. **procerus** (Greene) J. T. Howell. Rather common along tidal sloughs in salt marshes or in fresh water marshes near the coast: Rodeo Lagoon; Sausalito; Tiburon; Almonte; near McNears Landing; Bolinas Lagoon; Tomales Bay; Drakes Estero; Dillons Beach.

The calyx frequently becomes rose-tinged and in fruit the tall inflorescences may be very attractive. The basal leaves are quite large: one in which the petiole was nearly 9 inches long had a blade 1½ feet long and over a half foot broad.

4. **R. crispus** L. CURLY DOCK. In low places where water stands in winter, common and widespread: Angel Island, Sausalito, and Tiburon to Phoenix Lake, San Rafael Hills, and Tomales. Naturalized from Europe and Asia.

At Tiburon, a hybrid between *R. crispus* and *R. pulcher* L. has been found. It had the entire fruiting sepals of the former and the broader inflorescence of the latter.

5. **R. conglomeratus** Murr. Moist places, generally in the shade: Sausalito; Frank Valley; Mill Valley; Rock Spring, Mount Tamalpais; Stinson Beach; San Rafael Hills; Bolinas and Inverness, Point Reyes Peninsula. Introduced from Europe.

6. **R. obtusifolius** L. Occasional in low waste ground: Mill Valley; Stinson Beach; Lagunitas, acc. Stacey; Bolinas; Inverness. Introduced from Europe.

7. **R. pulcher** L. Rather common and widespread in waste places and in the hills: Sausalito, Tiburon, and Mount Tamalpais (Rock Spring) to San Rafael Hills

and Point Reyes Peninsula (Bolinas, Inverness, Point Reyes Lighthouse). Naturalized from the Mediterranean region.

8. **R. fueginus** Philippi. Wet soil bordering ponds or marshes in Marin County: Rodeo Lagoon; near Abbotts Lagoon, Point Reyes Peninsula.

3. Eriogonum

a. Plants annual; involucres in slender spikes ....3. *E. vimineum*
a. Plants perennial; involucres in heads at the end of peduncle-like stems or
    branches ....b
b. Base of plant compactly branched and densely leafy; stems tomentose, bearing
    1 or few heads ....1. *E. latifolium*
b. Base of plant generally loosely branched and less leafy; stems tomentose or
    glabrous, mostly with numerous heads ....2. *E. nudum*

1. **E. latifolium** Smith. Exposed mesas, dunes, and bluffs along the ocean: Angel Island; Sausalito; Tennessee Cove; Stinson Beach; Point Reyes; Tomales.

2. **E. nudum** Dougl. Shallow soil or rocky slopes, common and widespread from coastal hills to the interior: Tiburon and Mount Tamalpais (south side, Rock Spring, Phoenix Lake) to San Rafael Hills, San Antonio, and Inverness Ridge.

In Marin County the intergradation in habit between *E. nudum* and *E. latifolium* is such that a divisional line between the two is difficult to draw. Usually the stems of *E. nudum* are taller, more slender, more branching, and glabrous, but, at several stations towards the interior where typical forms of the species might be expected, plants with both glabrous and tomentose stems have been found together and the length of stem and extent of branching varies with the immediate environment.

3. **E. vimineum** Dougl. var. **caninum** Greene. Shallow soil of rocky slopes in shale or serpentine: Tiburon, the type locality; Mount Tamalpais (West Point, Bootjack, Rifle Camp, Camp Handy); Fairfax Hills; Carson country; San Rafael Hills.

This plant, with its graceful habit and bright rose flowers, is an attractive floral feature wherever it occurs throughout its long period of bloom from early summer until autumn. Greene gives no reason for the unusual varietal epithet; the connection between the plant and a dog must have been to him something personal or fanciful.

4. Chorizanthe

a. Involucral teeth nearly equal, bordered by a conspicuous rosy membrane
    ....1. *C. membranacea*
a. Involucral teeth unequal, not bordered by a membrane ....b
b. Involucre glabrous or nearly so, the teeth erect; perianth segments unequal,
    glabrous ....3. *C. valida*
b. Involucre hairy, the teeth spreading; perianth segments nearly equal, hairy ..c
c. Involucres mostly 1–3 in small axillary clusters; shorter involucral teeth much
    smaller than the longer teeth or lacking; plants becoming purplish-red
    ....2. *C. polygonoides*
c. Involucres usually many in rounded headlike clusters; shorter involucral teeth
    about half as long as the longer teeth; plants not becoming reddish ......d
d. Spines of involucral teeth straight or nearly so ....4. *C. villosa*

d. Spines of involucral teeth strongly hooked at the tip ....5. *C. cuspidata*

1. **C. membranacea** Benth. Shallow or rocky soil of grassy or brushy slopes, in Marin County known only from Mount Tamalpais near Bootjack and Rock Spring *(L. S. Rose)*.

2. **C. polygonoides** T. & G. Rare in thin soil of open rocky slopes: Tiburon Peninsula; near Rock Spring, Mount Tamalpais; head of Little Carson Canyon.

3. **C. valida** Wats. Sandy soil of coastal mesas in Sonoma and Marin counties; in Marin County known only from Point Reyes Peninsula where it was collected by Elmer.

4. **C. villosa** Eastw. Maritime dunes: Point Reyes Peninsula; Dillons Beach.

As with the preceding, *C. villosa*, is known only from Marin and Sonoma counties. It is one of a series of closely related species found along the California coast from Mendocino County to Santa Barbara County, the first of which to be named being *C. pungens* Benth.

5. **C. cuspidata** Wats. Dune hills near Dillons Beach.

The occurrence at Dillons Beach of both *C. cuspidata* and *C. villosa* suggests the possibility that these closely related plants may not be specifically distinct. Further studies in such a place where the two grow near each other may help decide the point.

5. Pterostegia

1. **P. drymarioides** F. & M. Wooded or brushy slopes, generally in shallow or rocky soil, occasionally on dunes: Angel Island; Sausalito; Corte Madera Ridge and Bootjack, Mount Tamalpais; San Rafael Hills; Little Carson; Dillons Beach; Inverness Ridge and McClure Beach, Point Reyes Peninsula.

The stems and leaves of this delicate annual are frequently tinged with pleasing shades of orange, rose, or red.

CHENOPODIACEAE. Goosefoot Family

a. Leaves opposite, reduced to inconspicuous bracts; upper stems succulent
....5. *Salicornia*

a. Leaves conspicuous, alternate or sometimes opposite; upper stems herbaceous, not succulent ....b

b. Flowers unisexual, the staminate with calyx, the pistillate without calyx but enclosed between two herbaceous bracts that enlarge and thicken in fruit
....4. *Atriplex*

b. Flowers perfect (or some pistillate in *Chenopodium*), calyx present c

c. Calyx lobes bearing a hooked spine; plants woolly-hairy ....1. *Bassia*

c. Calyx lobes not spiny; plants not conspicuously woolly-hairy (except sometimes in *Chenopodium ambrosioides*) ....d

d. Calyces becoming enlarged and hardened in fruit, cohering in clusters by their thickened bases ....2. *Beta*

d. Calyces not cohering in fruit ....3. *Chenopodium*

1. Bassia

1. **B. hyssopifolia** (Pall.) Kuntze. Along the railroad at Ignacio. Introduced from arid regions of the Old World.—*Echinopsilon hyssopifolius* (Pall.) Moq.

2. Beta. BEET

1. **B. vulgaris** L. Occasional escape from cultivation on subsaline flats bordering salt marshes: Sausalito; Marin City; Tiburon. Native of the Old World.

3. Chenopodium. GOOSEFOOT

a. Plants glandular-pubescent, at least on the lower side of the leaves, the herbage
    odorous ....b
a. Plants mealy or glabrous, not glandular, the herbage not especially odorous ..e
b. Calyx lobes united above the middle, the tube reticulate-veined; seeds vertical;
    leaves once or twice divided into narrow segments ....1. *C. multifidum*
b. Calyx lobes united to the middle or below, the tube not reticulate-veined;
    seeds horizontal; leaves not deeply divided into narrow segments c
c. Leaves glandular-hairy on the upper and lower sides ....2. *C. Botrys*
c. Leaves punctate-glandular on the lower side ....d
d. Stems finely hairy or glabrescent ....3. *C. ambrosioides*
d. Stems conspicuously woolly-hairy ....3*a. C. ambrosiodides* var. *vagans*
e. Calyx lobes united to the middle or above; seeds mostly vertical ....f
e. Calyx lobes united below the middle; seeds mostly horizontal ....g
f. Uppermost flower clusters in a nearly leafless spike; calyx herbaceous, loosely
    enclosing the fruit ....4. *C. californicum*
f. Flower clusters axillary; calyx membranaceous, closely enclosing the fruit
    ....5. *C. macrospermum*
g. Leaves dark green, shiny on upper side; seeds dull ....6. *C. murale*
g. Leaves light green, not shiny on upper side; seeds shiny ....7. *C. album*

1. **C. multifidum** L. A rare weed along roads and in waste places: Angel Island; Sausalito; San Rafael. Native of South America.—*Roubieva multifida* (L.) Moq.

2. **C. Botrys** L. Occasional weedy plant of roadside or gravelly stream bed: San Rafael; San Anselmo Creek above Fairfax; Alpine Dam; Point Reyes Station, acc. Stacey. Introduced from the Old World.

3. **C. ambrosioides** L. MEXICAN TEA. Weedy places about towns and in pastures in the hills: Sausalito; near Rock Spring, Mount Tamalpais; Stinson Beach; Ignacio. Indigenous in the warmer parts of North and South America.

3*a*. **C. ambrosioides** L. var. **vagans** (Standl.) J. T. Howell. Clay flats bordering salt marshes or in waste places about habitations: Angel Island; Manzanita; Ross; Ignacio; Stinson Beach; Bolinas; Inverness. Introduced from South America.—*C. vagans* Standl.

4. **C. californicum** (Wats.) Wats. Rare on brushy hills: Angel Island; Stinson Beach, acc. Stacey; McClure Beach, Point Reyes Peninsula.

5. **C. macrospermum** Hook. f. var. **farinosum** (Wats.) J. T. Howell. Abundant on the strand of Rodeo Lagoon as the water recedes in summer. This is the northernmost station for this plant which may be the same as *C. halophilum* Philippi of South America.—*C. farinosum* (Wats.) Standl.

6. **C. murale** L. A common weed about towns and along roads: Tiburon, Sausalito, and Mill Valley to Bolinas, San Rafael, and Point Reyes Peninsula (Inverness and McClure Beach). Native of the Old World.

7. **C. album** L. LAMBS QUARTERS. Occasional weed in disturbed ground along

roads, in fields, and in towns: Tamalpais Valley and Stinson Beach (acc. Stacey) to San Rafael, San Antonio Creek, and Inverness. Very widely distributed but perhaps not native.

4. Atriplex

a. Plants annual; leaves green above and below, little or not at all scurfy ....b
a. Plants perennial; leaves scurfy, at least on lower side c
b. Lower leaves lanceolate or oblong, not lobed or angled ....1. *A. patula*
b. Lower leaves triangular or rhomboidal with prominent angles or lobes
....1*a. A. patula* var. *hastata*
c. Leaves frequently dentate, green on upper side and pale on lower side; fruiting
    bracts nerved, becoming fleshy and reddish ....4. *A. semibaccata*
c. Leaves entire, alike on upper and lower sides, thinly or heavily scurfy; fruiting
    bracts not nerved, fleshy, or reddish ....d
d. Leaves narrowly oblong to oblanceolate, acute; fruiting bracts entire
....2. *A. californica*
d. Leaves broadly ovate to roundish, subacute or obtuse; fruiting bracts toothed
....3. *A. leucophylla*

1. **A. patula** L. Rare in salt marshes: Tiburon; near Black Point; Inverness.

1*a*. **A. patula** L. var. **hastata** (L.) Gray. Common in salt marshes and on bordering flats, occasional elsewhere in low places in clay soil: Sausalito, Tiburon, and Stinson Beach to Black Point, Inverness, and Drakes Estero.—*A. hastata* L.

Usually *A. patula* and its var. *hastata* can be readily distinguished but occasionally puzzling intermediates are found. One series of such intermediates was found near Inverness and since the plants grew with typical forms of the species and the variety, they may have originated by hybridization.

In the autumn, plants of *A. patula* var. *hastata* become tinged with rose and purple and add to the colorful effect of *Salicornia virginica* in the salt marshes.

2. **A. californica** Moq. Coastal dunes, borders of brackish marshes, and exposed maritime slopes, not common: Stinson Beach; Inverness; Point Reyes Lighthouse.

The relatively slender spreading stems arise each year from the crown of a much-enlarged fusiform root.

3. **A. leucophylla** (Moq.) Dietr. Sandy flats and dunes or borders of salt marshes: Angel Island; Lime Point, acc. Sutliffe; Rodeo Lagoon; Stinson Beach; Inverness, acc. Stacey; dunes and beaches, Point Reyes Peninsula.

4. **A. semibaccata** R. Br. Occasional near salt marshes bordering San Francisco Bay near Greenbrae and San Rafael. Introduced from Australia.

5. Salicornia. GLASSWORT

a. Plants perennial, shrubby at base ....1. *S. virginica*
a. Plants annual ....2. *S. Bigelovii*

1. **S. virginica** L. PICKLEWEED. The most abundant and characteristic flowering plant in Marin County salt marshes bordering San Francisco Bay and the ocean: Rodeo Lagoon, Tiburon, and Sausalito to Black Point, Drakes Bay, and Tomales Bay.—*S. ambigua* Michx.

In autumn the succulent stems and inflorescences become tinted red and purple and, since the plants form extensive colonies over much of the marshland, a beautiful and colorful effect is produced.

2. **S. Bigelovii** Torr. Rare in the salt marshes at the head of Richardson Bay below Mill Valley.

AMARANTHACEAE. AMARANTH FAMILY

1. Amaranthus. AMARANTH

a. Flowers terminal in dense compound spikes as well as axillary in smaller clusters ....b
a. Flowers only in axillary clusters ....d
b. Stems prostrate or decumbent; utricle indehiscent ....1. *A. deflexus*
b. Stems erect; utricle circumscissile c
c. Sepals of pistillate flowers lanceolate, acute or spinose-acuminate 2. *A. hybridus*
c. Sepals of pistillate flowers oblong to oblanceolate, obtuse or emarginate, generally mucronate ....3. *A. retroflexus*
d. Sepals of pistillate flowers 4 or 5, broad below and enclosing the lower part of the utricle; leaves lanceolate to oblong, acutish. Stems prostrate ....4. *A. graecizans*
d. Sepals of pistillate flowers 1–3, not enclosing base of the utricle; leaves narrowly obovate to nearly round, obtuse, or acute ....e
e. Stems erect; sepals of pistillate flowers 3, subequal; utricle readily dehiscent ....5. *A. albus*
e. Stems prostrate; sepals of pistillate flowers 1–3, one larger, the others (if present) rudimentary and scalelike; utricle tardily dehiscent 6. *A. californicus*

1. **A. deflexus** L. Common in towns as a garden or sidewalk weed: Sausalito; Mill Valley; Ross; San Rafael; Fairfax; Bolinas; Point Reyes Station. Widespread in tropical America and in the Old World, introduced in California.

2. **A. hybridus** L. Occasional weedy plant along streets or roads: Mill Valley; Frank Valley; Fairfax; Novato.

Specimens from Marin County are somewhat variable but none appears referable to *A. Powellii* Wats. which has been reported from western central California. European specimens of *A. hybridus* are extremely variable, and some phases of the species seem to approach *A. Powellii* too closely.

3. **A. retroflexus** L. Along roads and on the edge of pastures, rare in Marin County: Mill Valley, acc. Stacey; near San Rafael; Ignacio; Novato.

Near San Rafael, a plant with slender caudate inflorescences grew with the normal form but since its flowers were sterile it probably represented a hybrid between *A. retroflexus* and *A. hybridus*.

4. **A. graecizans** L. Rare in clay soil along roads: Novato; San Antonio Creek.—*A. blitoides* Wats.

5. **A. albus** L. Widespread but not common as a weedy plant of waste ground about habitations, in cultivated fields, and along roads: Mill Valley; Alpine Lake, Mount Tamalpais; Corte Madera; San Rafael; Novato; Fairfax; near Stinson Beach.—*A. graecizans* of authors, not L.

6. **A. californicus** (Moq.) Wats. Rare in summer-dried depressions that are covered with water in winter: strand of Alpine and Lagunitas lakes exposed by receding waters; floodbed of San Antonio Creek.

Growing with *A. californicus* at Alpine Lake was a second depressed or assurgent amaranth which superficially resembled it in general aspect but which differed in its 3 well-developed sepals and circumscissile utricle. Since these structures closely resemble those of *A. albus,* it is believed that the variant may have originated by hybridization.

NYCTAGINACEAE. FOUR-O'CLOCK FAMILY

1. Abronia. SAND-VERBENA

a. Flowers yellow ....1. *A. latifolia*
a. Flowers pink .... 2. *A. umbellata*

1. **A. latifolia** Esch. Dunes and beach sand above high tide: Rodeo Lagoon; Stinson Beach; McClure Beach and near the radio station, Point Reyes Peninsula; Dillons Beach.

2. **A. umbellata** Lamk. Dunes and beaches: Stinson Beach; McClure Beach, Point Reyes Peninsula; Dillons Beach.

Both of these Marin County sand-verbenas were described from material obtained on early-day exploring expeditions to the west coast of North America. *Abronia umbellata* was described from plants grown in Paris from seeds collected at Monterey in 1786 by members of the La Perouse Expedition. It was the first plant new to science to be described from California. Specimens of *A. latifolia* were collected at San Francisco in 1824 by Eschscholtz on the second Kotzebue expedition and were named and described in St. Petersburg in 1826.

AIZOACEAE. CARPETWEED FAMILY

a. Leaves alternate, broadly triangular-ovate. Plants annual ......4. *Tetragonia*
a. Leaves opposite or whorled, elongate ....b
b. Leaves thick with 3 nearly equal sides; plants perennial
....3. *Mesembryanthemum*
b. Leaves thin, plane or nearly so; plants annual c
c. Leaves whorled, stipules none or inconspicuous ....1. *Mollugo*
c. Leaves opposite, stipules present ....2. *Cypselea*

1. Mollugo

1. **M. verticillata** L. Generally found in low places that have been wet during the rainy period, in Marin County known only on the strand of Alpine Lake. In late summer and autumn, the mollugo and *Cypselea humifusa* produce a green, lawnlike effect over many acres of desiccated lake bottom.

2. Cypselea

1. **C. humifusa** Turpin. Depressions inundated during the winter: on the strand of Phoenix, Lagunitas, and Alpine lakes. Introduced from the West Indies.

In the autumn the plants have been observed to produce nodule-like subterranean cleistogamous flowers near the top of the taproot.

3. Mesembryanthemum

a. Leaves obtusely angled; flowers purplish-rose ....1. *M. chilense*
a. Leaves sharply angled; flowers yellowish, fading rose ....2. *M. edule*

1. **M. chilense** Mol. Forming broad mats on dunes near the coast: Rodeo La-

goon; Stinson Beach; Point Reyes dunes and lighthouse; Dillons Beach.—*M. ae-quilaterale* of California references.

2. **M. edule** L. Widely planted as a soil binder on road fills and cuts; occasionally naturalized on beaches and coastal bluffs: Angel Island; Bolinas sand spit; mouth of Rio Hondo near Bolinas; Dillons Beach. Introduced from South Africa.

4. Tetragonia

1. **T. expansa** Murr. NEW ZEALAND SPINACH. Occasional in sandy soil or on clay flats near salt marshes: Lime Point, acc. Sutliffe; Rodeo Lagoon; Angel Island; Manzanita; San Rafael; Stinson Beach; Dillons Beach; McClure Beach and the lighthouse, Point Reyes Peninsula. Native of New Zealand and Asia and widely dispersed as a maritime weed in both hemispheres.

PORTULACACEAE. PURSLANE FAMILY

a. Petals 12–18, about 2 cm. long; sepals 4–8 .... 4. *Lewisia*
a. Petals 2–6, 2–10 mm. long; sepals 2 .... b
b. Petals yellow; capsule circumscissile at the middle .... 1. *Portulaca*
b. Petals white, pink, or reddish; capsule 3-valved, dehiscing from apex c
c. Sepals sharply acute; petals generally rose or reddish; seeds numerous
  .... 2. *Calandrinia*
c. Sepals obtuse; petals white or pink; seeds 1–3 .... 3. *Montia*

1. Portulaca

1. **P. oleracea** L. COMMON PURSLANE. Occasional weed of gardens and waste ground: Mill Valley; Stinson Beach; Alpine Lake; Ross; San Rafael. Widely distributed in warm and temperate regions but not indigenous in Marin County.

2. Calandrinia

a. Sepals about equaling the capsule .... 1. *C. ciliata*
a. Sepals much shorter than the capsule .... 2. *C. Breweri*

1. **C. ciliata** (R. & P.) DC. var. **Menziesii** (Hook.) Macbr. RED MAIDS. Clay flats, gravelly hills, and sandy mesas, widespread but not very common: Sausalito, Tiburon, and Mount Tamalpais (near Rock Spring) to Ignacio, Point Reyes Peninsula (Inverness Ridge, Point Reyes), and Tomales.—*C. caulescens* H.B.K. var. *Menziesii* (Hook.) Gray.

The flowers are variable both in structure and size and even in a restricted area they may exhibit a considerable range of color. In the San Rafael Hills plants with both large magenta petals and with small mauve petals grew together, the two kinds undoubtedly being different genetic expressions of this variety. Also a mutant with white petals is sometimes seen growing with the usual type.

2. **C. Breweri** Wats. Rare in gravelly soil on brushy slopes, becoming locally abundant following fires: hills southeast of Muir Woods; Mount Tamalpais (Corte Madera Ridge, Berry Trail, Rifle Camp); Carson country; San Rafael Hills.

Vigorous plants in loose rich soil on burns characteristically produce several elongate, sparsely-leafy, prostrate branches up to 2 feet long. The flowers are medium-sized and the petals are usually rose color.

3. Montia

a. Leaves opposite .... b
a. Leaves alternate .... f
b. Stems branching, leafy .... 1. *M. fontana*
b. Stems simple, peduncle-like, with one pair of leaves subtending the inflorescence c
c. Cauline leaves separate, ovatish, usually more than 2 cm. long; racemes elongate with numerous bracts .... 2. *M. sibirica*
c. Cauline leaves indistinctly to conspicuously united; racemes bracteate only at the base .... d
d. Cauline leaves forming a symmetric or somewhat asymmetric disk .... 3. *M. perfoliata*
d. Cauline leaves not forming a disk, generally conspicuously united on one side, sometimes slightly united on the other side also .... e
e. Petals showy, 2–4 times as long as the sepals .... 4. *M. gypsophiloides*
e. Petals not showy, narrow and about twice as long as the sepals .... 5. *M. spathulata*
f. Basal leaves forming a rosette; cauline leaves mostly less than 1 cm. long. .... 6. *M. parvifolia*
f. Basal leaves not forming a rosette; cauline leaves mostly 1–2 cm. long .... 7. *M. diffusa*

1. **M. fontana** L. Wet soil of hills and low meadows where water seeps or stands in the spring, sometimes partly immersed in brooks or ponds: Tiburon; Mount Tamalpais (Rock Spring, Potrero Meadow, Lagunitas Meadows); Fairfax Hills; Tomales; Ledum Swamp, Point Reyes Peninsula.

Two forms of this variable plant occur in Marin County and should probably be recognized as distinct species. According to European workers who would reject entirely *M. fontana* L. as an ambiguous name, the Marin County plants would be *M. verna* Neck. (for the coarser one with turgid seeds nearly 1.5 mm. long) and *M. Hallii* (Gray) Greene (for the more delicate one with strongly flattened seeds 1 mm. long). The latter type is the more common, but the two grow together in Lagunitas Meadows.

2. **M. sibirica** (L.) Howell. Swamps and springy slopes near the coast: Sausalito; Point Reyes Peninsula (Ledum Swamp, dunes near the radio station). The plant on Point Reyes Peninsula was described as *Limnia bracteosa* Rydb.

3. **M. perfoliata** (Donn) Howell. MINERS LETTUCE. Common and widespread throughout the county in many forms.

Although *M. perfoliata* is one of the most easily recognized species, it is one of the most variable in Marin County. The plant is not only diverse in habit and in hue but the variability extends to the foliage, inflorescence, flowers, and fruits. Thus, the plants may be low and compact or tall and lax, the herbage may be green or reddish or copper color, the blades of the basal leaves may be broad or they may be scarcely wider than the petioles, the inflorescence may be abbreviated or elongate, stalked or sessile, the disk may be round or lobed, the flowers and fruits may be large or small. Numerous specific and varietal names have been applied to the plants that differ in one character or another. The following are the chief forms in Marin County:

M. perfoliata, the typical form, blades of basal leaves broad, green. Common in moist woods: Sausalito, Tiburon and Mount Tamalpais to San Rafael Hills and Tomales.

M. perfoliata f. **parviflora** (Dougl.) J. T. Howell, blades narrow, inflorescence elongate. Occurring with the typical form.

M. perfoliata f. **angustifolia** (Greene) J. T. Howell, blades narrow, inflorescence short. Frequently with the two preceding forms.

M. perfoliata f. **nubigena** (Greene) J. T. Howell, plants compact, corolla 4–6 mm. long, leaf-blades narrow or broad. Rocky slopes on the summit ridge of Mount Tamalpais, the type locality. This distinctive form is restricted to the rocky peaks of the Coast Ranges of central, coastal California.

M. perfoliata f. **cuprea** (Hel.) J. T. Howell, herbage copper color. Locally common on burn in the Carson country.

M. perfoliata f. **glauca** (Nutt.) J. T. Howell, plants low, compact, glaucous, disks small and lobed, flowers small. Shaly talus, Fairfax Hills.

4. **M. gypsophiloides** (F. & M.) Howell. Loose soil of shaly or gravelly detrital slopes: Fairfax Hills; San Geronimo Ridge.

This attractive plant with its slender stems and lax inflorescence is replaced on exposed serpentine slopes by var. *exigua* (T. & G.) J. T. Howell, in which the leaves are rosulately clustered, the stems are numerous and lower, and the petals are a little shorter: Rifle Camp and Camp Handy, Mount Tamalpais; Carson Ridge. After a fire the serpentine slopes may become rosy-pink from the many, full-flowered rosettes.—*M. exigua* (T. & G.) Jeps.; *M. spathulata* (Dougl.) Howell var. *exigua* (T. & G.) Robinson.

5. **M. spathulata** (Dougl.) Howell. Rocky slopes and ridges of shale, sandstone, or serpentine: Tiburon; Mount Tamalpais (Corte Madera Ridge, Rock Spring, Potrero Meadow); Fairfax Hills; Carson country, common in burned areas; Tomales.

In the usual typical form the cauline leaves are conspicuously joined on one side of the stem, but two variants, in which the cauline leaves are nearly distinct or only slightly joined near the base, may be noted as follows:

5*a*. **M. spathulata** (Dougl.) Howell var. **tenuifolia** (T. & G.) Munz. Basal and cauline leaves narrowly linear, elongate. Mount Tamalpais (on south side, *L. S. Rose*, Potrero Meadow); Carson Ridge.

5*b*. **M. spathulata** (Dougl.) Howell var. **rosulata** (Eastw.) J. T. Howell. Basal leaves oblong-linear; cauline leaves ovate-lanceolate, not elongate. Known only from the type locality, a serpentine outcrop near Rock Spring on Mount Tamalpais.

6. **M. parvifolia** (Moc.) Greene. Moist, moss-covered rocks, more or less shaded: Mill Valley, acc. Stacey; Cataract Gulch and Lagunitas Canyon, Mount Tamalpais; Big Carson Canyon, acc. Sutliffe; Camp Taylor.

7. **M. diffusa** (Nutt.) Greene. In the shade of redwoods, Cascade Canyon, Mill Valley. Not only is this the one known station for this rare species in Marin County but it is also the southernmost.

4. Lewisia

1. **L. rediviva** Pursh. BITTER ROOT. Shallow soil on rocks, Bald Mountain west of San Anselmo. According to Miss Eastwood the Bitter Root also grew on a rock outcrop now submerged in Alpine Lake.

CARYOPHYLLACEAE. PINK FAMILY

a. Stipules none ....b
a. Stipules present, sometimes inconspicuous ....f
b. Calyx tubular, the sepals united more than half their length ......5. *Silene*
b. Calyx not tubular, the sepals distinct nearly or quite to the base c
c. Stems and leaves conspicuously pubescent and more or less glandular; capsule
  elongate-cylindric ....1. *Cerastium*
c. Stems and leaves glabrous or inconspicuously pubescent (or rather conspicuously pubescent in *Stellaria littoralis*); capsule oblong, ovate, or globose ..d
d. Styles the same number as the sepals and alternate with them ....4. *Sagina*
d. Styles generally fewer than the sepals and opposite the outer ones ....e
e. Petals bifid, rarely none ....2. *Stellaria*
e. Petals entire or slightly emarginate ....3. *Arenaria*
f. Fruit several- to many-seeded; sepals not tipped with a bristle or spine; petals
  present ....g
f. Fruit 1-seeded; sepals tipped with a bristle or spine; petals none i
g. Leaves obovate, flat; sepals carinate ....6. *Polycarpon*
g. Leaves linear or linear-oblong, fleshy-thickened; sepals not carinate ....h
h. Leaves appearing verticillate; capsule with 5 valves ....7. *Spergula*
h. Leaves sometimes fascicled but not appearing verticillate; capsule with 3 valves
  ....8. *Spergularia*
i. Leaves hairy; sepals equal, bristle-tipped ....9. *Paronychia*
i. Leaves glabrous; sepals unequal, spine-tipped ....10. *Cardionema*

1. Cerastium

a. Petals showy, 2–3 times longer than the sepals; capsule about as long as the
  calyx. Plants perennial ....1. *C. arvense*
a. Petals inconspicuous, equaling or little exceeding the sepals; capsule much
  longer than the calyx ....b
b. Plants perennial; pedicels generally longer than the calyx .2. *C. holosteoides*
b. Plants annual; pedicels generally shorter than the calyx ....3. *C. glomeratum*

1. **C. arvense** L. Occasional on rocky slopes near the coast: Sausalito; Tomales; Dillons Beach; Point Reyes Lighthouse. The more vigorous plants with especially large attractive flowers along the northern California coast have been named var. *maximum* Holl. & Britt.

2. **C. holosteoides** Fries. In moist grassy places, introduced into Marin County as a weed in lawns: Mill Valley; San Anselmo.—*C. vulgatum* and *C. caespitosum* of California references.

3. **C. glomeratum** Thuill. MOUSE-EAR CHICKWEED. Widespread and rather common weed about habitations, along roads, and in woods and pastures: Sausalito, Tiburon, and Mount Tamalpais (Blithedale Canyon, Rock Spring) to San Rafael, Inverness, and Tomales.—*C. viscosum* of American references, *C. viscosum* L. in part.

2. Stellaria

a. Stems glabrous except for a longitudinal line of white hairs ....1. *S. media*
a. Stems glabrous or pubescent, hairs when present not in a line ....b
b. Annual with slender suberect stems; bracts of inflorescence hyaline ..2. *S. nitens*

b. Perennials with somewhat stouter assurgent or spreading stems (or erect when supported by other plants); flowers in the axils of leaves or foliaceous bracts c
c. Stems, leaves, and sepals finely glandular-pubescent ....5. *S. littoralis*
c. Stems, leaves, and sepals glabrous or nearly so ....d
d. Leaves ovate, membranaceous, the midrib not impressed above; flowers solitary in the axils of leaves ....3. *S. crispa*
d. Leaves lanceolate, somewhat thicker, the midrib impressed above; flowers in leafy-bracted cymes ....4. *S. sitchana*

1. **S. media** (L.) Cyr. CHICKWEED. Common weed in gardens and in moist woods and meadows: Sausalito, Tiburon, Mill Valley, and Mount Tamalpais (Phoenix Lake) to San Rafael, Inverness, and Dillons Beach. Native of Europe.

2. **S. nitens** Nutt. Occasional in shallow soil of rocky slopes and flats: Sausalito; Tiburon; Mount Tamalpais (Laurel Dell, Rock Spring, Camp Handy); Carson country; Big Rock Ridge; Tomales.

3. **S. crispa** C. & S. Rare near the coast on moist soil, generally shaded: Mud Lake; Bear Valley; swales of Point Reyes Peninsula. These are the southernmost known stations.

4. **S. sitchana** Steud. var. **Bongardiana** (Fern.) Hultén. Wet soil of coastal swamps: Olema Marshes; Point Reyes Peninsula (Mount Vision, Ledum Swamp, near the radio station); Dillons Beach. This species is not found south of the Golden Gate.—*S. borealis* Bigel. var. *Bongardiana* Fern.

5. **S. littoralis** Torr. Moist or wet places on coastal dunes: Point Reyes Peninsula, the type locality; Dillons Beach.

3. Arenaria

a. Plants perennial; leaves 3–6 mm. wide, longer than the internodes
....1. *A. macrophylla*
a. Plants annual; leaves 1 mm. wide or less, generally shorter than the internodes
....b
b. Petals shorter than the sepals or none ....4. *A. pusilla*
b. Petals showy, longer than the sepals c
c. Leaves needlelike, the lower about 1 cm. long or more ....2. *A. Douglasii*
c. Leaves narrowly ovate to lanceolate, the lower less than 5 mm. long
....3. *A. californica*

1. **A. macrophylla** Hook. Rare on shaded, rocky slopes: Sausalito, acc. Stacey; Mount Tamalpais (Rock Spring, *Eastwood, Leschke,* Potrero Meadow).

2. **A. Douglasii** Fenzl. Shallow soil of open rocky slopes, in Marin County commonly occurring on serpentine: Angel Island; Tiburon; Mount Tamalpais (West Point, Berry Trail, Rifle Camp); on sedimentary and volcanic rocks in Fairfax and San Rafael hills; Carson Ridge; head of Lucas Valley; on chert near Mud Lake, Point Reyes Peninsula.

Robust individuals growing on burns in the chaparral may be as much as 16 inches across and may bear several hundred blossoms. Usually the plants are sparsely branched and few-flowered.

3. **A. californica** (Gray) Brew. Rare in shallow soil overlying rocks that are

generally moistened by seeping water in the spring: Sausalito, acc. Lovegove; Point Reyes, acc. Stacey; Dillons Beach.

4. **A. pusilla** Wats. Rare in clay soil of open slope near Tomales.

4. Sagina. PEARLWORT

a. Leaf bases ciliolate; flowers apetalous, usually 4-parted ....1. *S. apetala*
a. Leaf bases not ciliolate; flowers petalous, usually 5-parted ....b
b. Stems spreading and rooting at nodes; flowers nodding in fruit
    ....2. *S. procumbens*
b. Stems not rooting at nodes; flowers erect at maturity c
c. Annual; stems slender, erect or rarely prostrate; sepals about 2 mm. long
    ....3. *S. occidentalis*
c. Perennial; stems fleshy, prostrate; sepals about 3 mm. long ....4. *S. crassicaulis*

1. **S. apetala** Ard. var. **barbata** Fenzl. Occasional inconspicuous weed in beaten ground about towns and along roads: Angel Island; Sausalito; Kentfield; San Rafael; Black Point; Little Carson, on burn; Inverness Ridge. Introduced from Europe.

2. **S. procumbens** L. Wet soil of mossy shaded banks: Inverness; Shell Beach. This plant which has every appearance of being indigenous on Point Reyes Peninsula is not known farther south along the Pacific coast.

3. **S. occidentalis** Wats. Rather common on wooded or brushy hills, occasional on sandy flats of coastal mesas: Sausalito, Tiburon, and Mount Tamalpais (Laurel Dell Trail) to San Rafael Hills, Dillons Beach, and Point Reyes Peninsula (Mud Lake, Inverness Ridge, and dunes).

Usually the slender stems of this annual are erect or slightly spreading, but on hard beaten ground on Point Reyes Peninsula plants have been found with prostrate stems. Even in these specimens there is no specific resemblance between *S. occidentalis* and *S. crassicaulis* in which the more fleshy-thickened stems are always prostrate.

4. **S. crassicaulis** Wats. Locally common in moist places on ocean bluffs and in swales among the dunes: Point Reyes Lighthouse and dunes; Bear Valley, acc. Stacey; Dillons Beach, the type locality.

5. Silene

a. Plants annual; petals only 3–5 mm. longer than the calyx ....b
a. Plants perennial; petals conspicuous, 7–15 mm. longer than the calyx .. ...d
b. Stems finely rough-pubescent below, the upper internodes with a broad
    glutinous band; calyx subglabrous ....3. *S. antirrhina*
b. Stems and calyx glandular-hairy c
c. Petals conspicuously exserted from the calyx tube; calyx with scattered long
    nonglandular hairs ....1. *S. gallica*
c. Petals scarcely or not at all exserted from the calyx tube; calyx without
    elongate nonglandular hairs ....2. *S. multinervia*
d. Leaves obtuse or acute; flowers generally numerous; petals rose ...4. *S. pacifica*
d. Leaves acute or acuminate; flowers solitary or few; petals red 5. *S. californica*

1. **S. gallica** L. WINDMILL PINK. Common weed of hillsides, roads, pastures, and

towns: Sausalito, Tiburon, and Mount Tamalpais to San Rafael Hills, Point Reyes Peninsula, and Dillons Beach. Introduced from Europe.

2. **S. multinervia** Wats. Rare in gravelly soil and usually common only after fires in the chaparral: south of Barths Retreat, Mount Tamalpais; Carson country. This species, which occurs locally from southern California northward in the Coast Ranges, is not known north of Marin County.

3. **S. antirrhina** L. Occasional in gravelly soil of hills and mountains on the edge of brush or chaparral: Matt Davis Trail and West Point, Mount Tamalpais; Little Carson; San Rafael Hills.

Usually the petals are visible and slightly exceed the calyx, but in a collection from Black Canyon in the San Rafael Hills the flowers were apetalous. The petals are white or pale pink at first but in age become purplish red.

4. **S. pacifica** Eastw. Occasional in shallow sandy soil on exposed rocky marine headlands: Fort Barry, acc. Sutliffe; Point Bonita, acc. Stacey; Rodeo Lagoon, the type locality; Point Reyes Lighthouse.—*S. grandis* Eastw. var. *pacifica* (Eastw.) Jeps.

Silene pacifica is one of several closely related entities found on coastal slopes of the western United States, *S. Scouleri* Hook. being the first named of the series. Marked by distinctive characters, these different plants may well be accorded specific recognition, as has been done for the maritime chorizanthes and erysimums.

5. **S. californica** Dur. INDIAN PINK. Occasional on brushy slopes in gravelly or rocky soil: Tiburon; Mount Tamalpais (Rattlesnake Camp, Bootjack, Potrero Meadow); Carson Ridge.

According to Gregory Lyon, *S. maritima* With., a glaucous, white-flowered perennial with inflated calyx, is naturalized on the bluffs overlooking Drakes Bay but no specimen has been seen from there.

6. Polycarpon

1. **P. tetraphyllum** (L.) L. Occasional in beaten clay soil of paths and roads about habitations: Sausalito; Angel Island; Mill Valley; Kentfield; Ross; San Rafael. Naturalized from Europe.

7. Spergula. SPURREY

1. **S. arvensis** L. Common and widespread in clay soil along roads, in pastures, and near salt marshes: Sausalito, Tiburon, and Mount Tamalpais (West Point to Santa Venetia, Point Reyes dunes, and Dillons Beach. Introduced from Europe.

8. Spergularia

a. Seeds not winged; plants annual (or *S. rubra* a short-lived perennial) ....b
a. Seeds winged and wingless in the same capsule; plants perennial with thickened tap-roots (or *S. media* an annual or short-lived perennial) ....d
b. Leaves becoming fascicled in the axils; stipules conspicuous, lanceolate, long-acuminate ....1. *S. rubra*
b. Leaves not fascicled in the axils; stipules not especially conspicuous, broad, acute or shortly acuminate c
c. Seeds mostly 0.5 mm. long or less, plump; stamens 6–10 ....2. *S. Bocconii*

c. Seeds mostly more than 0.5 mm. long, laterally flattened; stamens 2-5 (rarely 6)
....3. *S. marina*
d. Petals mauve ....6. *S. macrotheca*
d. Petals white ....e
e. Plants glabrous or subglabrous; leaves not fascicled; seeds usually about 1
mm. long ....4. *S. media*
e. Plants glandular-pubescent, heavily so above; leaves becoming fascicled; seeds
about 0.5 mm. long ....5. *S. villosa*

1. **S. rubra** (L.) J. & C. Presl. Common and widespread in hard soil along roads and paths and on clay flats of pastures and salt marsh borders: Sausalito, Tiburon, and Mount Tamalpais (Fern Canyon, Rifle Camp) to San Rafael, Black Point, and Inverness. Native of Europe.

The perennial phase of the species has been named var. *perennans* (Kindb.) Robins.

2. **S. Bocconii** (Scheele) Foucard. Not rare in packed soil of pastures and waste ground around towns: Sausalito; Angel Island; Tiburon; Mill Valley; Phoenix Lake; Fairfax; Santa Venetia; Stinson Beach; Bolinas. Introduced from the Old World.

3. **S. marina** (L.) Griseb. Common in low ground bordering salt marshes or in low alkaline pastures: Sausalito and Mill Valley to Ignacio, Stinson Beach, and Inverness.

4. **S. media** L. Locally abundant in salt marshes bordering the bay and ocean: Sausalito and Tiburon, acc. Rossbach; Almonte; Escalle; Drakes Estero. Introduced from Europe.

5. **S. villosa** (Pers.) Camb. Common along roads and on borders of salt marshes: Angel Island; Santa Venetia; Ignacio; Black Point. Introduced from South America.—*S. Clevelandii* (Greene) Robins.

6. **S. macrotheca** (Hornem.) Heynh. Maritime mesas and borders of salt marshes, occasionally in the marshes: Angel Island, Sausalito, and San Rafael to Stinson Beach, Inverness, and Point Reyes Lighthouse.

9. Paronychia

1. **P. franciscana** Eastw. Forming mats in shallow rocky soil on open hills: Sausalito, acc. Stacey; Elk Valley; Mountain Theater, Mount Tamalpais; San Rafael Hills; Dillons Beach; Point Reyes Lighthouse.

10. Cardionema

1. **C. ramosissimum** (Weinm.) Nels. & Macbr. Broad prostrate plants on marine bluffs and coastal dunes: Inverness; Point Reyes Lighthouse and dunes; Dillons Beach.—*Pentacaena ramosissima* (Weinm.) H. & A.

NYMPHAEACEAE. Water-lily Family

1. Nuphar. Cow-lily

1. **N. polysepalum** Engelm. Indian Pond-lily. Occasional in shallow ponds or marshy places: Lily Lake; Lagunitas, acc. Stacey; Olema, acc. Jepson; road to McClure Beach, Point Reyes Peninsula.—*Nymphaea polysepala* (Engelm.) Greene.

CERATOPHYLLACEAE. Hornwort Family

1. Ceratophyllum

1. **C. demersum** L. Hornwort. In Marin County this submerged aquatic is known only from Lake Lagunitas where it was collected by M. S. Jussel.

RANUNCULACEAE. Buttercup Family

a. Stems woody, climbing; leaves opposite ....1. *Clematis*
a. Stems herbaceous, not climbing; leaves alternate ....b
b. Flowers without spurs c
b. Flowers with spurs ....f
c. Flowers in a raceme; pistil 1; fruit a red berry ....2. *Actaea*
c. Flowers solitary or more or less paniculate; pistils more than 1, generally
    numerous; fruit an achene ....d
d. Flowers unisexual, the plants dioecious ....5. *Thalictrum*
d. Flowers perfect ....e
e. Flowers several; petals present, white or yellow, generally showy 3. *Ranunculus*
e. Flower solitary; petals none, the sepals petal-like, pale lavender ..4. *Anemone*
f. Flower regular, the 5 petals spurred ....6. *Aquilegia*
f. Flower irregular, the upper sepal spurred ....7. *Delphinium*

1. Clematis. Virgin's Bower

a. Flowers generally numerous and paniculate; pedicels less than 2 cm. long;
    sepals about 1 cm. long ....1. *C. ligusticifolia*
a. Flowers 1–3; pedicels 3–10 cm. long; sepals about 2 cm. long ..2. *C. lasiantha*

1. **C. ligusticifolia** Nutt. Occasional along streams where the woody stems climb high into the trees: Mill Valley; south side of Mount Tamalpais; San Anselmo Canyon; Tocaloma; Point Reyes Station.

2. **C. lasiantha** Nutt. Rare on dry slopes, frequently trailing over shrubs in the chaparral: Sausalito, acc. Lovegrove; Mount Tamalpais (Blithedale Canyon, acc. Sutliffe, Corte Madera Ridge, near Rattlesnake Camp, Lake Lagunitas); San Rafael, acc. Stacey.

2. Actaea. Baneberry

1. **A. arguta** Nutt. Moist shaded slopes of wooded canyons, not common: Mill Valley, acc. Stacey; Muir Woods; Steep Ravine, Mount Tamalpais; Black Forest and Inverness Park, Point Reyes Peninsula. In the late summer, the ample leaves set off to advantage the attractive bright red shiny berries.—*A. spicata* L. var. *arguta* (Nutt.) Torr.

3. Ranunculus. Buttercup

a. Plants aquatic, submerged leaves dissected into capillary segments; petals
    white ....b
a. Plants terrestrial; petals yellow c
b. Receptacle glabrous; upper leaves floating and not dissected ....1. *R. Lobbii*
b. Receptacle hairy; usually all leaves submerged and dissected .2. *R. aquatilis*
c. Leaves entire or crenate, not lobed or divided ....d
c. Leaves deeply cleft or divided ....e
d. Plant annual; stems erect; petals minute; achenes tuberculate ..3. *R. pusillus*

d. Plant perennial; stems widely creeping; petals small but showy; achenes smooth ....4. *R. Flammula*
e. Achenes bearing prickles or hairs ....f
e. Achenes not bearing prickles or hairs ....h
f. Stems stout; flowers large and rather showy; sides of achenes bearing stout prickles ....7. *R. muricatus*
f. Stems slender; flowers small, the petals inconspicuous; achenes hairy ....g
g. Plants perennial; hairs on achenes appressed-ascending ......5. *R. uncinatus*
g. Plants annual; hairs on achenes spreading, hooked ....6. *R. hebecarpus*
h. Stems widely creeping and rooting at the nodes ....8. *R. repens*
h. Stems erect or spreading, not creeping and rooting i
i. Beak of achene hooked, shorter than the achene ....9. *R. californicus*
i. Beak of achene straight, as long as the achene ....j
j. Plants more or less hirsutulous ....10. *R. orthorhynchus*
j. Plants glabrous ....11. *R. Bloomeri*

1. **R. Lobbii** (Hiern.) Gray. Pools and roadside hollows where water stands during the winter and spring, often in low fields adjoining the salt marshes: Mount Tamalpais (acc. Stacey) and Lily Lake (acc. Sutliffe) to Ignacio, Olema, and Point Reyes Peninsula (pond behind the dunes).

2. **R. aquatilis** L. var. **capillaceus** (Thuill.) DC. Occasional in pools and reservoirs: Phoenix Lake, Mount Tamalpais, and fields near China Camp to the laguna in Chileno Valley, Dillons Beach, and Point Reyes Peninsula (pond behind the dunes).

3. **R. pusillus** Poir. Low ground where water collects in the winter and along the edge of marshes: San Rafael, acc. Gray; Ignacio; Chileno Valley; near Aurora School; Olema Marshes.

4. **R. Flammula** L. var. **ovalis** (Bigel.) L. Benson. Among sedges and rushes in wet mountain meadows or coastal swales: Potrero Meadow and Hidden Lake, Mount Tamalpais; near Aurora School; Point Reyes Peninsula (Inverness, acc. Stacey, Ledum Swamp). These are the southernmost known stations for this creeping buttercup in the Coast Ranges.—*R. Flammula* L. var. *reptans* E. Mey.

5. **R. uncinatus** D. Don var. **parviflorus** (Torr.) L. Benson. Moist brushy places on Point Reyes Peninsula: Inverness; Shell Beach; head of Drakes Estero. This small-flowered buttercup has not been reported south of Marin County in the Coast Ranges.—*R. Bongardii* Greene.

6. **R. hebecarpus** H. & A. Occasional in moist soil on slopes shaded by brush or trees: Sausalito; Kentfield; Fairfax Hills; San Rafael Hills; Big Rock Ridge.

7. **R. muricatus** L. Rather common and ruderal in low fields and on grassy slopes: Angel Island, Sausalito, and Mount Tamalpais (Baltimore Canyon, acc. Sutliffe, Rock Spring) to San Rafael and Chileno Valley. Introduced from Europe.

8. **R. repens** L. Becoming locally common in marshy ground: Mill Valley; San Rafael; Bolinas Lagoon; Mud Lake and McClure Beach road, Point Reyes Peninsula. Native of Eurasia.

9. **R. californicus** Benth. CALIFORNIA BUTTERCUP. Common and widespread in moist soil of low fields, wooded slopes, and brushy hills: Sausalito, Tiburon, and Mount Tamalpais (Rock Spring, Cataract Gulch, Phoenix Lake) to San Rafael Hills and Tomales.

9a. **R. californicus** Benth. var. **cuneatus** Greene. Sandy soil of maritime slopes and mesas: Point Reyes Peninsula; Dillons Beach. In this coastal variety the stems are nearly prostrate and the roundish leaves are shallowly to deeply 3-parted.

10. **R. orthorhynchus** Hook. var. **platyphyllus** Gray. Wet places in fields and among dunes along the coast: Mill Valley and San Rafael to Point Reyes, acc. Stacey. In marshy ground on Point Reyes Peninsula and at Dillons Beach, the plants become several feet tall and produce golden flowers more than an inch across.

11. **R. Bloomeri** Wats. Rather rare in wet clay soil of boggy fields: between Sausalito and Rodeo Lagoon, acc. Sutliffe; Lagunitas Meadows; Kentfield; Ignacio; near Aurora School; Olema, acc. Stacey.

4. Anemone. WINDFLOWER

1. **A. quinquefolia** L. var. **Grayi** (Behr & Kell.) Jeps. Moist soil of wooded canyonsides, generally under redwoods: Sausalito, acc. Lovegrove; Mill Valley; Muir Woods; Lagunitas Canyon. *Anemone Grayi* Behr & Kell. was originally described from near Lagunitas.

5. Thalictrum. MEADOW RUE

1. **T. polycarpum** (Torr.) Wats. Brushy or grassy slopes and flats that are wet in the spring: Angel Island, Sausalito, Tiburon, and Mount Tamalpais (near Mountain Theater, Potrero Meadow, Phoenix Lake) to San Rafael Hills, San Anselmo Canyon, Tomales, and Point Reyes Peninsula (Mud Lake, Ledum Swamp).

6. Aquilegia. COLUMBINE

a. Stems glabrous or slightly glandular-puberulent ....1. *A. formosa*
a. Stems densely viscid-hairy ....2. *A. eximia*

1. **A. formosa** Fisch. var. **truncata** (F. & M.) Baker. Widespread and rather common on brushy slopes: Sausalito, Tiburon, and Mount Tamalpais (Rifle Camp, Phoenix Lake) to San Rafael, Tomales, and Point Reyes Peninsula (Mud Lake).—*A. truncata* F. & M.

2. **A. eximia** Van Houtte. A rare summer-blooming columbine in wet soil around springs on serpentine: Camp Handy, Mount Tamalpais; San Anselmo Canyon; Liberty Spring. An unusually handsome plant but disagreeably glandular.—*A. Tracyi* Jeps.

7. Delphinium. LARKSPUR

a. Flowers red or coral-color ....6. *D. nudicaule*
a. Flowers not red or coral-color ....b
b. Stems 1–2 m. tall; flowers sordid, whitish strongly tinged with lavender or
   violet ....5. *D. californicum*
b. Stems less than 1 m. tall; flowers deep violet or blue c
c. Follicles densely puberulent; stems stout; roots elongate, fusiform or fibrous
   ....4. *D. hesperium*
c. Follicles glabrous; stems slender; roots tuberous ....d
d. Leaves shallowly 5-lobed, the lobes broad, with several to many abruptly
   apiculate teeth ....1. *D. Bakeri*

d. Leaves deeply 3–5-lobed, the lobes usually narrow, entire or few-toothed ...e
e. Stems low, mostly less than 2 dm. tall; flowers 1–5 ....2. *D. decorum*
e. Stems taller, lax; flowers generally more than 5 ....3. *D. patens*

1. **D. Bakeri** Ewan. A rare larkspur known only from the type locality in Sonoma County and near Tomales in Marin County.

2. **D. decorum** F. & M. Coastal hills and dunes: Sausalito, acc. Stacey; Point Reyes Peninsula; Tomales.

3. **D. patens** Benth. Rocky soil of steep wooded or brushy slopes away from the coast: Phoenix Lake; Kentfield; San Anselmo Canyon; San Rafael Hills.— *D. decorum* F. & M. var. *patens* (Benth.) Gray.

4. **D. hesperium** Gray. Rather common in clay soil on drying, grassy hills: Tiburon and Mount Tamalpais (Bootjack) to San Rafael Hills and Tomales.

5. **D. californicum** T. & G. Occasional in brush on coastal hills: Sausalito; Muir Beach; Stinson Beach, acc. Stacey; Salmon Creek; Point Reyes Peninsula (Inverness, acc. Stacey, Ledum Swamp).

6. **D. nudicaule** T. & G. RED LARKSPUR. Widespread and sometimes locally common on rocky slopes in partial shade: Mount Tamalpais (above Muir Woods, Cataract Gulch, Fish Grade); San Anselmo Canyon; San Rafael Hills; Black Mountain; Tomales. Near Tomales *D. nudicaule* apparently hybridizes with *D. decorum* and the flowers vary in color from coral and yellowish to lavender and purple. The yellow-flowered forms have been reported from Marin County as *D. luteum* Hel., but that species appears to be restricted to the Bodega country in southern Sonoma County.

BERBERIDACEAE. BARBERRY FAMILY

a. Leaflets spiny-toothed; flowers yellow ....1. *Berberis*
a. Leaflets not spiny-toothed; flowers whitish ....2. *Vancouveria*

1. Berberis

a. Bud scales persistent, conspicuous; leaflets several-nerved from base
....1. *B. nervosa*
a. Bud scales deciduous, small; leaflets with 1 nerve from base ....2. *B. pinnata*

1. **B. nervosa** Pursh. OREGON GRAPE. Occasional on steep canyonsides in the deep shade of coniferous woods: Muir Woods; Mount Tamalpais (Bootjack Canyon, Cataract Gulch, north side West Peak); Big Carson Canyon.—*Mahonia nervosa* (Pursh) Nutt.

2. **B. pinnata** Lag. Hills near the coast, generally in exposed places about rocks: Sausalito; Tennessee Cove; southeast of Muir Woods; Stinson Beach, acc. Stacey; Little Carson Canyon; Tomales; McClure Beach road, Point Reyes Peninsula.—*Mahonia pinnata* (Lag.) Fedde.

2. Vancouveria

1. **V. planipetala** Calloni. INSIDE-OUT FLOWER. Occasional in moist places under redwoods or Douglas firs: Mill Valley; Cataract Gulch and Fish Grade, Mount Tamalpais; Lagunitas Canyon.—*V. parviflora* Greene.

Calloni based his species on a Bolander collection and gave as the type locality "paper mill red woods." Greene's species, described three years later than that of Calloni, was based on plants "from the Santa Cruz Mountains northward to and beyond Mt. Tamalpais."

LAURACEAE. LAUREL FAMILY

1. Umbellularia

1. **U. californica** (H. & A.) Nutt. CALIFORNIA LAUREL; CALIFORNIA BAY; OREGON PEPPERWOOD (plates 19, 20, 21). Widespread on shaded canyonsides, moist alluvial flats, exposed rocky summits, and along intermittent streams, one of the most common trees in Marin County: Sausalito, Angel Island, Tiburon, and Mount Tamalpais to San Rafael Hills, Black Point, and Point Reyes Peninsula.

In direct response to the diverse conditions under which it grows, the California laurel is extremely varied in habit. Along streams in protected canyons and on open valley lands it becomes a tall tree with dense rounded crown and gnarled massive trunk. In well-watered canyons, as in Steep Ravine on the seaward slope of Mount Tamalpais, groves of laurel are too dense to develop open woodland but instead form a pygmy forest in which the numerous slender moss- and lichen-covered trunks produce an eerie and fascinating effect. In the chaparral of open rocky slopes it is still further reduced by the heat and drought of summer to a twiggy shrub scarcely taller than a man. And on slopes exposed to the prevailing Westerlies, the wind-pruned crowns and wind-controlled branches not only bespeak the constancy and force of these summer winds but also show the laurel to be the tree most responsive to wind-shaping in our region.

The leaves of the California laurel may be substituted in cooking for those of *Laurus nobilis,* the bay-leaf of commerce, whence the common name, California bay. Caution must be used, however, for the leaves of our laurel contain an oil much more potent than do those of the classic laurel of the Mediterranean region.

PAPAVERACEAE. POPPY FAMILY

a. Plants shrubby. Leaves entire ....5. *Dendromecon*
a. Plants herbaceous ....b
b. Leaves entire, mostly opposite c
b. Leaves deeply divided or finely dissected, mostly alternate ....d
c. Leaves and sepals pilose ....1. *Platystemon*
c. Leaves and sepals glabrous ....2. *Meconella*
d. Sepals united; pod linear ....3. *Eschscholzia*
d. Sepals distinct; pod turbinate ....4. *Papaver*

1. Platystemon

1. **P. californicus** Benth. CREAM-CUP. Widely distributed and sometimes locally abundant on grassy hills and rocky ridges and also on sandy slopes near the ocean: Sausalito (acc. Lovegrove), Tiburon and Mount Tamalpais (serpentine knoll near Rock Spring, Bolinas Ridge) to Lucas Valley, Tomales, and Point Reyes Lighthouse.

Cream-cups are attractive wherever they grow. A notable wild-flower garden, graced by their pale beauty, is found among picturesque rocks on Tiburon.

The form of this variable species growing at San Rafael was named *P. communis* by Greene.

2. Meconella

1. **M. californica** Torr. Steep rocky slopes in hills near the coast, rare: Sausalito Hills, acc. Lovegrove; Tomales.—*M. oregana* Nutt. var. *californica* (Torr.) Jeps.

3. Eschscholzia

a. Plants perennial but blooming the first year; torus rim double 1. *E. californica*
a. Plants annual; torus rim simple ....2. *E. caespitosa*

1. **E. californica** Cham. CALIFORNIA POPPY. Widespread and frequently common on grassy flats, rocky slopes, and maritime dunes and bluffs: Sausalito, Angel Island, Tiburon, and Mount Tamalpais (Rock Spring, Lagunitas Meadow) to San Rafael, Tomales, and Point Reyes Peninsula (Bolinas, Inverness Ridge, dunes near radio station).

No poet has yet sung the full beauty of our poppy, no painter has successfully portrayed the satiny sheen of its lustrous petals, no scientist has satisfactorily diagnosed the vagaries of its variations and adaptability. In its abundance, this colorful plant should not be slighted: cherish it and be ever thankful that so rare a flower is common!

2. **E. caespitosa** Benth. var. **hypecoides** (Benth.) Gray. Widely distributed in the interior hills of California; in Marin County on Mount Tamalpais according to Jepson.

4. Papaver. POPPY

1. **P. californicum** Gray. This sole California representative of the true poppies of the Old World might well be called fire poppy, since it is rarely seen except in areas of brush or forest swept by fire. In such a place in Lagunitas Canyon after the conflagration of 1945 the plants were locally abundant, the flame-red petals of the medium-sized flowers recalling in their color the holocaust that awakened the long-dormant seeds. Many years ago Brandegee recorded the occurrence of the fire poppy in burns on Mount Tamalpais and Bolinas Ridge; and in 1930 it was found on the scorched front of the mountain above Mill Valley. Quite beyond burned areas, the poppy was found on Mount Tamalpais in 1948 along a freshly graded fire road above Phoenix Lake by Rimo Bacigalupi. The plant is restricted to the Coast Ranges from Marin County southward to San Diego County.

5. Dendromecon

1. **D. rigida** Benth. TREE POPPY. Rocky open slopes on the south side of Mount Tamalpais, a characteristic member of the chaparral.

FUMARIACEAE. FUMITORY FAMILY

1. Dicentra

1. **D. formosa** (Andr.) Walp. PACIFIC BLEEDING HEART. Occasional in deep soil in moist shaded places: Tocaloma, *Leschke;* Salmon Creek; Olema Marshes; Mud Lake and Ledum Swamp, Point Reyes Peninsula.

CRUCIFERAE. Mustard Family

a. Fruit conspicuously longer than wide, oblong to linear ....b
a. Fruit relatively broad, round or only a little longer than wide ....n
b. Fruit dehiscent c
b. Fruit indehiscent, the seeds in the modified beak ....m
c. Pods with a prominent beak. Flowers large, petals yellow ....4. *Brassica*
c. Pod without a prominent beak ....d
d. Cauline leaves entire or toothed ....e
d. Cauline leaves lobed or divided ....h
e. Pubescence of simple hairs or none ....f
e. Pubescence of forked or stellate hairs. (*Arabis glabra* sometimes with hairs
       only on lower part of stem) ....g
f. Petals with strongly undulate margin; cauline leaves sessile, auriculate-clasping;
       pods strongly flattened ....1. *Streptanthus*
f. Petals with plane margin; cauline leaves sessile or petioled, not auriculate;
       pods terete ....2. *Thelypodium*
g. Petals white or purplish rose, or if yellowish white then only about 5 mm.
       long; pods strictly erect ....11. *Arabis*
g. Petals orange or yellowish white, more than 1 cm. long; pods spreading or
       ascending ....12. *Erysimum*
h. Pods oblong, the seeds in 2 rows ....8. *Rorippa*
h. Pods linear, the seeds in 1 row i
i. Pods terete or 4-angled ....j
i. Pods flattened parallel to the partition l
j. Pods somewhat 4-angled, the valves with a prominent midvein; stems glabrous
       ....7. *Barbarea*
j. Pods terete; stems usually hirsute, at least near the base ....k
k. Petals creamy white; pods generally reflexed, rarely spreading or ascending
       ....2. *Thelypodium*
k. Petals yellow (fading ochroleucous in *Sisymbrium altissimum*); pods spread-
       ing or strictly erect ....3. *Sisymbrium*
l. Flowers large, petals more than 5 mm. long; plants perennial, with tuberous
       rootstocks ....9. *Dentaria*
l. Flowers small, petals less than 5 mm. long; plants annual ....10. *Cardamine*
m. Fruit elongate, several-seeded ....5. *Raphanus*
m. Fruit oblongish or elliptic, 1-seeded. Plants of sandy beaches ....6. *Cakile*
n. Pods 2- to many-seeded, generally dehiscent (indehiscent in *Cardaria* and the
       valves indehiscent in *Coronopus*) ....o
n. Fruits 1-seeded, indehiscent ....u
o. Pods small, broader than long and notched above and below, valves enclos-
       ing seeds like nutlets ....17. *Coronopus*
o. Pods round or longer than broad, not equally emarginate above and below,
       the valves not enclosing the seeds like nutlets ....p
p. Pods terete or inflated; plants perennial ....q
p. Pods strongly flattened, not inflated; plants annual (except *Lobularia*) .....r
q. Pods dehiscent, terete, several-seeded; lower cauline leaves pinnately divided
       ....8. *Rorippa*

q. Pods indehiscent, inflated, flattened slightly contrary to the partition, each cell 1-seeded; cauline leaves dentate ....16. *Cardaria*

r. Pods flattened contrary to the narrow partition ....s

r. Pods flattened parallel to the broad partition ....t

s. Pods round or nearly so, notched or emarginate, seed 1 in each cell
....15. *Lepidium*

s. Pods cuneiform, obcordate, seeds several in each cell ....18. *Capsella*

t. Petals rose; pods about 2.5 cm. long ....13. *Lunaria*

t. Petals white or lavender; pods about 3 mm. long ....14. *Lobularia*

u. Fruit bearing fine, hooked hairs ....19. *Athysanus*

u. Fruit glabrous or pubescent, hairs not hooked ....20. *Thysanocarpus*

1. Streptanthus

a. Lower stems and leaves hispid-hairy ....b

a. Lower stems and leaves glabrous or sometimes very sparsely hispid c

b. Flowers strongly tinged with reddish violet or purple ......1. *S. glandulosus*

b. Flowers whitish or pale yellowish, sometimes tinged with lavender
....2. *S. secundus*

c. Stems generally more than 1 dm. tall; flowers about 1 cm. long, the sepals deep brownish purple ....3. *S. niger*

c. Stems generally less than 1 dm. tall; flowers about 0.5 cm. long, the sepals purplish lavender ....4. *S. batrachopus*

1. **S. glandulosus** Hook. var. **pulchellus** (Greene) Jeps. TAMALPAIS JEWEL-FLOWER. Common on high exposed ridges, in serpentine or occasionally in shale or sandstone: Mount Tamalpais (East Peak, Rifle Camp, Rock Spring, head of Steep Ravine); Carson Ridge; Big Rock Ridge, *Robbins.* The individuals of this odd mustard are especially abundant after fires when the plants are much more robust and floriferous than usual. Mount Tamalpais is the type locality of the variety which is not known beyond Marin County.

2. **S. secundus** Greene. Localized on rocky ridges and slide areas of canyon-sides, on both serpentine and sedimentaries: Mount Tamalpais (Phoenix Lake, Lake Lagunitas, *Cantelow,* Rocky Ridge Fire Trail, Bill Williams Gulch, *Sutliffe*); Fairfax Hills; Carson Ridge; San Rafael Hills; east end of Lucas Valley. This species, which is very closely related to *S. glandulosus,* was first described from plants collected at the north base of Mount Tamalpais. It is known only from Marin and Sonoma counties.

3. **S. niger** Greene. Locally common on serpentine slopes at the south end of Tiburon Peninsula. This is the one known locality not only for the black jewel-flower but also for *Castilleja neglecta,* two of Marin County's most restricted endemics. The nearest relative of *S. niger* is *S. albidus* Greene, a plant almost equally rare, which is found only on serpentine slopes of the Santa Clara Valley south of San Francisco Bay.

4. **S. batrachopus** J. L. Morrison. High serpentine ridges, only three stations known: the type locality near Collier Spring, Mount Tamalpais; Simmons Trail, Mount Tamalpais; Carson Ridge. Usually the plants of this rare little jewel-flower are only 2 to 4 inches tall, but in an especially favorable situation plants

10 inches tall have been found. These larger plants somewhat resemble *S. Breweri* of the Inner Coast Ranges to which the Marin County plant is related.

2. Thelypodium

1. T. lasiophyllum (H. & A.) Greene. Widely distributed and sometimes locally common, especially after fires: Angel Island; Mount Tamalpais (Corte Madera Ridge, Fern Canyon, Rifle Camp); Carson Ridge; San Rafael Hills; Tomales; Point Reyes Lighthouse.

Usually the plants are robust (vigorous individuals on burned areas becoming 6 feet tall) and the pods are abruptly reflexed. There is a more delicate form with pods spreading or ascending which is found occasionally. This is var. *inalienum* Robins. and may be recorded from Tiburon, Nora Trail on Mount Tamalpais, Black Canyon in the San Rafael Hills, and Shell Beach on Point Reyes Peninsula.

3. Sisymbrium

a. Petals about 6 mm. long; pods widely spreading ....1. *S. altissimum*
a. Petals about 3 mm. long; pods closely appressed ....2. *S. officinale*

1. S. altissimum L. TUMBLE MUSTARD. A common and widespread European weed in the interior but in Marin County found only at Ignacio and near Ledum Swamp on Point Reyes Peninsula.

2. S. officinale (L.) Scop. HEDGE MUSTARD. Widespread and rather common in weedy places along roads and about habitations: Sausalito, Tiburon, and Mount Tamalpais (Blithedale Canyon, Phoenix Lake) to San Rafael, Chileno Valley, and Point Reyes Peninsula (Bolinas, Mud Lake, Inverness). Naturalized from Europe.

4. Brassica

a. Leaves subtending branches of the inflorescence broad; pods spreading or
   ascending ....b
a. Leaves subtending branches of the inflorescence narrow; pods appressed-ascending ....d
b. Upper cauline leaves not glaucous or auriculate, generally petiolate; pedicels
   shorter than the flowers, stout in fruit ....3. *B. Kaber*
b. Upper cauline leaves glaucous, auriculate-clasping; pedicels longer than the
   flowers, slender in fruit c
c. Basal leaves glabrous; buds exceeding the flowers at the top of the racemes
   ....1. *B. oleracea*
c. Basal leaves sparsely hirsute; flowers exceeding the buds ......2. *B. campestris*
d. Leaves of inflorescence flexuously recurved; valves of pod with prominent mid-
   nerve; beak slender, without seeds ....4. *B. nigra*
d. Leaves of inflorescence not curved; valves of pod longitudinally rugulose; beak
   stout, usually seed-bearing ....5. *B. geniculata*

1. B. oleracea L. CABBAGE. Locally naturalized on the rocky ocean bluffs at Point Bonita. Widely cultivated, growing wild on sea cliffs of western and southern Europe.

Escaping cultivation in the gardens behind the bold Marin headland, the Bonita cabbages have answered the call of the wild and have returned to that same sort of exposed maritime habitat from which their progenitors were weaned in prehistoric times in Europe. The centuries of domestication, it would seem, have never removed from the hereditary fabric of this plant its propensity for the difficult life on rocks swept by gales off stormy seas and wetted by flying scud from crashing surf. The sea, it would seem, has been a cabbage's one true and constant love!

In few groups of plants has man developed more numerous or varied forms than in the cabbages. With all the remarkable morphologic diversity and seeming genetic plasticity of this group of vegetables, however, the Bonita plants prove that cabbages have always remained essentially cabbages, a dramatic demonstration of the constancy of species.

2. **B. campestris** L. COMMON MUSTARD. Abundant and widespread in cultivated ground and along roads or trails: Sausalito, Angel Island, and Tiburon to San Rafael, Inverness, and Tomales. Native of Eurasia.

3. **B. Kaber** (DC.) L. C. Wheeler. CHARLOCK. Rather rare in weedy places around habitations and along roads: Sausalito, acc. Lovegrove; Tiburon; Mill Valley; San Rafael; Inverness. Native of Europe.—*B. arvensis* (L.) Rabenh.; *Sinapis arvensis* L.

4. **B. nigra** (L.) Koch. BLACK MUSTARD. Occasional on roadsides and in disturbed soil in the hills: Tiburon; between Mill Valley and Muir Woods; near Alpine Dam; Woodacre; south end of Tomales Bay; Inverness. Introduced from Europe.

5. **B. geniculata** (Desf.) J. Ball. MEDITERRANEAN MUSTARD; SUMMER MUSTARD. Widespread and abundant along roads and in waste places, the common summer-blooming species. In dry packed soil around habitations, the plants may be several times branched and only a foot or two high, but in favorable situations the plants may become 6 to 8 feet tall. This Old World weed is perhaps our most common mustard.—*B. adpressa* Boiss.; *B. incana* (L.) Meigen; *Sinapis incana* L.

5. Raphanus

1. **R. sativus** L. RADISH. Common and widespread in valleys and lower hills throughout the county: Sausalito, Angel Island, and Tiburon to Tomales and Point Reyes Peninsula (Bolinas, Inverness). Some of our most attractive springtime floral displays are produced by the lacy and colorful masses of this Old World introduction and garden escape.

6. Cakile. SEA ROCKET

a. Leaves crenate-dentate; petals not conspicuously exceeding the sepals; lower part of fruit without lateral spurs ....1. *C. edentula*
a. Leaves pinnately divided; petals conspicuous; lower part of the fruit with lateral spurs ....2. *C. maritima*

1. **C. edentula** (Bigel.) Hook. var. **californica** (Hel.) Fern. Occasional on beaches just above the high tide line: Sausalito, acc. Lovegrove; Stinson Beach; Bolinas and McClure Beach, Point Reyes Peninsula.

2. **C. maritima** Scop. Becoming common on beaches: Angel Island; Tiburon; Rodeo Lagoon; Muir Beach; Stinson Beach; McClure Beach, Point Reyes Peninsula; Dillons Beach. Introduced from the Old World. On Stinson Beach large plants form rounded mounds about 4 feet across and 2 feet tall.

7. **Barbarea.** WINTER-CRESS

a. Lateral divisions of the cauline leaves few, elliptic to narrowly oblong; style
gradually narrowed from a broad base ....1. *B. americana*

a. Lateral divisions of the cauline leaves in 2—5 pairs, mostly linear-oblong;
style abruptly narrowed ....2. *B. verna*

1. **B. americana** Rydb. Rocky or gravelly soil of bushy or wooded slopes in
partial shade: Sausalito, acc. Lovegrove; Mount Tamalpais (Corte Madera Ridge,
Berry Trail, Northside Trail); Carson Ridge, common in burned area; Lagunitas,
acc. Stacey; Tomales. Sometimes this plant is referred to *B. vulgaris* (L.) R. Br.
or *B. orthoceras* Ledeb., two closely related Old World species.

2. **B. verna** (Mill.) Aschers. European weed which has been found in Sausalito
and which may occur elsewhere in waste ground and along roads.

8. **Rorippa**

a. Petals white ....b

a. Petals yellow c

b. Stems erect from a thick tap-root; upper cauline leaves entire; pod globose or
a little elongate ....1. *R. Armoracia*

b. Stems floating, creeping, or ascending; upper cauline leaves pinnately divided;
pod linear-oblong ....2. *R. Nasturtium-aquaticum*

c. Pods linear, more or less curved; sepals generally less than 2 mm. long
....3. *R. curvisiliqua*

c. Pods oblong-linear, straight or nearly so; sepals 2 mm. long or more
....4. *R. islandica*

1. **R. Armoracia** (L.) Hitchc. HORSERADISH. In low ground along the highway
west of Point Reyes Station, where the plant has persisted for many years. Introduced from Europe.

2. **R. Nasturtium-aquaticum** (L.) Hayek. WATER-CRESS. In shallow water of
ponds and streams or on muddy flats, common and widespread: Rodeo Lagoon,
Tiburon, and Mount Tamalpais (Rock Spring, Alpine Dam) to Dillons Beach
and Point Reyes Peninsula (Bolinas, Inverness, Ledum Swamp). Although the
stems are usually low and spreading, in protected places they may become as
much as 3 feet tall.—*Radicula Nasturtium-aquaticum* (L.) Britt. & Rendle.

3. **R. curvisiliqua** (Hook.) Bessey. Occasional in moist canyons or on low wet
flats where water has stood: Lake Lagunitas; Alpine Lake; San Antonio Creek;
Chileno Valley; Bolinas; Point Reyes Station.—*Radicula curvisiliqua* (Hook.)
Greene.

4. **R. islandica** (Oeder) Borbás var. **occidentalis** (Wats.) Butters & Abbe. In
Marin County known only from the marshy borders of the laguna in Chileno
Valley.—*Rorippa palustris* (L.) Bess. subsp. *occidentalis* (Wats.) Abrams; *Radicula
palustris* (L.) Moench.

9. **Dentaria.** TOOTHWORT

1. **D. integrifolia** Nutt. RAIN-BELLS; MILK-MAIDS. Locally abundant in low fields
and meadows that are wet or marshy during the rainy season: Mill Valley,
Greenbrae, Ross, and Mount Tamalpais (Lagunitas Meadows) to Olema and
Point Reyes Station.—*D. californica* Nutt. var. *integrifolia* (Nutt.) Detling.

Much more widespread but rarely so abundant locally is the form *D. integrifolia* Nutt. var. *californica* (Nutt.) Jeps. In it the leaflets of the cauline leaves are ovate and more or less toothed or lobed, rather than oblong-oblanceolate and entire as they are in the species. It is found on rocky slopes of exposed ridges and deep canyons as well as in the partial shade of brush or woods: Sausalito, Tiburon, and Mount Tamalpais (Corte Madera Ridge, Bootjack Canyon, Cataract Gulch) to San Rafael, Tomales, and Point Reyes Peninsula (Drakes Estero). The flowers are usually white but on Mount Tamalpais they are frequently a deep and attractive rose.

10. **Cardamine.** BITTER-CRESS

1. **C. oligosperma** Nutt. Rather common in moist shaded places under brush or trees, occasionally in open meadows and on coastal dunes: Sausalito, Tiburon, and Mount Tamalpais (Phoenix Lake, Rock Spring, Cataract Gulch) to San Rafael, Dillons Beach, and Point Reyes Peninsula (Mud Lake, Inverness, dunes near the radio station). When the pods are mature, the valves roll up with great rapidity and expel the seeds forceably for a considerable distance.

11. **Arabis.** ROCK-CRESS

a. Petals purplish rose; cauline leaves not auriculate-clasping 1. *A. blepharophylla*
a. Petals white or yellowish white; cauline leaves auriculate-clasping . . . . .b
b. Cauline leaves generally conspicuously hairy; petals white; pods flattened
. . . . .2. *A. hirsuta*
b. Cauline leaves generally glabrous and glaucous; petals pale yellowish white; pods nearly terete . . . . .3. *A. glabra*

1. **A. blepharophylla** H. & A. In shallow soil on coastal slopes or occasionally on rocky outcrops inland: Sausalito; Mount Tamalpais, acc. Stacey; head of Lucas Valley, *Leschke;* Tomales; Point Reyes Lighthouse. This attractive plant, which frequently begins to bloom before the end of winter, adds much color to maritime rock gardens. Although the corolla is usually purplish rose, flowers with pale pink petals have been seen.
2. **A. hirsuta** (L.) Scop. Occasional on coastal flats or on moist shaded rocks: Salmon Creek; Point Reyes Peninsula (Inverness, *Linsdale,* Shell Beach, Abbots Lagoon).
3. **A. glabra** (L.) Bernh. TOWER MUSTARD. Occasional on grassy or brushy slopes in the hills: Sausalito; Angel Island; Tiburon; Laurel Dell Trail and Potrero Meadow, Mount Tamalpais; Lagunitas Canyon; San Rafael Hills; Tomales.

12. **Erysimum.** WALLFLOWER

a. Petals bright orange; pods markedly quadrangular . . . . .1. *E. capitatum*
a. Petals pale yellow or creamy-white; pods flattened . . . . .2. *E. concinnum*

1. **E. capitatum** (Dougl.) Greene. Occasional on rocky slopes of coastal hills in the northwestern part of the county: near Tocaloma; east of Point Reyes Station; Salmon Creek Canyon; Tomales. A single plant seen along a road in the Sausalito Hills was probably a waif.—*E. asperum* of Jepson, not *E. asperum* (Nutt.) DC.
2. **E. concinnum** Eastw. Rocky slopes of oceanic headlands and dune hills behind the beaches on Point Reyes Peninsula. In this handsome maritime plant,

the stems are low, the flowers are large, and the ascending pods are especially broad. It is very closely related to *E. Menziesii* (H. & A.) Wettst., a similar maritime plant found on the Monterey Peninsula. Among rocks on the ridge southeast of Muir Woods is another kind of plant belonging to this group of coastal wallflowers. It is taller and more slender and has smaller flowers and narrower pods. These and other variants belong to *E. capitatum* according to Jepson.

13. Lunaria. Moonwort

1. **L. annua** L. Honesty. An occasional garden escape in moist shaded woods: Sausalito; Mill Valley; Lagunitas; Inverness. Native of the Old World.

14. Lobularia

1. **L. maritima** (L.) Desv. Sweet Alyssum. Occasional in weedy places about habitations where it has escaped from cultivation: Point Bonita; Sausalito; Angel Island; Tiburon; Mill Valley. Native of southern Europe.—*Alyssum maritimum* (L.) Lamk.; *Koniga maritima* (L.) R. Br.

15. Lepidium. Peppergrass

a. Leaves markedly dimorphic, the lower finely bipinnatifid, the upper cauline entire and cordate-clasping; style evident ....1. *L. perfoliatum*
a. Leaves not dimorphic, the uppermost reduced; style very short or none ....b
b. Sepals persisting some time after anthesis ....2. *L. pubescens*
b. Sepals early deciduous c
c. Pedicels nearly terete ....3. *L. virginicum*
c. Pedicels strongly flattened ....d
d. Wings at apices of pods rounded, not widely divergent ....e
d. Wings at apices of pods acute and either elongate or divergent ....f
e. Pods glabrous, shining ....4. *L. nitidum*
e. Pods sparsely pubescent especially on the margin, dull ..5. *L. lasiocarpum*
f. Pods pubescent, the wings nearly parallel ....6. *L. latipes*
f. Pods glabrous, the wings widely divergent ....7. *L. oxycarpum*

1. **L. perfoliatum** L. Occasional along the railroad right-of-way near Manzanita and in the hills near San Anselmo. Native of Europe.

2. **L. pubescens** Desv. Widespread and rather common along roads and paths in hard packed soil: Sausalito and Mount Tamalpais (Blithedale Canyon, West Point) to San Rafael, Tomales, and Point Reyes Peninsula (Bolinas, Mud Lake, Inverness). Introduced from South America.

3. **L. virginicum** L. In Marin County known only from a collection made by Mary Courtright at Muir Beach.

4. **L. nitidum** Nutt. Common and widespread on valley flats, grassy hills, and rocky slopes: Tiburon, Sausalito, and Mount Tamalpais (Rock Spring) to San Rafael, Tomales, and Point Reyes Lighthouse. Both young folks and old find pleasure in munching the peppery fruits when they first appear in late winter and early spring.

5. **L. lasiocarpum** Nutt. Dunes of Point Reyes Peninsula and beach sand near Dillons Beach are the only stations known for this southern peppergrass in coastal northern California. There is no apparent reason to doubt that the plant is indigenous.

6. **L. latipes** Hook. Clay flats and slopes of interior hills, in Marin County known only from Ignacio.

7. **L. oxycarpum** T. & G. A rare peppergrass of alkaline valley floors and of saline flats adjacent to coastal salt marshes, in Marin County known only from low pastures bordering San Francisco Bay near Novato.

16. Cardaria

1. **C. Draba** (L.) Desv. HOARY CRESS. Locally established along a road north of San Rafael. Native of Europe.—*Lepidium Draba* L.

17. Coronopus

1. **C. didymus** (L.) Smith. WART-CRESS. Widespread weed of waste ground: Sausalito, Angel Island, and Mill Valley to San Rafael and Point Reyes Peninsula (Bolinas, Inverness, Point Reyes Lighthouse). Introduced from Europe.

18. Capsella

1. **C. Bursa-pastoris** (L.) Medic. SHEPHERD'S PURSE. Common vernal weed in hills and valleys: Sausalito and Tiburon to San Rafael and Point Reyes Station. Naturalized from Europe.

19. Athysanus

1. **A. pusillus** (Hook.) Greene. Occasional on hills in gravelly soil: Rock Spring and Bolinas Ridge, Mount Tamalpais; Carson Ridge; Lagunitas, *Stacey;* Big Rock Ridge; Tomales. The petals, if present, are usually quite inconspicuous, but in the plants from Rock Spring, the petals are large and relatively showy. In these same plants the sides of the fruits are generally glabrous although the sides usually bear both uncinate trichomes and arachnoid hairs.

20. Thysanocarpus

a. Cauline leaves scarcely or not at all auricled, the lower leaves glabrous. Petals about 2 mm. long, equaling or exceeding the stamens; fruit about 4 mm. broad . . . .**3.** *T. laciniatus*
a. Cauline leaves auriculate-clasping, the lower leaves more or less rosulate and hirsute . . . . .b
b. Petals not conspicuous, 1–1.5 mm. long, shorter than the stamens at time of anthesis; fruit 3–4 mm. broad . . . .**1.** *T. curvipes*
b. Petals relatively conspicuous, 2–3 mm. long, longer than the stamens; fruit 5–6 mm. broad . . . .**2.** *T. elegans*

1. **T. curvipes** Hook. FRINGE-POD. Gravel slopes on open hillsides: Sausalito; near Bootjack, Mount Tamalpais; Big Carson Canyon; Fairfax Hills; Big Rock Ridge; Black Mountain east of Point Reyes Station.

2. **T. elegans** F. & M. Open grassy hillsides: Fairfax Hills; Lagunitas Canyon. In the Marin County plants the wing on the fruit is sometimes perforate and sometimes not.—*T. curvipes* Hook. var. *elegans* (F. & M.) Robins.

3. **T. laciniatus** Nutt. var. **crenatus** Brew. Grassy opens on brush-covered slopes: San Rafael Hills; Big Rock Ridge. These are the northernmost known stations for this species which has not been reported heretofore north of San Francisco Bay.

RESEDACEAE. Mignonette Family

1. Reseda

a. Stems low, the branches spreading; pedicels equaling or longer than the flowers, 5–10 mm. long ....1. *R. odorata*

a. Stems tall, virgately branched; pedicels shorter than the flowers, less than 5 mm. long ....2. *R. Luteola*

1. **R. odorata** L. Mignonette. Occasional fugitive from cultivation in moist shady places: Marin County, acc. Jepson; Mill Valley, acc. Stacey. Native of North Africa.

2. **R. Luteola** L. Dyer's Mignonette. This European plant, cultivated for its yellow dye, is known in Marin County only from a Sausalito record by Stacey.

CRASSULACEAE. Stone-crop Family

a. Plants annual; leaves opposite; flowers 1–2 mm. long ....1. *Tillaea*

a. Plants perennial; leaves alternate, mostly forming a dense basal rosette; flowers 5 mm. long or more ....b

b. Petals spreading; basal leaves 3.5 cm. long or less ....2. *Sedum*

b. Petals erect; basal leaves generally more than 3 cm. long ....3. *Echeveria*

1. Tillaea

a. Petals 2–3 times longer than the sepals ....1. *T. aquatica*

a. Petals about equaling the sepals ....2. *T. erecta*

1. **T. aquatica** L. Occasional on muddy flats where water has stood: Phoenix and Lagunitas lakes; Lagunitas Meadows; Hamilton Field.—*Tillaeastrum aquaticum* (L.) Britt.

In plants collected in Lagunitas Meadows in May, the flower-stalks are shorter than the leaves, while in plants taken from the strand of Lake Lagunitas in September, the elongate flower-stalks are several times longer than the leaves. This matured aspect of the species has been called var. *Drummondii* by Jepson.

2. **T. erecta** H. & A. Widespread and sometimes common in loose well-drained gravelly or sandy soil on flats, slopes, and ridges: Sausalito, Angel Island, Tiburon, and Mount Tamalpais (Corte Madera Ridge) to San Rafael, Dillons Beach, and Point Reyes Peninsula (Drakes Estero and near the lighthouse).

This diminutive annual is one of the rather impressive number of species which are found both in Chile and in California.

2. Sedum

a. Leaves linear-oblongish, widest near the base ....1. *S. radiatum*

a. Leaves broadly obovate, widest near the apex ....2. *S. spathulifolium*

1. **S. radiatum** Wats. Rocky slopes in shallow soil, occurring both on sedimentaries and on serpentine: Tiburon; Cataract Gulch and Artura Trail, Mount Tamalpais; Lagunitas Canyon; Lucas Valley; Black Mountain east of Point Reyes Station; Tomales.

2. **S. spathulifolium** Hook. Rather common and widespread on steep rocky slopes, moist shaded bluffs, and exposed maritime headlands: Sausalito and Mount Tamalpais (East Peak, Cataract Gulch, Lake Lagunitas) to San Rafael Hills, Black Mountain, Dillons Beach, and Point Reyes Lighthouse.

Generally the herbage is glaucous but at times it is quite devoid of bloom. Plants from maritime habitats are sometimes conspicuously chalky and such a form was once named as a distinct species. Whether in shaded redwood canyon or on open ocean bluff, *S. spathulifolium* is one of our attractive rock plants.

3. Echeveria

a. Pedicels slender, about as long as the flower ....1. *E. laxa*
a. Pedicels stout, much shorter than the flower ....2. *E. caespitosa*

1. **E. laxa** Lindl. Rock Lettuce. Crevices of rocks on open ridges or southerly slopes, occasional: Mount Tamalpais (Corte Madera Ridge, Bootjack, Rock Spring, Fish Grade); Little Carson Canyon; Fairfax Hills; San Rafael Hills.— *Cotyledon laxa* (Lindl.) Brew. & Wats.; *Dudleya laxa* (Lindl.) Britt. & Rose.

The Marin County plants of this variable species were described as *Dudleya Sheldoni* Rose, the type of which came from the "north base of Mount Tamalpais."

2. **E. caespitosa** (Haw.) DC. Sea Lettuce (plate 11). Rocky slopes and ridges along the coast: Rodeo Lagoon; Sausalito; Tiburon; Angel Island; ridge southeast of Muir Woods; Stinson Beach; Rio Hondo near Bolinas; Point Reyes Lighthouse. —*Cotyledon caespitosa* Haw.; *Dudleya caespitosa* (Haw.) Britt. & Rose. The Marin County plant is *E. Cotyledon* of Jepson's Flora.

Plants with either green or glaucous leaves are commonly found growing together and seem identical except for the presence or absence of the chalky powder on the leaves. The form with densely glaucous leaves is treated by Abrams as *Dudleya farinosa*.

SAXIFRAGACEAE. Saxifrage Family

a. Stems herbaceous; leaves mostly basal, cauline leaves, if present, alternate and generally reduced ....b
a. Stems woody, leafy i
b. Flowers solitary on a slender scape, about 2 cm. across; leaves entire
,..1. *Parnassia*
b. Flowers several to numerous, usually 1 cm. across or less (except sometimes in *Lithophragma*); leaves denticulate to crenate or subentire c
c. Inflorescence paniculate; petals entire ....d
c. Inflorescence racemose; petals toothed, cleft, or pectinately pinnatifid ......g
d. Leaves ovate-oblong or subelliptic, cuneate at base ....2. *Saxifraga*
d. Leaves broadly ovate to round, truncate to cordate at base c
e. Petals conspicuous; stamens shorter than the calyx; ovary 2-celled 3. *Boykinia*
e. Petals relatively inconspicuous; stamens equaling to much longer than the calyx; ovary 1-celled ....f
f. Stamens 10, conspicuously longer than the calyx; petals filiform ..4. *Tiarella*
f. Stamens 5, equaling or little longer than the calyx; petals linear to oblong
....5. *Heuchera*
g. Flowers small, less than 5 mm. long; hypanthium saucer-shaped, stamens 5
....8. *Mitella*
g. Flowers relatively large, generally 5 mm. long or more; hypanthium turbinate or campanulate; stamens 10 ....h
h. Styles 3; petals white or pinkish, rather coarsely toothed or cleft; basal leaves 4 cm. long or less ....6. *Lithophragma*

h. Styles 2; petals greenish white becoming red, laciniate; basal leaves more than
   4 cm. long ....7. *Tellima*
i. Trailing shrub with opposite leaves; fruit a capsule ....9. *Whipplea*
i. Erect shrub with alternate leaves; fruit a berry ....10. *Ribes*

1. Parnassia. GRASS-OF-PARNASSUS

1. **P. californica** (Gray) Greene. Locally common in wet meadows on Mount
Tamalpais, blooming in late summer: Azalea Flat; Bootjack; Camp Handy.—
P. palustris L. var. *californica* Gray.

2. Saxifraga. SAXIFRAGE

a. Sepals not reflexed in flower and fruit; petals elongate, conspicuously exceeding the sepals ....1. *S. fallax*
a. Sepals reflexed in flower and fruit; petals roundish, only slightly longer than
   the sepals ....2. *S. californica*

1. **S. fallax** Greene. Steep northern slope in open woodland: Fairfax Hills
near the Meadows Club, the southernmost known station in the Coast Ranges.
2. **S. californica** Greene. Moist, partially shaded slopes in shallow soil or on
moss-covered rocks: Sausalito, acc. Lovegrove; Tiburon; Bootjack and Cataract
Gulch, Mount Tamalpais; Fairfax Hills; San Geronimo Ridge; San Rafael Hills;
Tomales.—*S. virginiensis* Michx. var. *californica* (Greene) Jeps.

The California saxifrage is sometimes an early bloomer: it has been found in
flower before the end of January, though usually it blooms in March and April.

3. Boykinia

1. **B. elata** (Nutt.) Greene. BROOK FOAM. Occasional on rocky stream beds and
on mossy rocks of deep canyons: Bootjack Canyon and Cataract Gulch, Mount
Tamalpais, *Leschke;* Lagunitas Canyon; Inverness and Ledum Swamp, Point
Reyes Peninsula.

4. Tiarella

1. **T. unifoliata** Hook. SUGAR-SCOOP. Listed from Mill Valley by Stacey but no
specimens from Marin County have been seen. This plant, which takes its
common name from the peculiar spoonlike fruit, is found in wooded canyons
near the coast from Santa Cruz County north to Alaska.

5. Heuchera. ALUM-ROOT

a. Hypanthium densely pilose ....1. *H. pilosissima*
a. Hypanthium thinly pubescent ....2. *H. micrantha*

1. **H. pilosissima** F. & M. Coastal bluffs and mesas; Point Reyes, acc. Stacey;
Tomales, *Leschke.*
2. **H. micrantha** Dougl. Common and widespread in deep canyons and on
shaded rocky slopes: Sausalito, Tiburon, and Mount Tamalpais (Cascade Canyon,
Cataract Gulch, Fish Grade) to Papermill Creek and Point Reyes Peninsula (Mud
Lake and Inverness Ridge).

The plants are variable and two forms have been recognized in the county:
var. *pacifica* Rosend., But., & Lak. in which the inflorescence and peduncle are

glandular but not hirsute, and var. *Hartwegii* (Wats.) Rosend. in which they are hirsute as well as glandular.

6. Lithophragma

a. Hypanthium truncate or broadly rounded at base; leaf axils frequently bearing bulblets ....1. *L. heterophyllum*
a. Hypanthium pointed at base; leaf axils without bulblets ....2. *L. affine*

1. **L. heterophyllum** (H. & A.) T. & G. Shaded slopes, generally in shallow rocky soil: Sausalito; Rifle Camp and Cataract Gulch, Mount Tamalpais; Carson country; San Rafael Hills; Big Rock Ridge.
2. **L. affine** Gray. WOODLAND STAR. Open, brushy, or wooded slopes and meadows: Sausalito (acc. Lovegrove), Tiburon, and Mount Tamalpais (near Rock Spring and Laurel Dell) to Fairfax Hills, San Rafael Hills, Big Rock Ridge (*Robbins*), and Tomales.

7. Tellima

1. **T. grandiflora** (Pursh) Dougl. FRINGE-CUPS. Moist rocky slopes in canyons: Sausalito; Cascade Canyon and Cataract Gulch, Mount Tamalpais; Lagunitas Creek, *Moore;* Mud Lake, Point Reyes Peninsula.

8. Mitella

1. **M. ovalis** Greene. Moist stream banks in deep shade: Lagunitas, *Eastwood;* Camp Taylor, acc. Bacigalupi. Not known south of Marin County.—*Pectiantia ovalis* (Greene) Rydb.

9. Whipplea

1. **W. modesta** Torr. MODESTY. Wooded or brushy slopes in shallow or rocky soil: Mount Tamalpais (Cascade and Blithedale canyons, Bootjack Ravine, Cataract Gulch); Lagunitas Canyon; San Rafael Hills; Point Reyes Station, acc. Stacey.

The original collection from which the genus and species were described was made in the redwoods of Marin County by John M. Bigelow in 1854.

10. Ribes. CURRANT; GOOSEBERRY

a. Stems not spiny ....b
a. Stems spiny c
b. Upper leaf surface not rough-hairy; style glabrous ....1. *R. sanguineum*
b. Upper leaf surface rough-hairy; style hairy ....2. *R. malvaceum*
c. Berry glabrous ....3. *R. divaricatum*
c. Berry spiny ....d
d. Leaves glabrous or nearly so except for hairs on margins; spines on fruit rarely gland-tipped ....4. *R. californicum*
d. Leaves more or less glandular-hairy; some spines on fruits gland-tipped ....e
e. Spines on fruits short and stout, only a little unequal; sepals pale greenish or frequently tinged with purple ....5*b. R. Menziesii* var. *Victoris*
e. Spines on fruits slender, noticeably unequal; sepals purplish ....f
f. Leaves thick and rugose, densely glandular-hairy beneath.....5. *R. Menziesii*
f. Leaves thinner and nearly plane, not densely hairy beneath
......................................5*a. R. Menziesii* var. *leptosmum*

1. **R. sanguineum** Pursh var. **glutinosum** (Benth.) Loud. FLOWERING CURRANT. Moist brushy slopes, generally in partial shade, not common: Rodeo Lagoon; Sausalito; Steep Ravine, Mount Tamalpais, *Lovegrove;* Ross Valley, acc. Stacey; Stinson Beach; Inverness and Ledum Swamp, Point Reyes Peninsula.—*R. glutinosum* Benth.

In some specimens the lower sides of the leaves are densely hairy as in typical *R. sanguineum* and in others the leaves are neither hairy nor glandular. No matter what the vesture of the leaves, however, our shrubs are among the choicest of our early spring flowering plants.

2. **R. malvaceum** Smith. Rare on rocky ridges on the edge of chaparral: Oceanview Trail, Mount Tamalpais, *Heggie;* Bolinas Ridge, acc. Greene.

3. **R. divaricatum** Dougl. Rather rare shrub in moist, wet, or marshy places, generally among willows or other shrubs: Willow Meadow, Mount Tamalpais; Olema, acc. Stacey; Salmon Creek Canyon; Mud Lake and Drakes Estero, Point Reyes Peninsula.—*Grossularia divaricata* (Dougl.) Cov. & Britt.

4. **R. californicum** H. & A. CALIFORNIA GOOSEBERRY. Occasional on wooded flats or open hills: Lake Lagunitas, *Sutliffe;* San Anselmo Canyon; Lucas Valley, *Leschke;* Tomales.—*Grossularia californica* (H. & A.) Cov. & Britt.

A gooseberry on Angel Island was listed as this species but it may have been *R. Menziesii* var. *leptosmum* which occurs on nearby Tiburon.

5. **R. Menziesii** Pursh. Occasional on Point Reyes Peninsula in coastal thickets on moist hills and mesas: Inverness; Inverness Ridge; Point Reyes road.—*Grossularia Menziesii* (Pursh) Cov. & Britt.

By reference to Jepson's Flora these stations may be regarded the southern distributional limit for the typical variety of the Menzies gooseberry; but, although it is not always easy to distinguish the typical variety from var. *leptosmum,* specimens nearly or quite representative of the Menzies gooseberry have been seen from the Santa Cruz Mountains to the south. The following plants, which are sometimes treated as distinct species, seem too closely related to the general *R. Menziesii* complex to be accorded more than varietal status:

5a. **R. Menziesii** Pursh var. **leptosmum** (Cov.) Jeps. Moist places on wooded or brushy slopes and along streams, occasional: Sausalito; Tiburon; Muir Woods; Lake Lagunitas, Mount Tamalpais; Big Carson Canyon; Salmon Creek Canyon; Tomales; Bear Valley, the type locality, *Eastwood;* Inverness Ridge.—*Grossularia leptosma* Cov.

With its thinner and slightly pubescent leaves, var. *leptosmum* may be the woodland homologue of typical *R. Menziesii* which is generally found in situations more exposed to wind and sun.

5b. **R. Menziesii** Pursh var. **Victoris** (Greene) Jancz. Rare on wooded flats and northerly slopes: Northside Trail and Lake Lagunitas, Mount Tamalpais; San Rafael Hills; Big Rock Ridge, *Robbins.*—*Grossularia Victoris* (Greene) Cov. & Britt.

Victor's gooseberry, with type locality "near the base of Mount Tamalpais," is known only from Marin County where it is not always easily distinguishable from var. *leptosmum.* In typical form, it has pale greenish flowers but generally these are more or less tinged with rose or purple. The fruits, densely covered with short stout yellowish spines, are perhaps the best mark of this local endemic.

ROSACEAE. Rose Family

a. Plants herbaceous, annuals or perennials ....b
a. Plants woody, shrubs or vines ....f
b. Petals present and showy c
b. Petals none ....d
c. Receptacle dry in fruit, not becoming red and juicy ....7. *Potentilla*
c. Receptacle becoming red and juicy in fruit ....8. *Fragaria*
d. Plants annual, less than 1 dm. tall ....13. *Alchemilla*
d. Plants perennial, more than 1 dm. tall ....e
e. Hypanthium angled and roughened but not spiny in fruit ....11. *Sanguisorba*
e. Hypanthium spiny in fruit ....12. *Acaena*
f. Ovary superior ....g
f. Ovary inferior ....n
g. Pistils several to numerous ....h
g. Pistil 1 l
h. Fruit fleshy, a drupe or drupelet i
h. Fruit dry, a follicle or achene ....j
i. Pistils numerous ....6. *Rubus*
i. Pistils 5. Plants dioecious; flowers in pendent racemes ....15. *Osmaronia*
j. Leaves compound; stems usually prickly; petals pink or rose ....14. *Rosa*
j. Leaves simple; stems not prickly; petals whitish ....k
k. Leaves palmately veined and lobed; flowers few in a rounded corymb
  ....1. *Physocarpus*
k. Leaves pinnately veined, coarsely serrate; flowers numerous in a panicle
  ....2. *Holodiscus*
l. Fruit a drupe. Leaves deciduous ....16. *Prunus*
l. Fruit an achene ....m
m. Petals lacking; leaves broadly cuneate, deciduous; achene with long plumose
  tail ....9. *Cercocarpus*
m. Petals present; mature leaves short, needle-like, mostly in fascicles, evergreen; achene not long-tailed ....10. *Adenostoma*
n. Leaves evergreen, elliptic-oblong; fruit red ....3. *Photinia*
n. Leaves deciduous; fruit dark purple or blackish ....o
o. Leaves roundish, serrate above the middle; flowers in a short raceme
  ....4. *Amelanchier*
o. Leaves ovate to obovate, serrate nearly to the base; flowers in a corymb
  ....5. *Crataegus*

1. Physocarpus

1. **P. capitatus** (Pursh) Ktze. Ninebark. Widely distributed but not common in moist, partially shaded places: Sausalito; Tiburon; Stinson Beach, acc. Stacey; Mount Tamalpais (Laurel Dell, Potrero Meadow, Lake Lagunitas); Ross; Fairfax; San Rafael Hills; Lagunitas Creek; Point Reyes Station; Salmon Creek; Tomales.

2. Holodiscus

1. **H. discolor** (Pursh) Maxim. Ocean-spray. Widespread on sheltered slopes and flats or occasionally in exposed rocky places: Sausalito, Tiburon, and Mount

Tamalpais (Matt Davis Trail, Cataract Gulch, Phoenix Lake) to San Rafael Hills, Tomales, and Point Reyes Peninsula (Mud Lake, Inverness).

A central California form with leaves less than 4.5 cm. long has been named as var. *franciscanus* (Rydb.) Jeps. but in Marin County it intergrades completely with the typical form which usually has larger leaves. On exposed wind-swept hills near Tomales shrubs of ocean-spray are dwarfed to brushy mats.

3. Photinia

1. **P. arbutifolia** (Ait.) Lindl. TOYON; CHRISTMAS BERRY. Common on brushy or openly wooded hills: Sausalito, Angel Island, Tiburon, and Mount Tamalpais (Blithedale Canyon, Azalea Flat, Phoenix Lake, Cataract Gulch) to San Rafael, Black Point, and Point Reyes Peninsula (Bolinas and Point Reyes road).—*Heteromeles arbutifolia* (Ait.) Roem.

4. Amelanchier. SERVICE BERRY

1. **A. pallida** Greene. Occasional on the edge of woods or rock outcrops of open hills: Sausalito; Tiburon; Mount Tamalpais (Rock Spring, Rifle Camp, Lake Lagunitas); San Anselmo Canyon; San Geronimo; San Rafael Hills; Salmon Creek School; Drakes Estero, Point Reyes Peninsula; Tomales.—*A. alnifolia* Nutt. var. *subintegra* (Greene) Jeps.

5. Crataegus

1. **C. Douglasii** Lindl. DOUGLAS HAWTHORN. Along streams or on wooded and brushy slopes that are wet in the spring, in Marin County known only from near Nicasio, *Malcolm Smith*. This is the southernmost known station.

The flowers would perhaps be more attractive to man if they did not emit a disagreeable fetid odor. This, however, is a strong lure to bees and flies, and these are much more important in the plant's life than man, anyway.

6. Rubus

a. Stems trailing, frequently forming shrubby mounds or low thickets; drupelets attached to receptacle in fruit .... b
a. Stems erect; drupelets separating from receptacle in fruit .... d
b. Leaflets 5 on vigorous shoots, palmate; flowers perfect c
b. Leaflets 3 on vigorous shoots, or very rarely 5 and then pinnate; flowers unisexual, the plants dioecious (except in cultivated forms) ......3. *R. ursinus*
c. Stems, petioles, and branches of the inflorescence bearing stout or slender prickles .... 1. *R. procerus*
c. Stems, petioles, and branches of the inflorescence without prickles
.... 2. *R. ulmifolius*
d. Leaves 3-foliolate; petals deep rose .... 4. *R. spectabilis*
d. Leaves simple; petals white .... 5. *R. parviflorus*

1. **R. procerus** P. J. Muell. HIMALAYA-BERRY. Common and widespread weed along roads and in waste places, frequently forming impenetrable thickets: Sausalito, Tiburon, and Mill Valley to San Rafael, Aurora School, and Point Reyes Peninsula (Bolinas, Mud Lake, Inverness). Objectionable as this European plant

is vegetatively, its blackberries are delicious and are sought after for jams and jellies.

2. **R. ulmifolius** Schott var. **inermis** (Willd.) Focke. Locally abundant on the edge of salt marshes below Mill Valley. Native of Europe.

3. **R. ursinus** C. & S. CALIFORNIA BLACKBERRY. Widespread and common in open or partly shaded places, especially on brushy or wooded hills: Sausalito, Angel Island, Tiburon, and Mount Tamalpais (Cascade Canyon, Matt Davis Trail) to San Rafael Hills, Tomales, and Point Reyes Peninsula (Bolinas, Inverness, Ledum Swamp).

In foliage and vestiture the California blackberry is one of the most variable plants in Marin County and several variants have been recognized as species. In typical *R. ursinus* the leaves are rather thick and more pubescent, the lower side being densely hairy. This is the most common form and is generally found in open sunny places. In shady places in canyons, a form in which the leaflets or lobes are about twice as long as broad is occasional. This is var. *sirbenus* (Bailey) J. T. Howell and has been collected in San Anselmo Canyon. Two other variants in which the leaves are only thinly pubescent or glabrous are found in moist, shaded places. The more usual form, in which the drupelets are thinly pubescent, is var. *glabratus* Presl and this corresponds to typical *R. vitifolius* C. & S. (the name that has been generally applied to the California blackberries). Much rarer and seemingly restricted in Marin County to azalea thickets on Mount Tamalpais, the type locality, is var. *Eastwoodianus* (Rydb.) J. T. Howell, a distinctive plant in which the fruit is glabrous and the stems and leaves are nearly glabrous.

Occasionally cultivated forms of the western blackberries escape and produce rampant brambles, as in Black Canyon in the San Rafael Hills. In this particular plant, which has been described as *R. titanus* Bailey, the flowers were perfect and the berries extra large. In the wild dioecious plants, the pistillate flowers are only about half the size of the staminate, and the berries, although not very large, are juicy and have a fine flavor.

4. **R. spectabilis** Pursh var. **franciscanus** (Rydb.) J. T. Howell. SALMON-BERRY. Occasional in well-watered canyons or in brushy thickets in wet soil near the coast: Sausalito; Steep Ravine, Mount Tamalpais; Tomales; Inverness and Ledum Swamp, Point Reyes Peninsula. North of Marin County the variety grades into the typical form in which the leaves are thinner and less hairy.—*R. spectabilis* Pursh var. *Menziesii* of authors, not *R. Menziesii* Hook.

5. **R. parviflorus** Nutt. var. **velutinus** (H. & A.) Greene. THIMBLE-BERRY. Common in thickets on moist or shaded hillsides near the coast: Sausalito, Angel Island, and Mount Tamalpais (Blithedale and Cascade canyons) to Tomales and Point Reyes Peninsula (Mud Lake, Inverness, Ledum Swamp).

Fassett has described two forms, f. *parahypomalacus* and f. *parvilosus*, from near Abbotts Lagoon, Point Reyes Peninsula.

7. Potentilla

a. Petals yellow; stamens attached to a disklike expansion of the receptacle....b
a. Petals white, sometimes tinged with pink or cream; stamens attached at top of a concave or cuplike hypanthium ....d

b. Stems widely creeping and rooting; leaflets silvery beneath ......1. *P. Egedii*
b. Stems erect or spreading; leaflets not silvery beneath c
c. Leaflets 3, not glandular-pubescent ....2. *P. millegrana*
c. Leaflets 5–9, more or less glandular-pubescent ....3. *P. glandulosa*
d. Hypanthium deeply cupshaped; bractlets 3–5-cleft or -toothed, rarely entire
....4. *P. californica*
d. Hypanthium shallowly cupshaped or saucershaped; bractlets entire ....e
e. Leaves viscidulous-glandular, scarcely, if at all, villous, leaflets in 10–20 pairs.
Pistils less than 30 ....7. *P. Micheneri*
e. Leaves densely silky or villous, leaflets in 5–12 pairs ....f
f. Leaves silky; pistils more than 30 ....5. *P. Lindleyi*
f. Leaves shaggy-villous; pistils generally less than 30 ....6. *P. marinensis*

1. **P. Egedii** Wormskj. var. **grandis** (T. & G.) J. T. Howell. Frequent in wet soil bordering salt marshes or coastal ponds: Tiburon; Escalle; Rodeo Lagoon; Muir Beach; Stinson Beach; Olema Marshes; Inverness; Dillons Beach.—*P. Anserina* L. var. *grandis* T. & G.; *P. pacifica* Howell.

Both the bicolored leaves and the bright yellow flowers of this marsh plant are attractive.

2. **P. millegrana** Engelm. Mount Tamalpais, acc. Stacey. A plant of low moist flats, it has a wide but sporadic distribution in California.—*P. leucocarpa* Rydb.

3. **P. glandulosa** Lindl. Common and widespread in grassland, brush, or woodland: Sausalito, Angel Island, Tiburon, and Mount Tamalpais (Corte Madera) to San Rafael, Tomales, and Point Reyes Peninsula (Bolinas, Mud Lake).

A robust variety, var. *Wrangelliana* Wolf (*P. Wrangelliana* Fisch. & Avé-Lall.), which is found along the coast of northern California, is listed from Mill Valley by Stacey but no plants referable to the variety have been seen from there.

4. **P. californica** (C. & S.) Greene. Common on coastal mesas and hills in open grassland or more commonly in brush: Rodeo Lagoon; Elk Valley; Muir Beach; Stinson Beach, acc. Stacey; Olema and Lagunitas, acc. Keck; Dillons Beach; Estero School; Point Reyes Peninsula (Bolinas, Point Reyes, acc. Jepson).—*Horkelia californica* C. & S.

5. **P. Lindleyi** Greene var. **sericea** (Gray) J. T. Howell. In Marin County known only from a collection made by Burtt Davy on Point Reyes Peninsula.—*P. Kelloggii* Greene; *Horkelia cuneata* Lindl. subsp. *sericea* (Gray) Keck.

6. **P. marinensis** (Elmer) J. T. Howell. Sandy flats near the coast, Point Reyes Peninsula, the type locality. This is a rare plant, known only from a few stations in Marin and Mendocino counties.—*P. Kelloggii* Greene var. *marinensis* (Elmer) Jeps.; *Horkelia marinensis* (Elmer) Crum.

7. **P. Micheneri** Greene. Occasional in clay soil in meadowy opens near springs and seepages on Mount Tamalpais, the type locality: old railroad grade above Cascade Canyon; Matt Davis Trail at Azalea Flat; Kent Trail at Willow Meadow. —*P. tenuiloba* (Torr.) Greene; *P. stenoloba* Greene; *Horkelia tenuiloba* (Torr.) Gray.

8. Fragaria. STRAWBERRY

a. Petals yellow ....3. *F. indica*
a. Petals white ....b

b. Leaflets thick, leathery, shiny on the upper side; plants of coastal flats and
dunes ....1. *F. chiloensis*

b. Leaflets thinner, not shiny above; plants of hills and meadows
....2. *F. californica*

1. **F. chiloensis** (L.) Duchesne. BEACH STRAWBERRY. Common in sandy places
along the ocean: Rodeo Lagoon; Stinson Beach, acc. Stacey; Tomales; Point
Reyes Lighthouse and dunes.

2. **F. californica** C. & S. Along sheltered borders of brush or woodland, wide-
spread: Sausalito, Tiburon, and Mount Tamalpais (Blithedale Canyon) to San
Rafael Hills, Tomales, and Point Reyes Peninsula (Bolinas, Inverness Ridge,
Shell Beach).

On exposed hilltops and ridges, the leaflets of the California strawberry become
thicker and more strongly veined, an ecologic form that has been described as
var. *franciscana* Rydb. It was described from plants collected in Marin County by
Alice Eastwood in 1896.

3. **F. indica** Andr. MOCK STRAWBERRY. An occasional escape from cultivation in
the coastal hills of California, the Indian strawberry is spontaneous and locally
common in the lower part of Steep Ravine. It is a native of India.—*Duchesnea indica* (Andr.) Focke.

9. Cercocarpus

1. **C. betuloides** Nutt. MOUNTAIN MAHOGANY. Occasional on brushy hillsides or
on dry rocky slopes in chaparral: Tiburon; Corte Madera Ridge and Phoenix
Lake, Mount Tamalpais; San Anselmo Canyon; south end of Carson Ridge; San
Rafael Hills.

10. Adenostoma

1. **A. fasciculatum** H. & A. CHAMISE. Widespread and frequently abundant on
dry rocky ridges, one of the commonest constituents of the chaparral, also occasional under pines on Inverness Ridge: Angel Island; Mount Tamalpais (very
common); Carson country; San Rafael Hills; Big Rock Ridge; Chileno Valley;
Salmon Creek; Inverness Ridge, Point Reyes Peninsula.

Chamise crown-sprouts after fire and at the same time seeds itself abundantly.
The leaves of these seedlings and often of the vigorous crown sprouts are very
different from the short needle-like or heather-like leaves that are so characteristic
of mature plants. These juvenile leaves are often 2 cm. long, 1.5 cm. wide, and
twice or thrice pinnatifid. The conspicuous stipules resemble leaf segments and,
like the segments, are somtimes lobed. At the bottom of the winged petiole-like
rachis and just above the stipules, the blade is jointed, a fact which indicates
that the blade is a part of a compound leaf. Higher on the stem the blades
gradually become narrower and less dissected until the mature leaf form is
reached, the xerophilous needle-like structure which actually represents the modified rachis of a leaflet.

11. Sanguisorba

1. **S. minor** (L.) Scop. Adventive along roads and in open woods: hills at Black
Point; Inverness Ridge, Point Reyes Peninsula. The California representative of

this Old World species seems to be the plant called subsp. *muricata* (Spach) Gams by European botanists.

12. Acaena

1. **A. californica** Bitt. Rather common on rocky open slopes near the coast, occasional on hills near the bay: Sausalito, Angel Island, Tiburon, and Mount Tamalpais (Rock Spring Trail) to San Rafael Hills, Tomales, and Point Reyes Peninsula (road to radio station).—*A. pinnatifida* R. & P. var. *californica* (Bitt.) Jeps.

13. Alchemilla

1. **A. occidentalis** Nutt. Widespread and generally common in shallow or rocky soil of hills and meadows: Sausalito; Tiburon; Rock Spring and Phoenix Lake, Mount Tamalpais; San Anselmo; Carson country; San Rafael; Tomales; Drakes Estero, Point Reyes Peninsula.—*A. arvensis* of California references.

14. Rosa. ROSE

a. Leaflets densely pubescent or glandular beneath. Usually stout shrubs more than 1 m. tall ....b
a. Leaflets glabrous beneath except for occasional glandular hairs or prickles along the veins c
b. Leaflets glandular beneath; sepals prominently lobed, densely glandular ....1. *R. rubiginosa*
b. Leaflets pubescent beneath, sometimes glandular-hairy; sepals entire, pubescent, rarely a little glandular ....2. *R. californica*
c. Hypanthium glandular, the upper part and the sepals persistent; stems low, 0.5 m. or less tall ....3. *R. spithamea*
c. Hypanthium glabrous, the upper part and the sepals deciduous; stems slender, 1 m. or more tall ....4. *R. gymnocarpa*

1. **R. rubiginosa** L. SWEETBRIER; EGLANTINE. A noxious weed in low coastal pastures near Dillons Beach, occasional shrubs occurring as far as Aurora School east of Tomales. The attractive flowers and fragrant foliage are scant recompense for the pastures invaded and captured by this rapacious European rose.

2. **R. californica** C. & S. Frequent and widely distributed along water courses in the hills and in low places along roads and in meadows: Rodeo Lagoon, Tiburon, Angel Island, and Mount Tamalpais (Potrero Meadow) to Bolinas, San Rafael Hills, Chileno Valley, and Point Reyes Station.

Forms with more numerous glands found at Stinson Beach and Lagunitas were listed by Stacey as *R. Aldersonii* Greene.

3. **R. spithamea** Wats. var. **sonomensis** (Greene) Jeps. Not uncommon on high rocky ridges and slopes in the chaparral: Mount Tamalpais (Matt Davis Trail, Rocky Ridge Fire Trail, Lake Lagunitas); Carson country; San Rafael Hills.—*R. sonomensis* Greene.

When, between fires, the chaparral shrubs grow tall and thick, the Sonoma rose is not conspicuous and rarely blooms; but in the ashes of a burn the little rose is attractively floriferous and its fragrance is something to be remembered.

4. **R. gymnocarpa** Nutt. WOOD ROSE. Occasional in the southern part of Marin

County in shaded canyons or on wooded slopes: Sausalito; Angel Island; Tiburon; Mount Tamalpais (Cascade Canyon, Cataract Gulch, Azalea Flat, Phoenix Lake); San Rafael Hills; Lagunitas, acc. Stacey.

15. Osmaronia

1. **O. cerasiformis** (T. & G.) Greene. Oso BERRY. Widespread but not frequent on moist brushy slopes and flats: Sausalito, Tiburon, and Mount Tamalpais (Steep Ravine, Camp Handy) to Bolinas, San Rafael, Tomales, and Olema Marshes. Shrubs in Steep Ravine have leaves as much as 14 cm. long and 6 cm. broad, though usually the leaves are only half as large.

16. Prunus

a. Flowers numerous in elongate racemes. Teeth on leaf margin sharp
....3. *P. demissa*
a. Flowers few in small rounded leafless clusters ....b
b. Flower cluster corymbose, pedunculate; teeth on leaf margin blunt
....1. *P. emarginata*
b. Flower cluster umbellate, sessile; teeth on leaf margin sharp ..2. *P. subcordata*

1. **P. emarginata** (Dougl.) Walp. BITTER CHERRY. Rocky northern slope near the summit of Mount Tamalpais on the Artura Trail; rocky bluffs south of Tomales. These are the only stations known in Marin County. The beautiful silver and gray mottled bark of the slender stems is one of the most attractive features of these shrubs.

2. **P. subcordata** Benth. PACIFIC PLUM. Rare on shaded hillsides: Mount Tamalpais, acc. Stacey; Manor, *Mexia;* Big Rock Ridge.

3. **P. demissa** (Nutt.) Walp. WESTERN CHOKE-CHERRY. Rare on the edge of woods along Papermill Creek at Lagunitas and Camp Taylor.—*P. virginiana* L. var. *demissa* (Nutt.) Torr.

LEGUMINOSAE. PEA FAMILY

a. Corolla regular, inconspicuous; stamens numerous, conspicuous ....b
a. Corolla irregular, papilionaceous, generally conspicuous; stamens 10, generally concealed in the keel formed by the two lower petals. (In *Amorpha* only the banner is present.) c
b. Stamens distinct; flowers in globose heads less than 1 cm. in diameter
....1. *Acacia*
b. Stamens united near the base; flowers in oblong spikes about 2.5 cm. across
....2. *Albizzia*
c. Leaflets more than 3 and palmate ....5. *Lupinus*
c. Leaflets none or 1 to many, pinnate if more than 3 ....d
d. Shrubs and trees ....e
d. Annual and perennial herbs ....k
e. Leaflets numerous (more than 5) ....f
e. Leaflets none or 1–5 ....g
f. Leaves punctate-glandular, odorous; corolla reduced to a purple banner; stipules not spiny ....14. *Amorpha*

f. Leaves not punctate-glandular or odorous; corolla white; stipules frequently
   spiny ....15. *Robinia*
g. Stems spinescent ....h
g. Stems not spinescent i
h. Leaves with 1–3 leaflets; corolla bright purplish rose ....4. *Pickeringia*
h. Leaves of mature plants reduced to a spiny petiole; corolla yellow ....7. *Ulex*
i. Flowers 1 cm. long or less, in sessile or shortly pedunculate umbels; legumes
   indehiscent ....12. *Lotus*
i. Flowers 1 cm. long or more, not in umbels; legumes dehiscent ....j
j. Branchlets strongly angled or pubescent; calyx 2-lipped ....6. *Cytisus*
j. Branchlets terete, subglabrous; calyx 1-lipped, split deeply above ..8. *Spartium*
k. Flowers solitary l
k. Flowers 2 to many ....n
l. Stipules reduced to small or inconspicuous glands ....12. *Lotus*
l. Stipules not reduced to glands ....m
m. Style tipped by a tuft or ring of hairs ....17. *Vicia*
m. Style hairy only on the upper side ....18. *Lathyrus*
n. Flowers in sessile or pedunculate umbels. Leaflets entire ....12. *Lotus*
n. Flowers in racemes, spikes, or heads, or if umbellate then the leaflets denticulate ....o
o. Leaflets 3 ....p
o. Leaflets more than 3, or in *Lathyrus* and *Pisum* sometimes only 2, the leaf then
   terminated by a tendril ....t
p. Herbage and calices glandular-punctate ....13. *Psoralea*
p. Herbage and calices not glandular-punctate ....q
q. Leaflets entire; stamens distinct. Stipules large and foliaceous; corolla bright
   yellow ....3. *Thermopsis*
q. Leaflets generally denticulate; stamens diadelphous ....r
r. Corolla withering persistent ....11. *Trifolium*
r. Corolla deciduous ....s
s. Inflorescences contracted or dense; style subulate ....9. *Medicago*
s. Inflorescences elongate, racemose; style filiform ....10. *Melilotus*
t. Leaves odd-pinnate ....16. *Astragalus*
t. Leaves pinnate, usually terminated by a tendril or the tendril rudimentary .u
u. Style tipped by a tuft or ring of hairs ....17. *Vicia*
u. Style hairy only on the upper side v
v. Stipules generally smaller than the leaflets (or large in *Lathyrus littoralis*);
   style not conspicuously flattened and folded ....18. *Lathyrus*
v. Stipules about as large as the leaflets; style conspicuously flattened and folded
   ....19. *Pisum*

1. Acacia

a. Leaves bipinnate, never reduced to phyllodes ....1. *A. decurrens*
a. Leaves generally reduced to phyllodes (or developing bipinnate blades on
   young plants or vigorous stump-sprouts in *A. melanoxylon*) ....b
b. Phyllodes 1-nerved ....2. *A. retinodes*
b. Phyllodes 1–4-nerved ....3. *A. melanoxylon*

1. **A. decurrens** Willd. GREEN WATTLE. Locally established and sometimes abundant in the vicinity of cultivated plants: Sausalito; Belvedere; Mill Valley; San Rafael. A native of Australia.

2. **A. retinodes** Schlecht. Naturalized on Belvedere and along Summit Avenue above Mill Valley, an escape from cultivation. Introduced from Australia.

3. **A. melanoxylon** R. Br. BLACK ACACIA. Locally and vigorously spontaneous in the vicinity of cultivated plants: Sausalito; Belvedere; Phoenix Lake; San Rafael Hills; Santa Venetia. A native of Australia.

2. Albizzia

1. **A. lophantha** (Willd.) Benth. Locally naturalized in Sausalito, Belvedere, Tiburon, and perhaps elsewhere near garden plants. A native of Australia.

3. Thermopsis

1. **T. macrophylla** H. & A. FALSE LUPINE. Open grass-covered hills or on grassy flats in the chaparral, sometimes locally abundant: Tiburon; Mill Valley; Corte Madera, acc. Stacey; Mount Tamalpais (Azalea Flat, Bootjack, Kent Trail, Lake Lagunitas); Carson country; Hicks Valley.

Thermopsis californica Wats. was based primarily on Bigelow's collection from Corte Madera.

4. Pickeringia

1. **P. montana** Nutt. CHAPARRAL PEA. Widespread and frequently common in the chaparral: Mount Tamalpais (Corte Madera Ridge, head of Cascade Canyon, Azalea Flat); Carson Country.

The chaparral pea in Marin County rarely sets fruit but the plants spread locally by underground stems from which vigorous rejuvenation takes place after chaparral fires.

5. Lupinus. LUPINE

a. Plants perennial ....b
a. Plants annual ....k
b. Plants herbaceous, aerial stems annual, not woody above the ground, arising
   from a simple or branched caudex at or below the ground c
b. Plants suffrutescent at base or shrubby and either subprostrate or erect, the
   stems woody above the ground ....g
c. Keel glabrous ....d
c. Keel hairy ....e
d. Stems erect, stout, very sparsely hairy; leaflets large, 10–20 cm. long, glabrous
   above ....1. *L. grandifolius*
d. Stems decumbent to erect, slender, sericeous or villous; leaflets smaller, generally less than 5 cm. long, sericeous above and below ......2. *L. formosus*
e. Keel hairy from the claws to the middle; plants of brushy or wooded slopes
   ....3. *L. latifolius*
e. Keel hairy from claws nearly to the tip; plants of coastal dunes and flats ...f
f. Corolla little longer than the calyx; upper calyx lip divided to the middle
   ....4. *L. Layneae*

f. Corolla showy, much exceeding the calyx; upper calyx lip entire or bidentate
....5. *L. variicolor*

g. Banner glabrous; upper calyx lip bidentate or entire ....h

g. Banner more or less hairy on the back near the top of the midrib; upper
calyx lip divided about to the middle i

h. Corolla usually yellow, sometimes lavender-tinged; seeds uniformly colored
brownish to blackish, broadly quadrate or oval ....6. *L. arboreus*

h. Corolla blue-violet to rose-purple; seeds mottled grayish and brownish, elliptic,
decidedly longer than broad ....7. *L. rivularis*

i. Petioles about as long as the leaflets; keel glabrous or with a few scattered hairs
....8. *L. Chamissonis*

i. Petioles of lower leaves much longer than the leaflets; keel hairy or sometimes
glabrous ....j

j. Floral bracts about 5 mm. long, shorter than the buds ....9. *L. albifrons*

j. Floral bracts more than 5 mm. long, conspicuously exceeding the buds at the
top of the raceme ....10. *L. Douglasii*

k. Pods oblong or linear-oblong, seeds more than 2 l

k. Pods broadly elliptic or roundish, seeds usually 2 ....o

l. Pedicels about 5 mm. long; flowers showy, generally 1–1.5 cm. long ....m

l. Pedicels 4 mm. or less long; flowers generally 1 cm. or less long ....n

m. Stems stout, fistulous; keel hairy near the claws above and below
....11. *L. succulentus*

m. Stems slender; keel hairy above or glabrous ....12. *L. nanus*

n. Keel elongate, slender-tipped; style long and slender ....13. *L. bicolor*

n. Keel short and wide, broad-tipped; style short and stout ....14. *L. micranthus*

o. Pedicels spreading to one side in fruit, the inflorescence becoming secund
....15. *L. densiflorus*

o. Pedicels erect in fruit, the inflorescence not secund ....16. *L. subvexus*

1. **L. grandifolius** Lindl. Wet soil of marshes and coastal swales: Mount Tamalpais, acc. Stacey; Olema Marshes; near the radio station, Point Reyes Peninsula.
This handsome plant (sometimes 5 feet tall) is very closely related to *L. polyphyllus* Lindl. and perhaps it should be regarded as merely a robust coastal
variety, var. *grandifolius* (Lindl.) C. P. Sm.

2. **L. formosus** Greene. Common and widespread in clay soil of valleys and on
open or brushy hills: Tiburon, Tamalpais Valley, and Mount Tamalpais (Lone
Tree) to San Rafael Hills, Black Point, Tomales, and Point Reyes Peninsula
(Inverness Ridge).

This is one of the most variable species, the plants differing in habit, vestiture.
foliage, and flowers. Specimens in which the flowers are large and the pubescence is spreading are probably referable to *L. Bridgesii* (Wats.) Heller, but there
are also plants with spreading hairs and smaller flowers. Generally the pubescence
is appressed but even then it varies in quantity, sometimes being sparse, sometimes dense. *Lupinus marinensis* Eastw., described from San Anselmo Canyon
above Fairfax, is a plant with dense satiny pubescence and narrow leaflets.
An undescribed form from Estero Americano is very pallid with velvety pubescence, while another form from Mud Lake is sparsely hairy and unusually green.
A distinctive plant from near Olema with appressed pubescence and with all the

petals conspicuously hairy has been named *L. punto-reyesensis* by C. P. Smith.

3. **L. latifolius** Agardh. Common in brushy or wooded places in valleys or hills: Sausalito, Angel Island, Tiburon, and Mount Tamalpais (Corte Madera Ridge, Steep Ravine, Alpine Dam) to San Rafael Hills and Tomales.—*L. rivularis* of Jepson's Flora of California.

4. **L. Layneae** Eastw. Known only from the dunes on Point Reyes Peninsula, the type locality. It is named in honor of Mary Katharine Layne Curran Brandegee who, as Mrs. Curran, collected the type specimen in 1886. This rare species is most closely related to *L. Tidestromii* Greene, an equally localized endemic restricted to the Monterey Peninsula.

5. **L. variicolor** Steud. As with *L. formosus,* this is part of a variable complex, the several aspects of which may generally be recognized by the character of the pubescence and size of flowers. Plants typical of the species with appressed pubescence and large flowers (1.3–1.8 cm. long) have been seen in Marin County only from the northern part near Aurora School and Tomales, *Leschke.* A similarly large-flowered plant but with pubescence villous-spreading is *L. eximius* Davy from Dillons Beach and Point Reyes Peninsula (Mount Vision, McClure Beach road, and Point Reyes). *Lupinus Micheneri* Greene with flowers about 1 cm. long is common on the sandy flats and dunes of Point Reyes Peninsula. All are related to *L. littoralis* Dougl., a coastal species found from northern California to British Columbia.

6. **L. arboreus** Sims. Common on coastal flats and dunes, occasionally in loose soil of road fills: Sausalito; Angel Island; Tiburon; Stinson Beach; Dillons Beach; Point Reyes Peninsula (common on the dunes and sandy flats back of the dunes).

In seasons of proper rainfall, the display of lupines along the road from Inverness to Point Reyes Lighthouse is spectacular and the plant that contributes most to the show is *L. arboreus.* It is usually yellow but frequently the flowers are vari-colored or particolored with lavender and violet. These color variations, as well as differences in the character of the pubescence, probably result from hybridization between *L. arboreus* and *L. variicolor* which is also a showy member in the lupine display.

7. **L. rivularis** Dougl. A widespread and sometimes common shrub of open or brushy hills, occasional in wooded canyons: Sausalito, Angel Island, and Mount Tamalpais (Corte Madera Ridge, Fern Canyon) to Bolinas, Inverness, and Point Reyes road.

In masses this shrubby lupine produces effects that are as colorful, if not quite so pleasing, as those of *L. arboreus* or *L. nanus.* As the plant occurs in Marin County it seems adequately distinct from *L. arboreus* to which it is related and with which it is usually combined in the floras. The application of the name *L. rivularis* is various: here Miss Eastwood, who applies it to the violet-flowered shrub of the California coast, is followed; C. P. Smith uses it for a more herbaceous but related species of the Pacific Northwest; and Jepson in the Flora of California associates it with the plant treated herein as *L. latifolius.*

8. **L. Chamissonis** Esch. In Marin County, restricted to Point Reyes Peninsula: shallow soil of rocky slope near Mud Lake; dunes near the radio station. The broad spreading shrubs on the Point Reyes dunes with their silvery foliage and fragrant blue flowers are among the choicest of all Marin County plants.

9. **L. albifrons** Benth. var. **collinus** Greene. Occasional in shallow rocky soil or in crevices of rocks: Sausalito; Angel Island; Mill Valley; Bootjack and Bolinas Ridge, Mount Tamalpais; Bald Mountain; Carson country; Salmon Creek.

10. **L. Douglasii** Agardh var. **fallax** (Greene) J. T. Howell. TAMALPAIS LUPINE. Common on Mount Tamalpais on rocky slopes in the chaparral or on the edge of woods: Corte Madera Ridge; Throckmorton Ridge; West Point; Bolinas Ridge; Alpine Dam.—*L. albifrons* or *L. albifrons* var. *Douglasii* as to Tamalpais plants.

The Tamalpais lupine, which was originally described by Greene as *L. fallax,* is closely related to *L. Douglasii* Agardh of the Monterey coast but differs in the wider leaflets and shorter floral bracts. It seems to be restricted to Mount Tamalpais where, with its abundant silvery foliage and fine racemes of blue flowers, it is one of the most beautiful flowering shrubs.

11. **L. succulentus** Dougl. Clay soil of low flats and steep slopes, often locally abundant: Angel Island; Tiburon; Stinson Beach, acc. Stacey; Baltimore Park; White Hill; Black Point.

12. **L. nanus** Dougl. SKY LUPINE. Widespread and frequently common on valley flats, grassy open hills, and brushy rocky slopes: Sausalito, Angel Island, Tiburon, and Mount Tamalpais (Corte Madera Ridge, Bootjack, Bolinas Ridge, Lagunitas Meadows) to San Rafael Hills, Big Rock Ridge, Olema, and Inverness Ridge. Robust plants with leaflets 3–5 cm. long are referable to var. *carnosulus* (Greene) C. P. Sm., the type locality of which is Olema.

It is this lupine which in the spring obscures the green of grassy hills and valleys with mantles and sheets of blue. And yet again it mingles with buttercups, poppies, popcorn-flowers, or tidy-tips to form flower gardens of surpassing beauty. The sky lupine is a favorite with everyone.

13. **L. bicolor** Lindl. Grassy hills and gravelly flats: Sausalito, acc. Lovegrove; Mill Valley, acc. Stacey; Fairfax Hills; San Rafael Hills.

An occasional form in which the flower-whorls are usually only 1 or 2 is var. *umbellatus* (Greene) C. P. Sm.: Angel Island; Rock Spring, Mount Tamalpais; Fairfax Hills; Dillons Beach; Point Reyes and Inverness Ridge, Point Reyes Peninsula.

14. **L. micranthus** Dougl. Widespread and rather frequent in valleys and hills: Angel Island, Mill Valley, and Mount Tamalpais (Fish Grade) to San Rafael, Black Point, and Point Reyes Peninsula (Point Reyes, *Eastwood*). This lupine may be confused with small-flowered forms of *L. bicolor* but the two are quite distinct and readily distinguishable by the difference in the keels.

15. **L. densiflorus** Benth. This is perhaps the most variable of all lupines in Marin County and although the plants may be low or tall, the vesture scant or dense, the flower-parts narrow or broad, and the color white, violet, or yellow, all the variants have in common the remarkable character of the horizontally spreading fruiting racemes along which the pods are distinctly secund. The typical form of the species, which may be recognized by the white, lavender, or pale yellow flowers, has been noted at Tiburon, Angel Island, Muir Beach, Fairfax and San Rafael hills, Tocaloma, and Point Reyes Station. A less common form with deeper yellow flowers that age brick-red is var. *Menziesii* (Agardh) C. P. Sm. which has been seen at Tiburon and Novato (*Leschke*), and in the Greenbrae and Fairfax hills. A rare coastal form with long-hairy stems and violet flowers is var. *crinitus*

Eastw. from the hills east of Dillons Beach. Jepson treats this variable complex as a part of the South American species, *L. microcarpus* Sims.

16. **L. subvexus** C. P. Sm. In Marin County known only from coastal hills between Muir and Stinson beaches.

6. Cytisus. BROOM

a. Stems sharply angled, not decidedly villous, generally leafless or nearly so; flowers solitary or paired ....1. *C. scoparius*
a. Stems obtusely ridged or angled, quite villous, leafy; flowers in small clusters terminating axillary branchlets ....b
b. Banner about 1 cm. long, glabrous on the back ......2. *C. monspessulanus*
b. Banner about 1.5 cm. long, hairy on the back ....3. *C. proliferus*

1. **C. scoparius** (L.) Link. SCOTCH BROOM. Common along roads and paths about towns, extensively naturalized in the grassland of open hills and invading the lower border of the chaparral: Sausalito; Mill Valley; Mount Tamalpais (Corte Madera Ridge, Troop 80 Trail); Inverness. An escape from cultivation, introduced from Europe.

2. **C. monspessulanus** L. FRENCH BROOM. A pernicious shrub weed exhibiting an aggressive vigor that the native vegetation cannot withstand, widely introduced and spreading: Sausalito; Angel Island; Tiburon; Mill Valley; Phoenix Lake, Mount Tamalpais; Ross; San Rafael; Lagunitas Creek; Chileno Valley; Inverness; Tomales. Native of southern Europe.

3. **C. proliferus** L. Occurring locally as an escape from cultivation in Sausalito (acc. Eastwood) and in the San Rafael Hills. Native of the Canary Islands.

7. Ulex

1. **U. europaeus** L. GORSE; FURZE. Locally common on open hills and along the borders of woods and brush: Sausalito, acc. Stacey; above Cascade Canyon, Mount Tamalpais; San Rafael; Tomales. Introduced from Europe.

8. Spartium

1. **S. junceum** L. SPANISH BROOM. Occasionally naturalized about towns on open rocky slopes: Belvedere; Mill Valley; San Anselmo; San Rafael; Fairfax. Native of southern Europe.

9. Medicago. MEDICK

a. Corolla violet or purplish; stems erect ....1. *M. sativa*
a. Corolla yellow; stems procumbent ....b
b. Pods recurved, reniform, 1-seeded; stems villous-pubescent ..2. *M. lupulina*
b. Pods spirally coiled, several-seeded; stems glabrous or nearly so c
c. Pods without prickles ....4b. *M. hispida* var. *confinis*
c. Pods with prickles ....d
d. Leaflets with dark brown spot; prickles curved from the base ...3. *M. arabica*
d. Leaflets without dark spot; prickles straight or commonly hooked at the tip .e
e. Prickles elongate, about equaling the radius of the fruit ....4. *M. hispida*
e. Prickles short, less than half the radius of the fruit

....4a. *M. hispida* var. *apiculata*

1. **M. sativa** L. ALFALFA. Rather widespread along roads but nowhere becoming common: Sausalito, Tiburon, and Mill Valley to San Rafael, Novato, and Point Reyes Peninsula (Bolinas, Inverness). Introduced from the Old World and widely cultivated as a forage plant.

2. **M. lupulina** L. BLACK MEDICK. Occasional weed in waste places and about habitations: Mill Valley, acc. Stacey; Ross; Mud Lake and Inverness, Point Reyes Peninsula. Native of the Old World.

3. **M. arabica** (L.) All. SPOTTED MEDICK. Occasional along roads and on grassy hills: Angel Island and Tiburon to San Rafael, Tomales, and Point Reyes Peninsula (Inverness, Point Reyes). Introduced from Europe.

4. **M. hispida** Gaertn. BUR CLOVER. Common and widespread as a weed about towns and as a valued forage plant in grassland of valleys and hills: Sausalito, Angel Island, and Tiburon to San Rafael, Tomales, and Point Reyes Peninsula (Bolinas, Mud Lake, Inverness). Native of the Old World.—*M. denticulata* Willd.

4*a*. **M. hispida** Gaertn. var. **apiculata** (Willd.) Burnat. Along roads: White Hill; Drakes Estero, Point Reyes Peninsula. Probably more common but mistaken for typical *M. hispida.—M. apiculata* Willd.

4*b*. **M. hispida** Gaertn. var. **confinis** (Koch) Burnat. Widespread but less common than the typical *M. hispida:* Sausalito, Angel Island, and Tiburon to San Rafael and Black Point.

This Old World plant is generally accepted as a species in American floras; but it is so closely allied to the bur clover that it seems preferable to follow the European botanists who treat it as a variety.

10. **Melilotus.** SWEET CLOVER; MELILOT

a. Corolla white ....1. *M. albus*
a. Corolla yellow ....b
b. Flowers 4–5 mm. long; plants becoming perennial ....2. *M. officinalis*
b. Flowers 2–3 mm. long; plants annual ....3. *M. indicus*

1. **M. albus** Desr. Occasional along roadsides and in weedy places: Sausalito and Tiburon to San Geronimo and Inverness. Native of Eurasia.

2. **M. officinalis** (L.) Lamk. Rare in waste ground along roads; in Marin County known only from Chileno Valley. Introduced from the Old World.

3. **M. indicus** (L.) All. Rather common along roads and paths in weedy places and in the hills: Sausalito, Angel Island, Tiburon, and Mount Tamalpais (Phoenix Lake) to Inverness. Native of the Old World.

11. **Trifolium.** CLOVER

a. Corollas becoming inflated in fruit. Heads involucrate, the involucre reduced
    to a narrow ring in *T. depauperatum* ....b
a. Corollas not becoming inflated in fruit, the limb of the banner sometimes
    becoming enlarged and shell-like ....g
b. Calyx teeth glabrous c
b. Calyx teeth plumose-hairy ....f
c. Corolla 1–2.5 cm. long, yellowish or greenish-yellow, tinged with rose or
    purple in age; stems stout ....d
c. Corolla 0.5–1 cm. long, yellowish or purplish; stems slender ....e

d. Lower calyx teeth subulate, longer than the calyx tube ......1. *T. flavulum*

d. Lower calyx teeth triangular-lanceolate, equaling the calyx tube or shorter
....2. *T. fucatum*

e. Involucre divided into oblong lobes ....3. *T. amplectens*

e. Involucre reduced to a narrow, entire ring ....4. *T. depauperatum*

f. Heads 1–1.5 cm. broad; corolla about equaling the calyx ....5. *T. barbigerum*

f. Heads 2–3 cm. broad; corolla much longer than the calyx ....6. *T. Grayi*

g. Heads involucrate ....h

g. Heads not involucrate. The uppermost leaves subtending the heads in *T. pratense* and *T. Macraei* ....o

h. Involucre bowl- or cup-shaped i

h. Involucre flat or nearly so ....j

i. Involucre glabrous ....7. *T. microdon*

i. Involucre hairy ....8. *T. microcephalum*

j. Plants perennial ....9. *T. Wormskjoldii*

j. Plants annual ....k

k. Plants glandular-pubescent, the glandular secretion pungently acid to taste
....14. *T. obtusiflorum*

k. Plants glabrous or nearly so, not glandular l

l. Calyx teeth abruptly narrowed to the awnlike tip, frequently bearing 1 or 2 prominent teeth on either side of the awn ....13. *T. tridentatum*

l. Calyx teeth gradually narrowed into a short awnlike tip ....m

m. Keel tipped with an elongate, linear appendage ......10. *T. appendiculatum*

m. Keel without an apical appendage ....n

n. Corolla showy, usually much longer than the calyx; calyx tube about 20-nerved, usually not conspicuously scarious ....11. *T. variegatum*

n. Corolla not showy, little longer than the calyx; calyx tube 10-nerved, conspicuously scarious ....12. *T. oliganthum*

o. Flowers on short pedicels, reflexed in age ....p

o. Flowers sessile or nearly so, not reflexed in age v

p. Corolla bright yellow in flower ....q

p. Corolla white, pink, or purplish ....r

q. Heads 5–7 mm. wide, loosely flowered; banner in fruit somewhat enlarged but not conspicuously striate ....15. *T. dubium*

q. Heads 8–10 mm. wide, compactly flowered; banner in fruit enlarged and shell-like, conspicuously striate ....16. *T. campestre*

r. Plants perennial ....s

r. Plants annual ....t

s. Stems creeping and rooting at the nodes; nerves of calyx tube prominent
....17. *T. repens*

s. Stems erect or spreading, not rooting; nerves of calyx tube inconspicuous
....18. *T. hybridum*

t. Peduncle pilose-hairy, especially near the top ....19. *T. bifidum*

t. Peduncle glabrous ....u

u. Calyx teeth entire ....20. *T. gracilentum*

u. Calyx teeth conspicuously denticulate ....21. *T. ciliolatum*

v. Plants perennial. Heads large, 2–3 cm. broad, generally subtended by the uppermost leaves ....22. *T. pratense*

v. Plants annual ....w
w. Heads, or most of them, closely subtended by the uppermost leaves
....23. *T. Macraei*
w. Heads not subtended by the uppermost leaves x
x. Corollas conspicuously longer than the calyx ....y
x. Corollas from much shorter to a little longer than the calyx ....z
y. Stems stout, generally simple at base; head 2–3 cm. broad ..24. *T. amoenum*
y. Stems slender, generally several from the base; head 1–2 cm. broad
....25. *T. dichotomum*
z. Corolla about equaling the calyx ....26. *T. albopurpureum*
z. Corolla much shorter than the calyx ....27. *T. olivaceum*

1. **T. flavulum** Greene. Widespread, sometimes locally abundant, in clay soil of low wet fields and seepages in the hills: Angel Island; Tiburon; Alpine Lake, Mount Tamalpais; San Anselmo Canyon; Carson Ridge; Tocaloma; Ignacio; between Salmon Creek and Chileno Valley.—*T. fucatum* Lindl. var. *flavulum* (Greene) Jeps.

A form with greenish corolla was named from "the hilly parts of Marin and Sonoma counties" as *T. virescens* Greene.

2. **T. fucatum** Lindl. SOUR CLOVER. Wet clay soil of valleys and hills, rather rare in Marin County: Lagunitas Meadows, Mount Tamalpais; Baltimore Park; Carson country.

3. **T. amplectens** T. & G. Common in meadows and on moist grassy slopes: Angel Island; Rock Spring and Lagunitas Meadows, Mount Tamalpais; San Anselmo Canyon; San Rafael; Tomales; Point Reyes Lighthouse. The Marin County plants are generally referable to var. *stenophyllum* (Nutt.) Jeps. in which the involucral lobes are not scarious-margined. The corollas are usually purplish-red but occasionally albinos with ivory-white corollas may be found.

4. **T. depauperatum** Desv. Occasional in Marin County on moist flats in the valleys or hills: Lagunitas, acc. Stacey; Fairfax Hills; Ignacio; Dillons Beach; Drakes Estero, Point Reyes Peninsula. The Lagunitas plant was referred by Stacey to var. *laciniatum* (Greene) Jeps., a form in which the leaflets are laciniately toothed. That from Ignacio and Fairfax Hills with narrow, entire or toothed leaflets, is var. *angustatum* (Greene) Jeps., while the one from Dillons Beach and Drakes Estero with broad leaflets is typical of the species.

5. **T. barbigerum** Torr. Common and widespread in low moist places: Sausalito, Tiburon, and Mount Tamalpais (Bootjack, Steep Ravine, Lagunitas Meadows) to Tomales and Point Reyes Peninsula (Mount Vision, radio station, lighthouse).

6. **T. Grayi** Loja. Rare, in Marin County known only from Point Reyes Peninsula where it has been collected by Brandegee and others.—*T. barbigerum* Torr. var. *Andrewsii* Gray.

7. **T. microdon** H. & A. Common and widespread in grassland of hills and valleys: Sausalito, Angel Island, Tiburon, and Mount Tamalpais (Corte Madera Ridge, Phoenix Lake, Rock Spring) to San Rafael Hills, Black Point, and Point Reyes Peninsula (Mount Vision and near the lighthouse).

Three Marin County clovers that occur also in Chile are *T. microdon, T. depauperatum,* and *T. Macraei*. Among the peas, there are species in *Lotus* and *Lupinus* also with this extreme north-south distribution.

8. **T. microcephalum** Pursh. Widespread but not generally common: Sausalito,

Angel Island, and Mount Tamalpais (Corte Madera Ridge, Summit Meadow, Phoenix Lake) to San Rafael Hills, Ignacio, and Tomales.

9. **T. Wormskjoldii** Lehm. An attractive and showy plant of wet hollows or marshy meadows as well as sandy coastal mesas: Sausalito Hills, acc. Sutliffe; Stinson Beach, acc. Stacey; Potrero Meadow, Mount Tamalpais; Dillons Beach; east of Aurora School; Point Reyes Peninsula (Bolinas, Inverness, dunes near radio station, lighthouse). *T. involucratum* Ort.; *T. fimbriatum* Lindl.; *T. Willdenovii* Spreng.

On dunes along the coast var. *Kennedianum* (McDerm.) Jeps. is to be expected but no plants have been seen with the reduced involucre that characterizes that variety.

10. **T. appendiculatum** Loja. Occasional in low wet fields and meadows: Lagunitas Meadows; Hamilton Field; Ignacio; Drakes Estero, Point Reyes Peninsula.

Jepson reports *T. appendiculatum* var. *rostratum* (Greene) Jeps. from "Marin Co." but the variety seems to be only an edaphic form of little moment.

11. **T. variegatum** Nutt. Common and widespread in low meadows or on moist hillsides: Angel Island, Tiburon, and Mount Tamalpais (Laurel Dell, Lagunitas Meadows) to Novato (*L. S. Rose*), Tocaloma, and Point Reyes Peninsula (Mount Vision).

Almost as widespread as the species is a low, small-headed variety, var. *pauciflorum* (Nutt.) McDerm. It is found in wet meadows and is particularly common on the gravelly or sandy beds of streams: Sausalito; Barths Retreat and Lagunitas Meadows, Mount Tamalpais; Greenbrae Hills; San Rafael Hills; Tomales.

12. **T. oliganthum** Steud. Occasional in clayey or gravelly soil of slopes and ridges: Sausalito, acc. Stacey; Corte Madera Ridge, Mount Tamalpais; Fish Grade; Fairfax Hills; San Anselmo Canyon; San Rafael Hills.

13. **T. tridentatum** Lindl. Common and widespread in low fields and meadows, on grassy slopes, or on rocky ridges: Sausalito, Angel Island, Tiburon, and Mount Tamalpais (Corte Madera Ridge, Rock Spring, Phoenix Lake) to San Rafael Hills, Dillons Beach, and Point Reyes Peninsula (Mount Vision).

14. **T. obtusiflorum** Hook. Moist hillsides and gravelly streambeds, occasional: in burn on Rocky Ridge, Mount Tamalpais; Fairfax Hills; San Rafael Hills; Point Reyes Post Office, *Elmer*.

15. **T. dubium** Sibth. SHAMROCK. Occasional along roads and paths, the plants sometimes abundant and forming a loose turfy mat: Alpine Lake, Mount Tamalpais; Fairfax Hills; San Rafael; Stinson Beach; Bolinas; Inverness. It is now pretty well agreed by botanists and others interested that the original Celtic shamrock is this lowly European clover and not *Oxalis Acetosella* L. which is the shamrock of English literature.

16. **T. campestre** Schreb. HOP CLOVER. Occasional in California, in Marin County known only from Fairfax. The name to be assigned this European clover varies and according to some authorities should be *T. procumbens* L.

17. **T. repens** L. WHITE CLOVER. Escaping from cultivation and established in moist places along roads and in meadows: Muir Beach, Stinson Beach, and Mount Tamalpais (Blithedale Canyon) to San Rafael, Hamilton Field, and Point Reyes Peninsula (Bolinas, Mud Lake, road to lighthouse). Naturalized from Europe.

18. **T. hybridum** L. ALSIKE CLOVER. Rare in Marin County: Mill Valley, acc. Stacey; Bolinas. Introduced from Europe.

19. **T. bifidum** Gray. Widely distributed but not common on grassy hills and flats: Sausalito, Angel Island, Tiburon, and Mount Tamalpais (Corte Madera Ridge, Bootjack, Phoenix Lake, Berry Trail) to San Rafael Hills, Black Point, Tomales, and Point Reyes Peninsula (Inverness Ridge). Two leaf-forms are about equally well represented in Marin County: the typical form in which the leaflets are narrow and more deeply notched, and var. *decipiens* Greene in which the leaflets are somewhat broader and entire or merely emarginate at the apex.

20. **T. gracilentum** T. & G. Occasional in hills and valleys: Angel Island; Rock Spring, Mount Tamalpais; Baltimore Park; San Rafael Hills; Ignacio; Black Point; Carson country; Mount Vision and Point Reyes, Point Reyes Peninsula.

21. **T. ciliolatum** Benth. Rather rare on grassy hills and flats: Tiburon; Phoenix Lake, Mount Tamalpais; San Rafael; San Anselmo Canyon; Ignacio; Black Point. —*T. ciliatum* Nutt.

22. **T. pratense** L. RED CLOVER. Rare, in Marin County known only from Mill Valley (acc. Stacey) and Inverness Park. Usually the heads are sessile between the uppermost leaves but in the plant from Inverness Park the heads are borne on slender peduncles. In Europe, where the species is indigenous, some varieties are described as having stalked heads.

23. **T. Macraei** H. & A. Widely distributed in Marin County but not common, generally in open places in sandy or gravelly soil of well-drained slopes: Sausalito, Angel Island, and Mount Tamalpais (Corte Madera Ridge, near Rock Spring) to San Rafael Hills, Dillons Beach, and Point Reyes Peninsula (Inverness Ridge). Near the Point Reyes Lighthouse occurs a more robust maritime form which has been described as *T. mercedense* Kennedy.

24. **T. amoenum** Greene. A robust showy species of open flats or low hills in clay soil, in Marin County reported from Olema by Jepson. Known only from the counties immediately bordering San Francisco Bay.

25. **T. dichotomum** H. & A. Clay soil of steep slopes and rocky ridges, occasional: Rock Spring and Phoenix Lake, Mount Tamalpais; Carson country; Salmon Creek School. All the Marin County plants seem referable to var. *turbinatum* Jeps., the type locality of which is Ross Valley. A form with large heads 2.5 cm. broad and flowers 1 to 1.3 cm. long has been found near Point Reyes Station. These dimensions approach those of *T. amoenum* but the plants lack the simple robust base of that remarkable species.

26. **T. albopurpureum** T. & G. In loose or rocky clayey soil on grassy open slopes: Bootjack and Phoenix Lake, Mount Tamalpais; Fairfax Hills; Deer Park. In these plants the corolla is a little shorter than the calyx. Much more common in Marin County is a form in which the corolla is a little longer than the calyx, var. *neolagopus* (Loja.) McDerm., which has been found in the following places: Tiburon; Phoenix Lake; Carson country; San Rafael Hills; Black Point.

27. **T. olivaceum** Greene. This species was reported from "Marin County" by McDermott. A related plant with corollas nearly half as long as the calyx has been collected near Deer Park in the Fairfax Hills. It resembles specimens of *T. olivaceum* var. *griseum* Jeps. from the South Coast Ranges in which the corollas are a little longer and more evident than in typical *T. olivaceum*.

12. Lotus

a. Stipules membranaceous or foliaceous, attached at the node by a broad base .b
a. Stipules reduced to small or inconspicuous glands. If the leaves are sessile or

subsessile, the lowest leaflets may simulate foliaceous stipules but they are attached by a short, distinct petiolule ....d

b. Bract attached at top of peduncle; banner bright yellow, wings and keel rosy ....1. *L. formosissimus*

b. Bract attached below top of peduncle; banner not bright yellow c

c. Plant glabrous or nearly so; bract 3-foliolate, petiolate ......2. *L. crassifolius*

c. Plant villous-pubescent; bract 4- or 5-foliolate, subsessile ......3. *L. stipularis*

d. Plants annual. Pods straight, dehiscent, not slender-beaked ....e

d. Plants perennial i

e. Peduncles generally bracteate and more than 5 mm. long (or if occasionally less, the leaves with more than 1 leaflet on each side of the rachis and the seeds roughened) ....f

e. Peduncles not bracteate and less than 5 mm. long. Leaves with only 1 leaflet on side of the rachis; seeds smooth ....h

f. Flowers yellow, becoming reddish; seeds roughened, the ends truncate ....6. *L. strigosus*

f. Flowers whitish or pinkish when first opened, sometimes becoming reddish later; seeds smooth, the ends not truncate ....g

g. Bract 1-foliolate; teeth of calyx longer than the tube ....4. *L. Purshianus*

g. Bract 3-foliolate; teeth of calyx shorter than the tube ......5. *L. micranthus*

h. Teeth of calyx longer than the tube ....7. *L. humistratus*

h. Teeth of calyx equaling the tube or shorter ....8. *L. subpinnatus*

i. Pods dehiscent, not narrowed into a long slender beak; flowers generally 1 cm. or more long, if shorter, then the flowers borne on a much-elongate peduncle; stems erect, herbaceous ....j

i. Pods indehiscent, narrowed into a long slender beak; flowers 1 cm. or less long; umbels pedunculate or subsessile; stems erect or prostrate, more or less woody, especially towards the base l

j. Leaves subsessile or petiolate, leaflets 5–9, the lateral frequently alternate; herbage canescent with short subappressed hairs ....11. *L. grandiflorus*

j. Leaves sessile, leaflets 5, the lateral opposite or nearly so; herbage glabrous or sparsely long-pilose ....k

k. Leaflets broadly obovate, veins conspicuous; umbel mostly 8–12-flowered ....9. *L. uliginosus*

k. Leaflets lanceolate, veins inconspicuous; umbel mostly 3–6-flowered ....10. *L. corniculatus*

l. Stems and leaves loosely tomentulose ....12. *L. eriophorus*

l. Stems and leaves glabrous or thinly appressed-pubescent ....m

m. Calyx teeth broadly triangular ....13. *L. junceus*

m. Calyx teeth narrow, linear, reduced to the midrib ....14. *L. scoparius*

1. **L. formosissimus** Greene. Locally common in low wet meadows: Sausalito Hills, acc. Lovegrove; Potrero and Lagunitas meadows, Mount Tamalpais; Corte Madera, acc. Stacey; Hicks and Chileno valleys, abundant; Dillons Beach; Ledum Swamp and Drakes Bay, Point Reyes Peninsula.—*Hosackia gracilis* Benth.

2. **L. crassifolius** (Benth.) Greene. In Marin County known only near Corte Madera, the type locality of *Hosackia stolonifera* Lindl. var. *pubescens* Torr.—*H. crassifolia* Benth.

3. **L. stipularis** (Benth.) Greene. A rare plant of clay flats on the borders of chaparral: Azalea Flat and Bolinas Ridge, Mount Tamalpais; Lagunitas, acc.

Stacey. Beneath the villous hairs of the Marin County plants are glands that impart a yellow stain. This glandular variant of the species is *L. balsamiferus* (Kell.) Greene or *Hosackia stipularis* subsp. *balsamifera* (Kell.) Abrams.

4. **L. Purshianus** (Benth.) Clements & Clements. SPANISH CLOVER. Widespread on dry slopes and flats in clayey or sandy soil: Sausalito, Tiburon, and Mount Tamalpais (Rock Spring) to Fairfax Hills, San Rafael, and Point Reyes Peninsula (Bolinas, Point Reyes road).—*L. americanus* (Nutt.) Bisch.; *Hosackia americana* (Nutt.) Piper.

5. **L. micranthus** Benth. Common and widespread on hills in grassy and brushy places: Sausalito, Angel Island, Tiburon, and Mount Tamalpais (Cascade Canyon, Rock Spring) to San Rafael, Dillons Beach, and Point Reyes Peninsula (Mud Lake, Point Reyes road).—*Hosackia parviflora* Benth.

6. **L. strigosus** (Nutt.) Greene. Rare on warm brushy slopes: Angel Island; Cascade Canyon and East Peak, Mount Tamalpais. These are the northernmost known stations for this species.—*Hosackia strigosa* Nutt.

7. **L. humistratus** Greene. Frequently common on warm slopes and flats in clay, sand, or gravel: Tiburon; Mount Tamalpais (Corte Madera Ridge, Rock Spring, Rifle Camp); Carson country; Fairfax and San Rafael hills.—*Hosackia brachycarpa* Benth.

8. **L. subpinnatus** Lag. Widespread and often common on warm slopes in both brush and grassland: Angel Island, Tiburon, and Mount Tamalpais (Rock Spring, Phoenix Lake) to San Rafael Hills, Lucas Valley, and Tomales.—*Hosackia subpinnata* (Lag.) T. & G.

9. **L. uliginosus** Schkuhr. In Marin County known only from low moist ground between Mill Valley and Almonte, the station no longer extant. Introduced from the Old World, occasional in the Coast Ranges of northwestern California.—*L. trifoliolatus* Eastw.

10. **L. corniculatus** L. Becoming widely established in California; in Marin County found only near Stinson Beach and on the road to Point Reyes. Native of the Old World.

11. **L. grandiflorus** (Benth.) Greene. A rare plant on wooded or brushy slopes in rocky soil: Mount Tamalpais, acc. Stacey; Bolinas Ridge; Lily Lake. The Marin County plant has been reported as *L. grandiflorus* var. *anthyloides* Gray or *Hosackia leucophaea* (Greene) Abrams but in that plant the pubescence is spreading while in ours the pubescence is subappressed.

The close resemblance in technical floral and fruiting characters between *L. grandiflorus* and the two preceding Old World species is indicative of an intrageneric relationship scarcely susceptible of sectional separation in spite of the continental hiatus between the regions where the plants are indigenous.

12. **L. eriophorus** Greene. Dune hills and sandy mesas along the coast: Stinson Beach, acc. Stacey; Point Reyes Peninsula (Ledum Swamp, dunes near the life-saving station, and the lighthouse).—*Hosackia tomentosa* H. & A.

13. **L. junceus** (Benth.) Greene. Shallow soil of rocky slopes, rare: Corte Madera Ridge and West Point, Mount Tamalpais; Point Reyes Lighthouse.—*Hosackia juncea* Benth.

Generally in our region the stems are more slender and the peduncles longer, a form described from Mount Tamalpais as *L. Biolettii* Greene (*L. junceus* var.

Biolettii Ottley). It is occasional on the south side of Mount Tamalpais and under pines on Inverness Ridge and at Shell Beach.

14. **L. scoparius** (Nutt.) Ottley. DEERWEED. Common and widespread on coastal dunes and mesas, brushy slopes, and wooded hills: Sausalito, Angel Island, Tiburon, and Mount Tamalpais (common on south side, Northside Trail, Phoenix Lake) to San Rafael, Salmon Creek, and Point Reyes Peninsula (Bolinas, Point Reyes road).—*Hosackia glabra* (Vogel) Torr.

Generally deerweed forms an erect twiggy or broomlike shrub but along the coast the stems are nearly or quite prostrate and spread in broad circular mats a meter or two across. Away from the coast in shaded canyons, a slender trailing habit, quite unlike the normal form, may develop.

13. Psoralea

a. Shrub; leaflets spiny-tipped ....1. *P. fruticans*
a. Herbs; leaflets not spiny-tipped ....b
b. Stems prostrate, the petioles and peduncles erect ....2. *P. orbicularis*
b. Stems above ground erect c
c. Calyx teeth nearly equal; corolla whitish with purple-tipped keel; stems less
    than 1 m. tall ....3. *P. physodes*
c. Calyx teeth unequal; corolla violet or purplish; stems generally more than 1
    m. tall ....4. *P. macrostachya*

1. **P. fruticans** (L.) Rydb. A specimen of this South African plant collected in 1873 along "Streams of Tamelpais" was described by Kellogg as *P. fruticosa*. Apparently it has not been found again.

2. **P. orbicularis** Lindl. Occasional in wet ground in low meadows or on springy slopes: Rodeo Lagoon; Potrero Meadow, Mount Tamalpais; Ross; San Anselmo Canyon; Salmon Creek; east of Aurora School; near Abbotts Lagoon, Point Reyes Peninsula.

3. **P. physodes** Dougl. CALIFORNIA TEA. Occasional on brushy or wooded canyonsides in the southern part of the county: Blithedale Canyon and Phoenix Lake, Mount Tamalpais; Fairfax Hills; Lagunitas Canyon.

The leaves have been brewed as a tea, hence the common name.

4. **P. macrostachya** DC. CALIFORNIA HEMP. Occasional around springs or in the rocky beds of streams: Mount Tamalpais (West Point, Bootjack, Collier Spring, Blithedale Canyon); San Rafael Hills; Point Reyes, acc. Stacey.

A variant of the California hemp has been recognized as *P. Douglasii* Greene and Jepson refers to it a collection from Fairfax. In *P. Douglasii* the pubescence is less dense, the leaflets are broader, and the calyx scarcely equals (instead of exceeding) the corolla.

14. Amorpha

1. **A. californica** Nutt. var. **hispidula** (Greene) Palmer. MOCK LOCUST. Widespread but usually not common in the Tamalpais area on sunny slopes or in deep shade of woods: Mount Tamalpais (Cascade Canyon, Murray Trail, Corte Madera Ridge, Northside Trail); between Ross and Phoenix Lake; Lagunitas Canyon. The mock locust is one of the numerous shrubs of the region that stump sprout following fire.

15. Robinia

1. **R. Pseudo-Acacia** L. BLACK LOCUST. Escaping from cultivation and becoming locally established: Sausalito; Belvedere; Ross; San Rafael. Native of eastern and central United States.

16. Astragalus. LOCO WEED

a. Plants annual, less than 3 dm. tall; flowers in rounded heads or short oblong racemes ....b
a. Plants perennial, more than 4 dm. tall, usually 1 m. tall; flowers in elongate oblong racemes c
b. Flowers 3–5 mm. long; fruit 3 mm. long ....1. *A. Gambellianus*
b. Flowers about 10 mm. long; fruit, including long slender beak, 1–1.5 cm. long ....2. *A. Breweri*
c. Young leaves densely white-hairy; calyx teeth about as long as the tube; flowers 8–10 mm. long, in dense racemes ....3. *A. pycnostachyus*
c. Young leaves sparsely white-hairy or subglabrous; calyx teeth much shorter than the tube; flowers 10–14 mm. long, in loose racemes ....d
d. Stems erect, about 1 m. tall; stipules distinct ....4. *A. franciscanus*
d. Stems low and spreading, subprostrate; stipules united on the side of the stem opposite the leaf attachment ....5. *A. Nuttallii*

1. **A. Gambellianus** Sheld. Widespread and rather common in gravelly or clayey soil of open hills and meadows: Angel Island; Tiburon; Bootjack and Rock Spring, Mount Tamalpais; Carson country; Lagunitas, acc. Jepson; San Rafael Hills; Lucas Valley.—*A. nigrescens* Nutt.

In typical *A. Gambellianus* the corolla is only about 3 mm. long and is about twice as long as the calyx. In meadowy places on the north side of Mount Tamalpais is var. *Elmeri* (Greene) J. T. Howell in which the corolla is 4–6 mm. long and relatively showy. This larger-flowered form was originally described as *A. Elmeri* Greene from Ross Valley.

2. **A. Breweri** Gray. Locally abundant in gravelly soil on serpentine slopes, above Bootjack and near Rock Spring, Mount Tamalpais. This attractive plant is not known south of Marin County.

3. **A. pycnostachyus** Gray. Along the upper edges of salt marshes in sandy soil: Stinson Beach; Drakes Estero. This species was originally described from specimens collected by Bolander at Bolinas Bay.

4. **A. franciscanus** Sheld. In Marin County known only from Angel Island where it is rare on southern slopes.—*A. Menziesii* Gray subsp. *virgatus* (Gray) Abrams.

5. **A. Nuttallii** (T. & G.) J. T. Howell. Exposed maritime bluff, McClure Beach, Point Reyes Peninsula.—*A. Menziesii* Gray; *A. vestitus* (Benth.) Wats. var. *Menziesii* (Gray) Jones.

17. Vicia. VETCH

a. Flowers borne near the end of an elongate peduncle ....b
a. Flowers 1–4 in the axils of the leaves, sessile or very shortly stalked i
b. Flowers small, 1 cm. long or less c
b. Flowers more than 1 cm. long ....d

c. Pods hairy, 2-seeded ....1. *V. hirsuta*

c. Pods glabrous, 3–8-seeded ....2. *V. exigua*

d. Flowers yellowish tinged with red or brown, or purple-red fading dark purple ....e

d. Flowers lavender, violet, or bluish purple, rarely whitish ....f

e. Native perennial; leaves and stems weakly pubescent; pods glabrous ....3. *V. gigantea*

e. Introduced annual; leaves and stems silky-villous; pods pubescent ....4. *V. benghalensis*

f. Flowers numerous, generally 10 or more; introduced plants ....g

f. Flowers few, mostly 2–6; native plants ....h

g. Pubescence villous ....5. *V. villosa*

g. Pubescence sparse, subappressed ....6. *V. dasycarpa*

h. Stems glabrous or minutely puberulent ....7. *V. americana*

h. Stems finely villous ....8. *V. californica*

i. Leaflets 4–7 cm. long, coriaceous, tendril reduced to a filiform nontwining tip or obsolete ....9. *V. Faba*

i. Leaflets 3 cm. long or less, thin, tendrils of upper leaves well developed and branched ....j

j. Flowers yellowish, fading rosy; hairs on pods with pustulate base ....10. *V. lutea*

j. Flowers rosy- or violet-purple; hairs, if present on pods, not pustulate at base.k

k. Leaflets linear to oblanceolate-elliptic; flowers 1–2 cm. long; pods blackish at maturity; seeds round ....11. *V. angustifolia*

k. Leaflets narrowly to broadly obovate, rarely narrower; flowers 2–3 cm. long; pods light to dark brown; seeds more or less compressed-angled.12. *V. sativa*

1. **V. hirsuta** (L.) S. F. Gray. Becoming locally common along roads: San Rafael, *O'Conor;* Inverness. Native of the Old World.

2. **V. exigua** Nutt. Gravelly soil of brushy slopes or rarely in clay soil of open woods, widespread but not very common: Angel Island; Tiburon; Mount Tamalpais, acc. Stacey; Carson country; San Rafael Hills; Tomales.

A variant, var. *Hassei* (Wats.) Jeps., in which the leaflets are broader and more or less notched at the apex is occasional: Rocky Ridge, Mount Tamalpais; Carson Ridge; Black Point. This variety, frequently common in the South Coast Ranges and southern California, is not known north of Marin County.

3. **V. gigantea** Hook. Rampant climber in sheltered canyons near the coast, occasionally on moist open hillsides: Rodeo Lagoon, Sausalito, Tiburon, and Mount Tamalpais (Cascade and Blithedale canyons) to Bolinas and Inverness.

4. **V. benghalensis** L. One of the most common and colorful of the several vetches introduced along roadsides: Sausalito, Tamalpais Valley, and Tiburon to Fairfax, Novato, Lucas Valley, and Point Reyes Peninsula (Bolinas, Inverness Ridge). Native of the Mediterranean region.—*V. atropurpurea* Desf.

5. **V. villosa** Roth. Rather widespread and often locally abundant along roads and trails: Baltimore Park; Fairfax; Inverness Ridge. Native of the Old World.

6. **V. dasycarpa** Ten. Not so common as the preceding species which it resembles in its masses of attractive purple-blue flowers: Olema; Inverness. Native of Europe and North Africa.

At several places in central California a vetch has been collected which is quite like *V. dasycarpa* in general appearance but which differs from both *V. dasycarpa* and *V. villosa* in having a pubescent rather than glabrous pod. In these characters it corresponds to the description of the Old World *V. salaminia* Heldr. & Sart. and it may belong to that species. On the other hand, our California plant may represent a hybrid between *V. dasycarpa* and *V. benghalensis,* and it is so accepted here until it is better understood.

7. **V. americana** Muhl. var. **oregana** (Nutt.) Nels. Widespread through the hills on open grassy slopes, among brush, or in woods: Sausalito, Angel Island, Tiburon, and Mount Tamalpais (Cascade and Blithedale canyons, Rock Spring) to San Rafael Hills and Tomales.

The American vetch is extremely variable in leaflet shape and three distinctive forms are recognizable. Var. *oregana* has entire, ovate or oval leaflets but also in Marin County is var. *truncata* (Nutt.) Brew. with leaflets coarsely 2- or 3-toothed near the apex. The latter occurs near Dillons Beach and Jepson reports it from Ross Valley. In contrast to these forms with broad leaflets is var. *linearis* (Nutt.) Wats. in which the leaflets are narrowly oblong to linear. It is occasional on wooded or open slopes from Sausalito, Tiburon, and Mount Tamalpais to Carson Ridge and Point Reyes (acc. Stacey).

8. **V. californica** Greene. Certain plants in Marin County seem definitely referable to the California vetch, although as a species it is nowhere too well separated from the variable American vetch. As long ago as 1898 Miss Eastwood reported it from the "road between Fairfax and Cataract Gulch" and recently it has been found on Corte Madera Ridge, Mount Tamalpais, and in Lagunitas Canyon.

9. **V. Faba** L. BROAD BEAN; HORSE BEAN. Occasionally escaped from cultivation, as at Stinson Beach, but probably not persisting. Native of the Old World.

10. **V. lutea** L. Locally and abundantly established as a roadside weed in Salmon Creek Canyon. Native of Europe and North Africa.

11. **V. angustifolia** L. Occasional weed along roads and in waste places: Bootjack, Mount Tamalpais; Larkspur; Olema; Bolinas; Inverness. Introduced from the Old World.

Much more common than the species is the var. *segetalis* (Thuill.) Koch, a ranker plant with larger flowers and fruits. In the species the flowers are usually 1–1.5 cm. long and the fruits 4–5 mm. wide, while the leaflets are linear or linear-lanceolate. In the variety, the flowers may be 1.8 cm. long and the fruits 6 mm. wide, while the leaflets are oblanceolate to elliptic. Through the variety *V. angustifolia* approaches *V. sativa* closely, and both in the United States and in the Old World the plants are not always separated as species. In Marin County, var. *segetalis* is widely distributed through the hills and valleys from Tiburon, Mill Valley, and Stinson Beach to Black Point, Chileno Valley, and Inverness.

12. **V. sativa** L. In fields and along roads, widespread but not so common as *V. angustifolia* var. *segetalis:* Sausalito, Tiburon, and San Rafael to Black Point, Chileno Valley, and Inverness. Native of Europe and North Africa.

18. Lathyrus

a. Tendril reduced to a short, nontwining, unbranched rudiment ....b
a. Tendril well developed, branched and twining c
b. Leaves and stems silvery-silky; stipules larger than the leaflets 1. *L. littoralis*

b. Leaves and stems sparsely villous, green; stipules very narrow, smaller than the leaflets ....2. *L. Torreyi*

c. Leaflets more than 5 ....d

c. Leaflets 2. Stems winged ....f

d. Plants glabrous; stipules about as large as the leaflets ......5. *L. polyphyllus*

d. Plants more or less hairy; stipules generally smaller than the leaflets ....e

e. Stem angled but not winged ....3. *L. Bolanderi*

e. Stem with narrow wings on the angles ....4. *L. Watsonii*

f. Petiole broadly winged; plants perennial ....6. *L. latifolius*

f. Petiole narrowly winged or only angled; plants annual ....g

g. Stems, inflorescence, and pods glabrous ....7. *L. tingitanus*

g. Stems, inflorescence, and pods villous-hirsute ....8. *L. odoratus*

1. **L. littoralis** (Nutt.) Endl. BEACH PEA. Dunes and sandy flats along the ocean: Stinson Beach, *Leschke;* Point Reyes Peninsula near the radio station.

2. **L. Torreyi** Gray. REDWOOD PEA. Deep soil in shady woods, occasional: Mill Valley and Corte Madera, acc. Stacey; Phoenix Lake, Mount Tamalpais; Little Carson.

The herbage has a delicate fragrance of vanilla when dry.

3. **L. Bolanderi** Wats. Common and widespread in wooded canyons, on brushy hills, and on maritime slopes: Sausalito, Angel Island, Tiburon, and Mount Tamalpais (Cascade and Blithedale canyons, Bolinas Ridge) to San Rafael, Tomales, and Point Reyes Peninsula (Inverness, Point Reyes). The plant collected at Tomales Bay and Corte Madera by Bigelow was described as *L. vestitus* Nutt. var. *multiflorus* Torr.

This wild sweet pea is one of the first plants to bloom after the beginning of the rainy season, flowers having been observed as early as October. A form with few flowers and small leaflets has been found on Mount Tamalpais, Carson Ridge, and San Rafael Hills and was quite common after the burn of 1945.

4. **L. Watsonii** White. Dry brushy slopes and flats away from the coast, rare in Marin County: Lagunitas Meadows, Mount Tamalpais; White Hill, acc. Stacey; Novato, *Leschke.—L. californicus* Wats.

5. **L. polyphyllus** Nutt. Rare in Marin County where it is represented by var. *insecundus* Jeps., Olema, the type locality.

6. **L. latifolius** L. Occasionally escaping from cultivation and becoming established around habitations: Angel Island; Belvedere; Mill Valley; Corte Madera; San Anselmo; Fairfax; Lagunitas; Bolinas; Point Reyes Station. Native of Europe.

7. **L. tingitanus** L. TANGIER PEA. Occasionally cultivated and becoming well established around towns and along roads: Sausalito; Tiburon; Mill Valley; Corte Madera; San Rafael; Lagunitas; San Geronimo; Aurora School; Tomales; Inverness Park. Introduced from the Mediterranean region.

8. **L. odoratus** L. SWEET PEA. Commonly cultivated but only rarely escaping to waste ground and then perhaps fugitive: Sausalito; Tiburon; Fairfax; Chileno Valley road. Native of Italy.

19. Pisum. PEA

1. **P. sativum** L. Two forms of the pea have been found growing spontaneously in Marin County. The common garden pea with white flowers occurred as a waif at Black Point and may be expected as a fugitive from cultivation elsewhere. The

second form is var. *arvense* (L.) Gams, the field pea, marked by its violet banner and purplish wings. This is locally established in Fairfax. Both forms are introduced from the Old World.

GERANIACEAE. Geranium Family

a. Upper sepal spurred, the spur adnate to the pedicel which appears jointed. Leaves roundish in outline, odorous ....1. *Pelargonium*
a. Upper sepal not spurred ....b
b. Leaves roundish in outline, palmately lobed or divided; stamens with anthers 10 ....2. *Geranium*
b. Leaves ovate to elliptic or oblong in outline, pinnate or pinnately lobed or divided; stamens with anthers 5 ....3. *Erodium*

1. Pelargonium

1. **P. grossulariaefolium** Ait. In waste ground, of rare occurrence or perhaps overlooked: San Rafael; Inverness. A native of South Africa.

The leaves have a pungent odor remotely suggesting turpentine.

2. Geranium

a. Plants annual ....b
a. Plants perennial, the stems from top of a thick, carrotlike root ....e
b. Sepals not prominently awn-tipped; carpel body glabrous, transversely wrinkled ....1. *G. molle*
b. Sepals prominently awn-tipped; carpel body hairy, not wrinkled c
c. Pedicels in fruit much longer than the calyx; beak on style column elongate ....4. *G. Bicknellii*
c. Pedicels in fruit about equaling the calyx or shorter; beak on style column short ....d
d. Inflorescence glandular-hairy; leaf lobes acute; petals purplish rose; anthers lavender; stigma purplish; seeds marked with deep roundish pits ....2. *G. dissectum*
d. Inflorescence hirsutulous, scarcely if at all glandular; leaf lobes somewhat rounded; petals pink; anthers and stigmas yellow; seeds marked with very shallow elongate pits ....3. *G. carolinianum*
e. Hairs on the stems retrorse, appressed, whitish, dull; petals conspicuously longer than the sepals; seeds finely reticulate ....5. *G. retrorsum*
e. Hairs on the stems retrorse-spreading, hispid, translucent, shining; petals little longer than the sepals; seeds coarsely reticulate ....6. *G. pilosum*

1. **G. molle** L. Cranesbill. Widespread and sometimes abundant on grassy or brushy hills: Sausalito, Tiburon, and Mount Tamalpais (Blithedale Canyon, Rock Spring, Phoenix Lake) to San Rafael, Tomales, and Inverness. Introduced from Europe.

2. **G. dissectum** L. Occasional throughout the county in grassland or chaparral: Sausalito, Angel Island, Tiburon, and Mount Tamalpais (Blithedale Canyon, Potrero Meadow, Phoenix Lake) to San Rafael, Tomales, and Inverness. Native of Europe.

3. **G. carolinianum** L. Rather rare in more or less shaded places on open or

wooded slopes: Sausalito, acc. Lovegrove; Camp Taylor, *Cannon;* San Rafael Hills; Black Point; Tomales.

4. **G. Bicknellii** Britt. Locally common on steep slopes in loose gravelly soil in a chaparral burn: north of West Peak, Mount Tamalpais; Little Carson. These are the southernmost known stations in California.—*G. carolinianum* L. var. *longipes* Wats.

5. **G. retrorsum** L'Hér. Occasional in grassy places along roads or on the edge of brush: Baltimore Park; San Rafael; near Stinson Beach; pond south of Olema; Bolinas and Drakes Estero, Point Reyes Peninsula. Native of Australasia.—*G. pilosum* Forst. f. var. *grandiflorum* Knuth; *G. pilosum* Forst. f. var. *retrorsum* (L'Hér.) Jeps.

6. **G. pilosum** Forst. f. In moist partially shaded places near the south end of Tomales Bay: Inverness Park; Olema Marshes. Native of Australasia. The plant reported in 1898 from the vicinity of Olema as *G. sibiricum* is apparently a shade form of *G. pilosum* in which the penduncles are mostly 1-flowered.

3. Erodium

a. Leaves simple, deeply lobed or divided; beak of fruit 5 cm. long or more ...b
a. Leaves pinnate; beak of fruit less than 5 cm. long c
b. Concavities (foveolae) at top of fruit subtended by a single fold (plica), this upper part of the fruit more or less hairy; beak of the fruit 5–9 cm. long; sepals with a short, green mucro ....1. *E. obtusiplicatum*
b. Concavities at top of fruit subtended by 2 folds, the upper part of the fruit glabrous; beak of the fruit 9–12 cm. long; sepals with a prominent, reddish mucro ....2. *E. Botrys*
c. Leaflets broad, coarsely toothed or serrate; claws of petals glabrous
    ....3. *E. moschatum*
c. Leaflets pinnately lobed or divided; claws of petals ciliate-hairy
    ....4. *E. cicutarium*

1. **E. obtusiplicatum** (Maire, Weiller, & Wilczek) J. T. Howell. Widely distributed and rather common in weedy places about towns, along roads, and on open grassy hillsides: Tiburon; Phoenix and Alpine lakes, Mount Tamalpais; Ross; Fairfax; San Rafael Hills; Drakes Estero, Point Reyes Peninsula. Native of North Africa.

2. **E. Botrys** (Cav.) Bertol. Broad-leaf Filaree. Common in clayey or gravelly soil in grassland or brush: Sausalito, Angel Island, Tiburon, and Mount Tamalpais (Bootjack, Bolinas Ridge, Phoenix Lake) to the Carson country, San Rafael Hills, and San Geronimo. Introduced from the Mediterranean region.

This species is the most common filaree on the Marin County hills. So abundant is it in some places that its rosy-mauve flowers impart to the landscape color that can be perceived at a considerable distance. The bowl-shaped flowers are usually erect but on rainy days they nod and so keep dry inside.

3. **E. moschatum** (L.) L'Hér. White-stem Filaree. Occasional in relatively moist places in deep soil: Sausalito, Angel Island, and Tiburon to San Rafael, Dillons Beach, and Point Reyes Peninsula (Drakes Estero). Naturalized from the Mediterranean region.

The pale mauve petals of this filaree do not have the more pleasing rosy hues of the other filarees of this area.

4. **E. cicutarium** (L.) L'Hér. RED-STEM FILAREE. Common and widespread on grassy flats, slopes, and ridges from waste places about towns to coastal mesas and summit meadows: Sausalito, Angel Island, Tiburon, and Mount Tamalpais (Blithedale Canyon, Bolinas Ridge) to San Rafael, Tomales, and Point Reyes Peninsula (Inverness Ridge, Drakes Estero). Introduced from the Old World.

OXALIDACEAE. OXALIS FAMILY

1. Oxalis

a. Petals white, pink, or purplish rose. Leaves basal ....b
a. Petals yellow ....d
b. Flowers solitary. Leaves and peduncles arising from slender stoloniferous stems
....3. *O. oregana*
b. Flowers in a simple or branched umbel c
c. Pedicels, calyx, and petals more or less pubescent; leaves and peduncles arising from the crown of a thickened tuberiform stem ....1. *O. rubra*
c. Pedicels, calyx, and petals glabrous; leaves and peduncles arising from scaly bulbs ....2. *O. Martiana*
d. Leaves and peduncles basal; flowers showy, usually more than 1.5 cm. long
....4. *O. cernua*
d. Leaves and peduncles cauline; flowers smaller, usually 1 cm. or less long ....e
e. Inflorescence forked, the rather numerous flowers racemosely arranged along the branches; annual with simple stems ....5. *O. laxa*
e. Inflorescence umbellate with few flowers; perennial with branching stems ....f
f. Hairs on pedicels scant and appressed-ascending, on stems upwardly appressed
....6. *O. corniculata*
f. Hairs on pedicels dense and spreading, on stems reflexed ....7. *O. pilosa*

1. **O. rubra** St. Hil. Escaping from cultivation and locally naturalized in towns: Sausalito; Tiburon; Mill Valley; San Rafael; Inverness. Introduced from Brazil.

The petals, which are usually bright purplish rose or occasionally white, have been described as glabrous but in the Marin County plants they are lightly pubescent on the lower side. In this character they resemble *O. lasiopetala* Zucc., another South American species.

2. **O. Martiana** Zucc. A neat plant for beds and borders but one likely to become weedy and obnoxious. Together with *O. corniculata* L., it grows without cultivation in the park at Ross. Although it is a native of South America, it is now widely naturalized through tropical and subtropical regions.

3. **O. oregana** Nutt. REDWOOD SORREL. Common and attractive ground-cover under redwoods: Mill Valley; Muir Woods; Fern Canyon and Cataract Gulch, Mount Tamalpais.

4. **O. cernua** Thunb. Escaping from cultivation and becoming established in weedy patches and along roads and paths: Sausalito; Angel Island; Tiburon; Mill Valley; Dipsea Trail southeast of Muir Woods.

Although it is neither Bermudan nor ranunculaceous, this bright South African flower is generally known as Bermuda buttercup, a truly excellent example of nonbotanical discrimination in a common name.

5. **O. laxa** H. & A. This distinctive Chilean herb has been found at several stations in the foothills of the Sierra Nevada but the only known Coast Range record is one from Stinson Beach.

6. **O. corniculata** L. Both in its typical green form and in its purple-leaved form (var. *purpurea* Parl.), this little sorrel is not uncommon around towns where it has escaped from cultivation: Sausalito, Mill Valley, and Ross to Fairfax, San Rafael, and Inverness. Our garden plant is probably of European origin.

7. **O. pilosa** Nutt. Widespread and not uncommon on brushy canyonsides and open coastal hills, sometimes becoming ruderal in towns: Sausalito, Tiburon, and Mount Tamalpais (Blithedale Canyon, West Point, Steep Ravine) to San Rafael, Tomales, and Point Reyes Peninsula (Bolinas, Mud Lake, Inverness, Ledum Swamp).

Recently Jepson confused this indigenous sorrel with the introduced *O. corniculata,* but, although the two are more closely related than some of the diverse elements of this large and difficult genus occurring in California, they are adequately distinct. Expressing a much more natural and closer relationship is the name *O. Wrightii* Gray var. *pilosa* (Nutt.) Wiegand, but our coastal plant seems sufficiently distinct from that more xerophilous plant of the arid southwest.

TROPAEOLACEAE. Tropaeolum Family

1. Tropaeolum

1. **T. majus** L. Garden Nasturtium. Occasionally fugitive from cultivation and sometimes becoming locally abundant in coastal thickets: Tiburon; Belvedere; Bolinas Lagoon; Stinson Beach; Drakes Estero, Point Reyes Peninsula. Native of South America.

LINACEAE. Flax Family

1. Linum

a. Petals 1–1.5 cm. long, blue; styles 5; capsule 5–10 mm. long ....b
a. Petals less than 1 cm. long, white or rose; styles 3; capsule 2–4 mm. long c
b. Petals about 1.5 cm. long; fruit 7–10 mm. long; seeds 4 mm. long, brown, shiny
    but a little roughened; leaves narrowly lanceolate ......1. *L. usitatissimum*
b. Petals about 1 cm. long; fruit 5 mm. long; seeds 3 mm. long, brownish green,
    shiny and smooth; leaves linear ....2. *L. angustifolium*
c. Petals less than 5 mm. long. Sepals glabrous; stems puberulent just above the
    nodes ....3. *L. micranthum*
c. Petals 5–8 mm. long ....d
d. Sepals glabrous ....4. *L. californicum*
d. Sepals pubescent ....5. *L. congestum*

1. **L. usitatissimum** L. Flax. Occasional in weedy places along roads, not persisting: Mill Valley; Ignacio. Introduced through cultivation, the species unknown in the wild state.

2. **L. angustifolium** Huds. Open grassy hills and valleys, in places becoming locally common: San Rafael; Tocaloma; Olema; Inverness Park; Tomales; Chileno Valley. Introduced from the Old World.—*L. usitatissimum* of Jepson's Flora in part.

3. **L. micranthum** Gray. Gravelly soil of open or brushy slopes, in our region commonly found on serpentine: Mount Tamalpais (West Point, Barths Retreat, Berry Trail); Little Carson Falls; Carson Ridge.

4. **L. californicum** Benth. This species, usually reported from the Inner Coast Ranges and farther east, is listed by Stacey as occurring near San Rafael.

5. **L. congestum** Gray. In Marin County seemingly confined with other rare plants to the serpentine island on Tiburon Peninsula. Marin County is the type locality of this flax which is otherwise known only from the San Francisco Peninsula.—*L. californicum* Benth. var. *congestum* (Gray) Jeps.

ZYGOPHYLLACEAE. CALTROP FAMILY

1. Tribulus

1. **T. terrestris** L. PUNCTURE VINE. Widespread and troublesome weed in the warmer and drier valleys of California; reported from northeastern Marin County (Weeds of California, p. 254).

SIMAROUBACEAE. SIMAROUBA FAMILY

1. Ailanthus

1. **A. altissima** (Mill.) Swingle. TREE OF HEAVEN. Locally naturalized about habitations: Belvedere; Ross; San Rafael; Santa Venetia; Ignacio; Black Point. Introduced from China.—*A. glandulosa* Desf.

POLYGALACEAE. MILKWORT FAMILY

1. Polygala. MILKWORT

1. **P. californica** Nutt. Not uncommon on rocky ridges and slopes in chaparral or open woods: Mount Tamalpais (Throckmorton Ridge, Blithedale Canyon, Mountain Theater); Lagunitas Canyon; Carson Ridge; San Rafael Hills; Inverness Ridge, Point Reyes Peninsula.

The flowers are usually an attractive bright rose but occasionally they are paler and rarely even an albino is found.

EUPHORBIACEAE. SPURGE FAMILY

a. Stems and leaves densely covered with stellate hairs; flowers unisexual, the staminate with calyx, the pistillate without calyx ....1. *Eremocarpus*
a. Stems and leaves glabrous or hairy, the hairs not stellate; flowers unisexual, both the staminate and pistillate without calyx and borne in a campanulate involucre, the whole appearing like a single perfect flower ....2. *Euphorbia*

1. Eremocarpus

1. **E. setigerus** Benth. TURKEY MULLEIN. Occasional in dry soil of valleys and open hillsides, maturing in summer and autumn: Mount Tamalpais (West Point, Rock Spring, Lagunitas Meadows); San Rafael Hills; Ignacio; Point Reyes Station, acc. Stacey.

2. Euphorbia. SPURGE

a. Stems erect; lower leaves alternate, the upper opposite or whorled ....b
a. Stems generally prostrate, rarely ascending or erect; leaves opposite ....e

b. Glands of the involucre entire; capsule warty ....1. *E. spathulata*
b. Glands of the involucre with horns or slender lobes c
c. Stems stout; lower leaves linear-lanceolate, sessile ....4. *E. Lathyrus*
c. Stems slender; lower leaves obovate to roundish, petiolate or cuneate at base .d
d. Capsule valves smooth on back ....2. *E. crenulata*
d. Capsule valves with thin longitudinal ridges on back ....3. *E. Peplus*
e. Stems, leaves, and capsules glabrous ....5. *E. serpyllifolia*
e. Stems, leaves, and capsules hairy ....f
f. Transverse wrinkles on the seeds rounded and usually broader than the spaces
    between them ....6. *E. maculata*
f. Transverse wrinkles on the seeds sharp and narrower than the spaces between
    them ....7. *E. prostrata*

1. **E. spathulata** Lamk. On grassy or open brushy slopes moist in the spring, not common: Angel Island; Tiburon; Mount Tamalpais, acc. Stacey; San Anselmo Canyon; Tomales.—*E. dictyosperma* F. & M.

2. **E. crenulata** Engelm. Rare on grassy slopes in partial shade of brush or trees: Sausalito; Tiburon; Mount Tamalpais, acc. Stacey; Nicasio; Inverness Ridge.

3. **E. Peplus** L. Garden weed, sometimes naturalized along shaded paths or roads: Sausalito; Angel Island; Tiburon; Mill Valley; Ross; San Rafael. Introduced from the Old World.

4. **E. Lathyrus** L. COMPASS PLANT; CAPER SPURGE; GOPHER PLANT. Occasional in waste places around towns and on moist flats at the upper edge of salt marshes: Tamalpais Valley; Mill Valley; Stinson Beach, acc. Stacey; Bolinas; Inverness. Native of the Mediterranean region.

5. **E. serpyllifolia** Pers. Widespread but not very common in low places swampy or submerged in the spring: Alpine, Lagunitas, and Phoenix lakes; San Anselmo; San Rafael; Burdell; San Antonio Creek; east of Aurora School.

6. **E. maculata** L. Weed in lawns and gardens, also occurring along roads in dry packed ground: Phoenix Lake; Ross; San Rafael. Naturalized from the eastern United States and Canada.—*E. supina* Raf.

7. **E. prostrata** Ait. Garden weed in Mill Valley. Widely distributed in the southern United States and southward but introduced in California.—*E. Chamaesyce* of some references.

CALLITRICHACEAE. WATER STARWORT FAMILY

1. Callitriche. WATER STARWORT

a. Adjacent carpels separated nearly to the axis, the fruit broader than long;
    leaves linear-oblong, all alike ....1. *C. hermaphroditica*
a. Adjacent carpels united except for a shallow groove; leaves oblanceolate to
    roundish, or if submerged, then linear ....b
b. Fruit pedunculate, the peduncle usually exceeding the fruit in length
    ....2. *C. marginata*
b. Fruit sessile or very shortly pedunculate c
c. Fruit deeply notched, back of carpels conspicuously winged ..3. *C. stenocarpa*
c. Fruit shallowly notched; back of carpels obtuse, acute, or narrowly winged ..d
d. Styles shorter than the fruit; back of carpels acute or narrowly winged
    ....4. *C. palustris*

d. Styles equaling the fruit or longer, back of carpels acute or obtuse
....5. *C. Bolanderi*

1. **C. hermaphroditica** L. Submerged aquatic, in Marin County known only from a roadside pond east of Aurora School.—*C. autumnalis* L.

2. **C. marginata** Torr. Occasional on wet ground or where water has stood: Larkspur, acc. Stacey; near McNears Landing; Ignacio; Ledum Swamp, Point Reyes Peninsula.

3. **C. stenocarpa** Hegelm. Wet ground recently under water, east of Aurora School.—*C. palustris* L. var. *stenocarpa* (Hegelm.) Jeps.

4. **C. palustris** L. Shallow ponds or emergent on muddy flats: Lagunitas road, *Moore in 1879;* Olema Marshes; laguna in Chileno Valley.

5. **C. Bolanderi** Hegelm. Submerged in shallow ponds and streams or terrestrial on wet ground: Mount Tamalpais (Hidden Lake, Potrero and Lagunitas meadows); San Anselmo Canyon, *Eastwood;* near McNears Landing; east of Aurora School; Point Reyes Peninsula (Point Reyes, *Eastwood,* Mud Lake).—*C. palustris* L. var. *Bolanderi* (Hegelm.) Jeps.

The fruits of *C. Bolanderi* are rounded on the margins, and near the base the carpels are a little ventricose. The markings on the nearly round fruits are finer than in the slightly obovate fruits of *C. palustris*. In *C. stenocarpa* the fruits are also round or nearly so but there the markings are even coarser than in *C. palustris* and the carpels are conspicuously winged along the entire back.

LIMNANTHACEAE. Meadow Foam Family

1. Limnanthes

1. **L. Douglasii** R. Br. Meadow Foam. Low fields and meadows where water collects in the spring or around seepages on open slopes in the hills, frequently abundant and coloring broad areas: near Phoenix Lake; Fairfax; Novato; Chileno Valley; east of Aurora School.

The usual form growing at the stations cited above has petals white with a yellow base but on the downs of Point Reyes Peninsula an equally attractive form occurs in which the petals are golden-yellow throughout.

ANACARDIACEAE. Sumac Family

1. Rhus

1. **R. diversiloba** T. & G. Poison Oak. Throughout Marin County, inhabiting valley thickets, brushy hillsides, dense or open woodland, and ocean headlands: Sausalito, Angel Island, Tiburon, and Mount Tamalpais to San Rafael, Black Point, Tomales, and Point Reyes Peninsula.

Not only is poison oak our most common and widespread shrub but, in its adaptations to its environments, it is also the most variable. On open hillsides and in the chaparral it develops a dense twiggy habit and is stiffly erect. In open woodland it is also erect but much less twiggy, while in sheltered canyons the stems are gracefully diffuse and the long pliant branches are horizontally spreading. In woods the plants frequently become vinelike and clamber up the trunks to bush out high above the ground. In marked contrast are the shrubby mats on wind-swept hills along the ocean where this most variable plant creeps along the ground and is only an inch or two tall.

Because it is so variable, poison oak is not always readily recognized by those who would avoid its baneful touch, and, at times, even botanists may be puzzled. However, those who are not immune to the poisonous sap do not have to touch the plant to develop the painful dermatitis, but they may contract it from the smoke of fires in which the plant is burned or even from the roadside dust which it has contaminated.

In spring its ivory flowers perfume the sunny hill or sheltered glade, in summer its fine green leaves contrast refreshingly with dried and tawny grassland, in autumn its colors flame more brilliantly than in any other native: but one great fault, its poisonous juice, nullifies its every other virtue and renders this beautiful shrub the most disparaged of all within our region.

CELASTRACEAE. Burning Bush Family

a. Stems erect, more than 1 m. tall; leaves deciduous, petiolate ....1. *Euonymus*
a. Stems low and spreading, mostly less than 1 m. tall; leaves evergreen, nearly
   sessile ....2. *Paxistima*

1. Euonymus

1. **E. occidentalis** Nutt. Western Burning Bush; Pawnbroker Bush. Moist shaded streambanks in canyons or under trees, rare: Muir Woods; Devils Gulch, Mount Tamalpais, acc. Sutliffe; Big Carson, *Leschke;* Lagunitas, acc. Stacey; Second Canyon, Inverness, *Leschke.*

2. Paxistima

1. **P. Myrsinites** (Pursh) Raf. Oregon Boxwood. This northern species was first discovered on Mount Tamalpais southeast of Laurel Dell by Robert H. Menzies in 1941 and since then it has been found along the trail between Laurel Dell and Barths Retreat. These are the southernmost known stations.

ACERACEAE. Maple Family
1. Acer

a. Leaves simple, 5-lobed or -parted; flowers perfect and staminate in same cluster;
   petals present ....1. *A. macrophyllum*
a. Leaves 3-foliolate; flowers unisexual; petals none ....2. *A. Negundo*

1. **A. macrophyllum** Pursh. Big-leaf Maple (plate 24). Occasional along streams and in well-watered canyons near the coast: Sausalito, acc. Lovegrove; Tiburon; Mount Tamalpais (Blithedale and Cascade canyons, Cataract Gulch, Lagunitas Meadows); San Anselmo Canyon; Papermill Creek; Point Reyes Station, acc. Stacey.

Most of the color in our autumnal woods comes from poison oak and big-leaf maple. Flecks of blue sky seen between the bright yellow leaves of a sun-lit maple crown produces an effect approximating fine stained glass.

2. **A. Negundo** L. var. **californicum** (T. & G.) Sarg. California Box Elder. Occasional along streams in valleys: Larkspur; Ross; Ignacio; Tocaloma; Point Reyes Station; Olema Marshes.

If the big-leaf maple is most colorful with its leaves in the fall, the box elder is most colorful without its leaves in the spring, when the abundant rosy inflorescences hang in fringes and tassels from every branch. And again the blue of the sky may provide a breathtaking color contrast for background.

HIPPOCASTANACEAE. Horse-chestnut Family

1. Aesculus

1. **A. californica** (Spach) Nutt. California Buckeye (plates 6, 7). Widespread almost throughout Marin County, in places common on hills: Sausalito, Tiburon, and Mount Tamalpais (Cataract Gulch, Phoenix Lake) to San Rafael Hills, Black Point, and Point Reyes Peninsula (Bolinas, Mud Lake, Inverness).

In several aspects and at all seasons the buckeye is attractive, whether it is the tracery of bare branches etched against the winter sky, or the vernal opulence of leaf and flower conformed into a huge bouquet, or the dull rich tone of colored foliage bronzing the autumnal canyonside. At times the flowers are more rosy-tinged than at others, but always they emit a pleasant fragrance. The gradually increasing development of leaflets on successive bud-scales in the spring demonstrates admirably the protective function assumed by modified petioles. The large glossy chestnut-colored seeds have beauty, and also has the seedling with its delicate foliage, whether germination is in the woods or indoors in a bowl of water.

RHAMNACEAE. Buckthorn Family

a. Flowers greenish, the petals, if present, tiny and not clawed; fruit fleshy
....1. *Rhamnus*
a. Flowers white, bluish, or purplish, the petals rather small but conspicuous, long-clawed; fruit a capsule ....2. *Ceanothus*

1. Rhamnus

a. Leaves oblong or broadly lanceolate, usually more than 2 cm. long, fruit black
....1. *R. californica*
a. Leaves elliptic to roundish, less than 2 cm. long; fruit red ......2. *R. crocea*

1. **R. californica** Esch. Coffee Berry. Common shrub of open woodland, brushy canyonsides, and chaparral: Sausalito, Tiburon, and Mount Tamalpais (Cascade Canyon, Azalea Flat, Rifle Camp, Phoenix Lake) to San Rafael Hills, Black Point, Tomales, and Point Reyes Peninsula (Bolinas, Mud Lake, Ledum Swamp).

2. **R. crocea** Nutt. Red-berry. Rather widespread but not common on rocky slopes in shallow soil: Angel Island; Muir Beach; Mill Valley, acc. Stacey; Phoenix Lake, Mount Tamalpais; San Anselmo Canyon; San Rafael Hills; Big Rock Ridge; Tomales.

2. Ceanothus

a. Leaves alternate, the stipules not becoming corky-thickened ....b
a. Leaves opposite, the stipules becoming corky-thickened ....e
b. Branchlets ridged or angular; leaves pale or yellowish green beneath, the margin inconspicuously glandular. Leaves prominently 3-nerved from the base; flowers bluish ....4. *C. thyrsiflorus*
b. Branchlets round; leaves frequently pale or whitish beneath, the margin rather conspicuously glandular-toothed c

c. Leaves mostly more than 3.5 cm. long, conspicuously 3-nerved from the base; flowers white ....1. *C. velutinus*

c. Leaves mostly less than 3.5 cm. long, not conspicuously 3-nerved; flowers lavender to deep blue ....d

d. Stems usually low and spreading, the branchlets flexible; leaves 0.5–1.5 (or 2.5) cm. long, undulate-crisped; flowers indigo-blue ....2. *C. foliosus*

d. Stems tall and erect, the branchlets rigid; leaves mostly 1–4 cm. long, plane; flowers pale lavender to light blue ....3. *C. sorediatus*

e. Leaves generally entire ....5. *C. ramulosus*

e. Leaves generally toothed ....f

f. Leaves hollylike with 4 or 5 coarse subspinose teeth on each side; fruit with prominent crests between the thick wrinkled horns ....8. *C. Jepsonii*

f. Leaves with few to many teeth, not hollylike; fruit with low wrinkled crests between the stubby or slender horns ....g

g. Leaves obovate to roundish, usually more than 1 cm. broad, with 6–15 teeth on the sides ....h

g. Leaves oblong to obovate, usually less than 1 cm. broad, with 2–8 teeth on the sides i

h. Stems prostrate ....6. *C. gloriosus*

h. Stems erect, 1–2 m. tall ....6a. *C. gloriosus* var. *exaltatus*

i. Stems low and spreading, the branchlets slender and not very rigid ....6b. *C. gloriosus* var. *porrectus*

i. Stems erect, 1–2 m. tall, the branchlets rigid ....7. *C. Masonii*

1. **C. velutinus** Dougl. var. **laevigatus** (Hook.) T. & G. Rare and local in the chaparral: Corte Madera Ridge, Mount Tamalpais; Mount Vision, Point Reyes Peninsula, acc. Jepson. According to McMinn, the variety is found usually near the coast from Washington south to the Santa Cruz Mountains, California.

2. **C. foliosus** Parry. INDIGO BRUSH. Common in open rocky places in the chaparral or in partial shade of open woodland: Mount Tamalpais (Corte Madera Ridge, Throckmorton Ridge, Azalea Flat, Bolinas Ridge); San Anselmo Canyon; San Geronimo Ridge.

3. **C. sorediatus** H. & A. JIM BRUSH. Occasional in the chaparral on Mount Tamalpais (Corte Madera Ridge, West Point, Northside Trail, Phoenix Lake), in the Carson country (San Anselmo Canyon), and in Lucas Valley, *Leschke*.

A distinctive feature of this plant is the acrid odor emitted by its foliage, an odor frequently discernible along trails even before the shrub is seen. This smell is not noticeable in a few leaves (as in a bouquet, for instance) but apparently only where there is an entire plant or thicket. In full bloom this species is one of the most attractive in *Ceanothus*. In 1889, before *C. sorediatus* was properly understood, Parry named the plant on Mount Tamalpais *C. intricatus.*

4. **C. thyrsiflorus** Esch. BLUE BLOSSOM. Common and widespread in woodland, borderland brush, and chaparral and on exposed maritime slopes: Sausalito, Angel Island, and Mount Tamalpais (Blithedale Canyon, Bolinas Ridge, Phoenix Lake) to San Rafael Hills, Tomales, and Point Reyes Peninsula (Bolinas, Inverness Ridge, Point Reyes).

This species, which is the most abundant in our region, is the one commonly referred to as wild lilac or mountain lilac. It is extremely variable in habit: on well-watered slopes in deep soil it frequently attains the size of a small tree. Such arborescent individuals on San Geronimo Ridge and near Mud Lake are 25 feet tall and have trunks to 9 inches in diameter covered with a nearly smooth brownish gray bark. In marked contrast, plants on exposed ocean bluffs are prostrate with only short leafy twigs arising from the ground. This form is var. *repens* McMinn, the type locality of which is Point Reyes.

Suspected hybrids between *C. thyrsiflorus* and two other species have been recorded from Mount Tamalpais. One, between this species and *C. foliosus* in upper Cascade Canyon, is represented by tall shrubs with leaves intermediate between the two species. Another between *C. thyrsiflorus* and *C. sorediatus* has been reported by McMinn.

5. **C. ramulosus** (Greene) McMinn. BLUE BUCK-BRUSH. Rocky and gravelly slopes in the chaparral, often gregarious and coloring broad areas on steep canyonsides: Sausalito; Mount Tamalpais (Corte Madera Ridge, above Muir Woods, Potrero Meadow, Phoenix Lake); San Anselmo Canyon; San Geronimo Ridge.—*C. cuneatus* (Hook.) Nutt. var. *ramulosus* Greene.

From field observations, *C. ramulosus* has been observed to hybridize with several other species on Mount Tamalpais. Plants apparently derived from a cross between this species and *C. Jepsonii* have been seen near Bootjack Camp and southeast of Laurel Dell. Plants combining the thick leaves of *C. ramulosus* and the alternate leaf arrangement of *C. foliosus* have been found on the south side of the mountain above Cascade Canyon. An apparent cross between *C. ramulosus* and *C. gloriosus* var. *exaltatus* has become well established on Bolinas Ridge and has been described as *C. Masonii*.

6. **C. gloriosus** J. T. Howell. GLORY MAT. Sandy flats near the radio station and exposed ridges on McClure Beach road, Point Reyes Peninsula.

This prostrate coastal plant and the following erect variety occur northward to Mendocino County but south of the Golden Gate at Monterey they are replaced by the closely related and highly restricted *C. rigidus* Nutt. Specimens of *C. gloriosus* collected by Bigelow on "Punto de los Reyes" were described as *C. rigidus* Nutt. var. *grandifolius* Torr.

6a. **C. gloriosus** var. **exaltatus** J. T. Howell. GLORY BRUSH. Rocky slopes in the chaparral on Bolinas Ridge, the only locality known in Marin County. A handsome plant when in full bloom.

6b. **C. gloriosus** var. **porrectus** J. T. Howell. MOUNT VISION CEANOTHUS. Known only from the bishop pine forest on Inverness Ridge: slope of Mount Vision; Ledum Swamp; near summit of Point Reyes road, the type locality.

7. **C. Masonii** McMinn. Restricted to the chaparral on the rocky slopes of Bolinas Ridge.

In character, *C. Masonii* is intermediate between *C. gloriosus* var. *exaltatus* and *C. ramulosus* and field evidence would indicate that it is a hybrid derivative. It grows with the former on Bolinas Ridge and the other probable parent is found in the vicinity to the east and on Mount Tamalpais. *Ceanothus Masonii* is the Marin County plant referred to *C. rigidus* by Jepson.

8. **C. Jepsonii** Greene. MUSK BUSH. Serpentine slopes of high open ridges: Mount Tamalpais (Bootjack, Barths Retreat, Northside Trail); Carson Ridge; San Geronimo Ridge, the type locality.

The flowers are few in the clusters but they are among the largest in

Ceanothus and of a pleasing violet-blue color. Just before the fruits are ripe, they are frequently reddish-tinged and attractive, too. The flowers and the fruits, or just the leaves, make this plant one of the most interesting and attractive on the Marin County serpentine barrens.

VITACEAE. Grape Family

1. **Vitis.** Grape; Vine

1. **V. californica** Benth. The California grape has been seen in Marin County only at Ignacio where it occurred near the railroad and where it was probably introduced in stream gravels for road embankments, just as *Salix melanopsis, Mentzelia laevicaulis,* and *Chrysopsis oregona* have been introduced. The grape is very common and characteristic along streams in Sonoma County, so it may yet be found as indigenous in northern Marin County.

MALVACEAE. Mallow Family

a. Corolla brick-red; carpels 15 or more, dehiscent, bearing 2 short slender horns on the back, cavity divided by a horizontal partition ......4. *Modiola*
a. Corolla white, pink, or mauve; carpels 5–15, indehiscent, rounded on the back and not horned, cavity not divided ....b
b. Stamens in 2 whorls at the top of the stamen column; bractlets none (or 1–3 in *Sidalcea Hickmanii*) ....3. *Sidalcea*
b. Stamens not in 2 whorls at the top of the stamen column; bractlets 1–3, attached near the base of the calyx c
c. Stamens clustered in a single whorl at top of stamen column; bractlets 1–3, distinct from each other at the base ....1. *Malva*
c. Stamens scattered below the top of the stamen column; bractlets 3, united at the base ....2. *Lavatera*

1. **Malva.** Mallow

a. Bractlets ovate or elliptic-ovate; corolla about twice as long as the calyx in flower ....1. *M. nicaeensis*
a. Bractlets linear to lanceolate; corolla about equaling the calyx in flower ....b
b. Calyx lobes acute, much broader than long; carpels pubescent, the edges toothed ....2. *M. parviflora*
b. Calyx lobes shortly acuminate, about as long as broad; carpels glabrous, the edges rounded ....3. *M. verticillata*

1. **M. nicaeensis** All. Common weed along roads and paths and in waste places around towns: Sausalito, Tiburon, and Mill Valley to San Rafael, Dillons Beach, and Point Reyes. Introduced from the Mediterranean region.—*M. borealis* of California references.

2. **M. parviflora** L. Cheese-weed. Weed of roadsides and waste ground, scarcely as common as the preceding: Sausalito (acc. Lovegrove), Angel Island, Tiburon, and Mill Valley to San Rafael, Bolinas, and Dillons Beach. Native of the Mediterranean region.

3. **M. verticillata** L. var. **crispa** L. The only California collection of this Old World curly-leaved mallow that has been seen is one made by Stacey in Mill Valley in 1933.

2. Lavatera

a. Bractlets longer than the calyx and tending to conceal it; leaves densely
   pubescent above and below ....3. *L. arborea*
a. Bractlets shorter than the calyx, not tending to conceal it; leaves much less
   pubescent above than below ....b
b. Shrub; petals 3–4 cm. long, long-clawed ....1. *L. assurgentiflora*
b. Annual herb; petals 1–2 cm. long, short-clawed ....2. *L. cretica*

1. **L. assurgentiflora** Kell. This plant, native on the islands off southern California, is used extensively as a windbreak in the central coastal part of the state
where occasionally it is spontaneous, as in Sausalito (acc. Stacey).

2. **L. cretica** L. This Mediterranean malva-like plant is common enough as a
weed in waste ground in San Francisco but in Marin County, it is known only
from very localized occurrences at Point Bonita, Sausalito, and Angel Island. When
once established, it can be expected to spread widely through the county.

3. **L. arborea** L. Locally established at several stations along the central California coast, this Mediterranean species has been seen in Marin County as a
fugitive from cultivation at Point Bonita, Sausalito, and Dillons Beach.

3. Sidalcea

a. Stipules (at least above middle of plant) divided into linear segments. Plants
   annual ....1. *S. diploscypha*
a. Stipules lanceolate to elliptic or ovate, simple ....b
b. Bractlets 1–3, rarely none; upper and lower leaves nearly alike, semicircular
   or round-reniform, crenate or shallowly crenately lobed. Plants perennial
   ....5. *S. Hickmanii*
b. Bractlets none; upper leaves usually much more deeply lobed or divided than
   the lower c
c. Leaves and stems conspicuously pubescent, more or less stellate-hirsute; carpels
   reticulately veined, the transverse veins more prominent than the longitudinal. Plants perennial ....4. *S. malvaeflora*
c. Leaves and stems sparsely hairy or glabrous; carpels longitudinally striate on
   the back ....d
d. Plants annual; stipules and stipular bracts mostly 0.5–1 cm. long, acute
   ....2. *S. calycosa*
d. Plants perennial with long-creeping stoloniferous base; stipules and stipular
   bracts mostly 1–1.5 cm. long, obtusish ....3. *S. rhizomata*

1. **S. diploscypha** (T. & G.) Gray. Clay soil of open slopes or valley flats, not
common: Fairfax, *Sutliffe;* San Anselmo Canyon; San Rafael, *Jackson;* Chileno
Valley. A most attractive species with its large rosy-pink flowers.

2. **S. calycosa** Jones. Rare in clay soil of low wet meadows: Lagunitas Meadows;
Point Reyes Station, acc. Stacey.

3. **S. rhizomata** Jeps. Not uncommon in swales and marshes on Point Reyes
Peninsula, the type locality: near the radio station; McClure Beach road. In
habit and habitat as well as in several technical characters, this flaccid perennial
that is at home among coastal tussoks of sedge and rush need not be confused
with the preceding with which it is sometimes combined.

4. **S. malvaeflora** (DC.) Gray. CHECKER-BLOOM. Widespread and sometimes abundant and colorful on open grassy hills and maritime mesas: Sausalito, Angel Island, Tiburon, and Mount Tamalpais (Rock Spring, Lagunitas Meadows) to San Rafael Hills, Tomales, and Point Reyes Peninsula (road to lighthouse). The coastal plants with most of the leaves conspicuously round-reniform are probably a distinct entity deserving recognition.

At its best on coastal hills, the checker-bloom is attractive enough but its flowers of a rosy-mauve just miss being a really good color. The flowers are of two distinct sizes: a large flower of pale mauve that is perfect or staminate, and a small flower of deep rose that is pistillate.

5. **S. Hickmanii** Greene. One of the rare plants of California, in Marin County occurring on Carson Ridge in disturbed soil or in burned areas. Following the fire in 1945, seedlings bloomed in one year. Besides the Marin County station, this rare mallow is known in Monterey County and, as var. *Parishii* Robins., in the San Bernardino Mountains.

4. Modiola

1. **M. caroliniana** (L.) G. Don. Weed of lawns, paths, and waste ground, rare in Marin County: Point Reyes Station, acc. Stacey; Inverness. Native in eastern North America, South America, and South Africa.

GUTTIFERAE. ST. JOHN'S WORT FAMILY

1. Hypericum. ST. JOHN'S WORT

a. Petals about equaling the sepals, without black dots on the margin
 . . . . .1. *H. anagalloides*
a. Petals much longer than the sepals, with black dots on the margin . . . . . . .b
b. Sepals linear-deltoid, acuminate, mostly without black dots; lower leaf axils with conspicuous sterile branches . . . . .4. *H. perforatum*
b. Sepals ovate, oblong, or obovate; lower leaf axils without elongate sterile branches c
c. Leaves oblong to ovate, mostly flat; sepals obtuse or subacute, mostly without black dots . . . . .2. *H. Scouleri*
c. Leaves linear to lanceolate, mostly folded; sepals acute with numerous black dots on the margin . . . . .3. *H. concinnum*

1. **H. anagalloides** C. & S. TINKER'S PENNY. Common in wet ground of coastal marshes and meadows: Rodeo Lagoon; Elk Valley; Azalea Flat and Rock Spring, Mount Tamalpais; Lagunitas, acc. Stacey; Bolinas, Shell Beach, and Ledum Swamp, Point Reyes Peninsula.

2. **H. Scouleri** Hook. Wet meadows on Mount Tamalpais: near Bootjack; Laurel Dell Trail; Potrero Meadow; Willow Meadow.—*H. formosum* H.B.K. var. *Scouleri* (Hook.) Coulter.

3. **H. concinnum** Benth. GOLD WIRE. Dry rocky ridges in the chaparral: Mount Tamalpais (Corte Madera and Throckmorton ridges, Potrero Meadow); San Anselmo Canyon. This species, so attractive with its sparse foliage and abundant golden blossoms, is not known south of Marin County in the Coast Ranges.

4. **H. perforatum** L. KLAMATH WEED. This evil Old World weed is known to be locally established at four stations in Marin County: Sausalito; south of

Muir Woods; Cascade Drive, Fairfax; Ridgecrest Road, Fairfax. Unfortunately, while it might still be easily eradicated, it is allowed to flourish and propagate.

ELATINACEAE. Waterwort Family

1. Elatine. Waterwort

1. **E. brachysperma** Gray. Small prostrate annual of muddy flats or drying ponds, in Marin County known only from the strand of Lake Lagunitas and from the laguna in Chileno Valley.

FRANKENIACEAE. Frankenia Family

1. Frankenia

1. **F. grandifolia** C. & S. Common in the salt marshes and in subsaline soil along their borders: Sausalito, Tiburon, Burdell, and Black Point to Rodeo Lagoon, Stinson Beach, Tomales Bay, and Drakes Estero.

The low bushy plants, sometimes hoary with crystals of salt, are not unattractive with their rosy-mauve flowers.

CISTACEAE. Rock-rose Family

a. Leaves alternate; petals 0.3–1 cm. long, yellow ....1. *Helianthemum*
a. Leaves opposite; petals 2–3 cm. long, mauve or purplish ....2. *Cistus*

1. Helianthemum

1. **H. scoparium** Nutt. Broom-rose. Open pine woods on Inverness Ridge.

Much more common and widespread on rocky or gravelly slopes in the chaparral is var. *vulgare* Jeps. in which the flowers are only about half as large as in the typical pineland variety: Mount Tamalpais (Corte Madera Ridge, Throckmorton Ridge, Barths Retreat); Carson Ridge; San Geronimo Ridge; San Rafael Hills. After a fire in the chaparral, seedlings come up by the thousands, very leafy and quite different in appearance from the nearly leafless, broomlike, mature plants.

2. Cistus. Rock-rose

1. **C. villosus** L. One of the forms of this variable species, which is so common and widespread in the Mediterranean region, has become abundantly naturalized on the openly wooded hills near Black Point. The variant occurring here is var. *corsicus* (Lois.) Gross., a glandular plant lacking the villous stems and leaves that characterize the typical variety.

VIOLACEAE. Violet Family

1. Viola. Violet

a. Petals violet or white and purplish ....b
a. Petals yellow c
b. Petals violet, somewhat paler at the base ....1. *V. adunca*
b. Petals white, the back of the upper two, purple ....2. *V. ocellata*
c. Leaves cordate; petals not purplish- or brownish-tinged on the back ....d
c. Leaves narrowed or truncate at base, generally not cordate; petals tinged with

purplish or brownish ....e
d. Aerial stems erect; leaves thin, not evergreen ....3. *V. glabella*
d. Aerial stems creeping, stoloniferous; leaves subcoriaceous, evergreen
....4. *V. sempervirens*
e. Upper and lower leaves broadly ovate; flowers 1-2 cm. long ..5. *V. pedunculata*
e. Upper leaves elongate, ovate-lanceolate, the lower generally broader; flowers
generally 1 cm. long or less ....6. *V. quercetorum*

1. **V. adunca** Smith. On open grassy slopes, sandy flats back of the dunes, and on the edge of brush under pines, rather common near the coast: Sausalito; Bolinas Ridge; Lagunitas, acc. Stacey; Tomales; Ledum Swamp and Point Reyes dunes, Point Reyes Peninsula.

Viola Howellii Gray, a closely related northern species with a short blunt spur, has been reported from Point Reyes Peninsula by Brainerd and by Mason but no plants differing from *V. adunca* have been seen from there.

2. **V. ocellata** T. & G. WESTERN HEART'S EASE. Rocky slopes and grassy flats, occasional: Kent Trail and Lagunitas Meadows, Mount Tamalpais; Big Carson Canyon; Lagunitas.

3. **V. glabella** Nutt. Moist shaded places along streams or in woods, rare in Marin County: Lagunitas Canyon; Papermill Creek; Nicasio, *Leschke.*

4. **V. sempervirens** Greene. Deep shade of wooded flats and canyons, particularly common under redwoods with *Oxalis oregana:* Mill Valley; Muir Woods; Bolinas Ridge; Big Carson Canyon, acc. Sutliffe; Inverness Ridge.—*V. sarmentosa* Dougl.

5. **V. pedunculata** T. & G. JOHNNY-JUMP-UP; YELLOW PANSY. Not widespread but locally common on grassy hills: Sausalito; Angel Island; Tiburon; Mill Valley, acc. Orr; Black Point.

6. **V. quercetorum** Baker & Clausen. Rare in grassy places near the top of Mount Tamalpais: Mountain Theater; Potrero Meadow, acc. Sutliffe. Perhaps also on San Geronimo Ridge.—*V. purpurea* of authors in part.

LOASACEAE. LOASA FAMILY

1. Mentzelia

a. Upper leaves saliently toothed or acutely serrate-lobed; petals oblanceolate
....1. *M. laevicaulis*
a. Upper leaves pinnately parted or divided; petals obovate ......2. *M. Lindleyi*

1. **M. laevicaulis** (Dougl.) T. & G. BLAZING STAR. Although widespread in California in loose rocky soil, this plant was undoubtedly introduced into Marin County with river gravels used for railroad embankments: San Anselmo, acc. Pollard; common along railroad north of San Rafael; Ignacio, acc. Stacey.

2. **M. Lindleyi** T. & G. Native of the South Coast Ranges and Sierra Nevada foothills, occasionally escaping from cultivation in the North Coast Ranges: north of San Rafael, *Pollard.*

DATISCACEAE. DATISCA FAMILY

1. Datisca

1. **D. glomerata** (Presl) Baill. DURANGO ROOT. Rank erect herb in rocky beds of summer-dried streams, reported from Mount Tamalpais by Jepson.

THYMELAEACEAE. DAPHNE FAMILY

1. Dirca

1. D. occidentalis Gray. WESTERN LEATHERWOOD. Moist hillsides in partial shade, rare: Lagunitas, *Eastwood;* Devils Gulch, Samuel P. Taylor State Park, *Paul Wilson;* near confluence of Nicasio and Lagunitas creeks, *Menzies;* Inverness Ridge south of Inverness, *Lockhart.*

This, one of the rarest shrubs in California, is found only in hills around San Francisco Bay. It is also one of the most beautiful, when, in the spring, its neat dark brown stems are decked with small clusters of bright yellow fragrant flowers. Otherwise it is an inconspicuous shrub with its soft green leaves and yellowish green fruits.

LYTHRACEAE. LOOSE-STRIFE FAMILY

a. Leaves alternate; calyx cylindric; capsule elongate, cylindric ......1. *Lythrum*
a. Leaves opposite; calyx campanulate; capsule globose ....2. *Ammannia*

1. Lythrum. LOOSE-STRIFE

a. Stems erect, usually 1 m. or more tall; petals about 5 mm. long
....1. *L. californicum*
a. Stems low and erect or longer and adsurgent; petals 1–3 mm. long ....b
b. Plant annual, the stems not stoloniferous ....2. *L. Hyssopifolia*
b. Plant perennial, the stems stoloniferous ....3. *L. adsurgens*

1. L. californicum T. & G. Widespread through California in wet ground around springs or marshes, in Marin County reported from Mount Tamalpais by Stacey.

2. L. Hyssopifolia L. Common in low places in fields and meadows and in roadside hollows that are wet in the spring: Rodeo Lagoon, Tiburon, and Mount Tamalpais (Lagunitas Meadows, Phoenix Lake, Rock Spring) to San Rafael Hills, Black Point, and Point Reyes Peninsula (Inverness, Ledum Swamp).

3. L. adsurgens Greene. Low places that are wet in the spring, perhaps only a rare perennial phase of *L. Hyssopifolia:* Lake Lagunitas, Mount Tamalpais; Stinson Beach; Bolinas; Lagunitas, acc. Jepson.

2. Ammannia

1. A. coccinea Rottb. In Marin County known only from the summer-dried strand of Lake Lagunitas below the high-water line.

MYRTACEAE. MYRTLE FAMILY

1. Eucalyptus

a. Buds roughly wrinkled and ridged, 1.5–2 cm. long; flowers 3–4 cm. broad
....1. *E. globulus*
a. Buds smooth, 0.5–1 cm. long; flowers 1–1.5 cm. broad ....2. *E. viminalis*

1. E. globulus Labill. BLUE GUM. Abundantly spontaneous around old cultivated specimens, reproduction particularly prolific on cleared or burned ground: Sausalito, Angel Island, Tiburon, Belvedere, and Mount Tamalpais (Cascade

Canyon, Lagunitas Meadows) to San Rafael, Chileno Valley, Tomales, and Bolinas. This tree is by far the most aggressive tree weed in Marin County. Introduced from Australia.

2. **E. viminalis** Labill. Commonly cultivated, only occasionally spontaneous, as in Tamalpais Valley. Native of Australia.

ONAGRACEAE. EVENING PRIMROSE FAMILY

a. Flowers without a distinct hypanthium, the sepals persisting on the ovary after flowering ....b
a. Flowers with a free hypanthium, the hypanthium and attached sepals deciduous after flowering c
b. Leaves alternate; sepals 5; petals showy ....1. *Jussiaea*
b. Leaves opposite; sepals 4; petals minute or lacking ....2. *Ludwigia*
c. Sepals and petals 2, the petals white; fruit 1- or 2-seeded, indehiscent and bearing hooked hairs ....9. *Circaea*
c. Sepals and petals 4; fruit a many-seeded capsule ....d
d. Seeds with a tuft of hairs at one end ....e
d. Seeds without a tuft of hairs ....f
e. Flowers red, the hypanthium usually more than 2 cm. long ....3. *Zauschneria*
e. Flowers white to mauve or deep rose, the hypanthium less than 2 cm. long ....4. *Epilobium*
f. Petals yellow or turning reddish in age ....8. *Oenothera*
f. Petals white, pink, or reddish purple ....g
g. Sepals distinct, erect; petals generally small and not showy ....5. *Boisduvalia*
g. Sepals reflexed, sometimes remaining united at the tip; petals generally large and showy ....h
h. Petals distinctly clawed, the claws equaling the blade or a little shorter ....6. *Clarkia*
h. Petals scarcely, if at all, clawed ....7. *Godetia*

1. Jussiaea

1. **J. uruguayensis** Camb. Low wet ground at the south end of Tiburon Peninsula, the only station known in California. The plants exhibit two habits—a low creeping form on the muddy strand of a pond, and a robust erect form 3 or 4 feet tall on somewhat higher ground among willows. It is the latter form that blooms and fruits. Native of South America.

2. Ludwigia

1. **L. palustris** (L.) Ell. var. **pacifica** Fern. & Griscom. Muddy ground where water has stood: the laguna in Chileno Valley, the only locality known in Marin County and the southernmost in the Coast Ranges.

3. Zauschneria

1. **Z. californica** Presl. CALIFORNIA FUCHSIA. Occasional in shallow soil on dry rocky slopes: Angel Island; Belvedere; Bootjack, Mount Tamalpais; Fairfax Hills; Little Carson Canyon; Lagunitas, acc. Stacey.
On the rocks above Bootjack the scarlet tubular corollas of *Penstemon corym-*

bosus of early summer are replaced in late summer and early fall by the equally brilliant red flowers of the zauschneria.

4. Epilobium. WILLOW-HERB

a. Plants perennial, generally growing in wet places; epidermis of stems not exfoliating ....b
a. Plants annual, generally growing in dry ground; epidermis of stems exfoliating ....f
b. Rootstocks producing small bulblike winter buds (i.e., turions); stems slender, generally 3 dm. tall or less and unbranched ....1. *E. Halleanum*
b. Rootstocks not producing turions; stems stouter, generally more than 3 dm. tall and branched above c
c. Petals 2–6 mm. long, white to purplish rose; upper leaves generally alternate .d
c. Petals 6–10 mm. long, purplish rose; leaves generally opposite ....e
d. Upper stems and inflorescence glandular ....2. *E. adenocaulon*
d. Upper stems and inflorescence pubescent but not glandular 3. *E. californicum*
e. Stems and leaves gray-pubescent ....4. *E. Watsonii*
e. Stems and leaves glabrous or nearly so ....4a. *E. Watsonii* var. *franciscanum*
f. Stems generally less than 3 dm. tall, pubescent to the base; hypanthium about 1 mm. long ....5. *E. minutum*
f. Stems generally more than 3 dm. tall, glabrous below, pubescent above; hypanthium 2–15 mm. long ....g
g. Hypanthium 6 mm. long or more ....6c. *E. paniculatum* var. *jucundum*
g. Hypanthium 2–6 mm. long ....h
h. Hypanthium 2–3 mm. long i
h. Hypanthium 4–6 mm. long. Pedicels and capsules thinly to densely glandular-puberulent ....6b. *E. paniculatum* f. *laevicaule*
i. Pedicels or capsules or both slightly glandular-puberulent ..6. *E. paniculatum*
i. Pedicels and capsules more densely glandular-puberulent

....................................6a. *E. paniculatum* f. *adenocladon*

1. **E. Halleanum** Hausskn. Marshy ground, Potrero Meadow, Mount Tamalpais. This is the southernmost station for any of the members of the *E. brevistylum* group in the Coast Ranges.

2. **E. adenocaulon** Hausskn. var. **occidentale** Trel. Occasional in wet springy places or along streams: Elk Valley; Stinson Beach; above Muir Woods; Mill Valley; Point Reyes, acc. Stacey.—*E. californicum* Hausskn. var. *occidentale* (Trel.) Jeps.

3. **E. californicum** Hausskn. Rather common in low wet ground: Tiburon: Frank Valley; Mill Valley; West Point, Mount Tamalpais; San Rafael Hills; Santa Venetia.

The conspicuously gray-hairy form, var. *holosericeum* (Trel.) Jeps., has been collected by Leschke in Gallinas Valley.

4. **E. Watsonii** Barbey. A rare plant of marshy places, in Marin County being known only from Dillons Beach and near the radio station on Point Reyes Peninsula. In both places it grew with the common glabrous plant which is here treated as a variety, though the nearness of the relationship of the two is only surmised.

4a. **E. Watsonii** Barbey var. **franciscanum** (Barbey) Jeps. Common in wet ground of coastal slopes and valleys: Sausalito, Angel Island, Tiburon, and Mill Valley to Dillons Beach and Point Reyes Peninsula (Bolinas, Mud Lake, Ledum Swamp).

5. **E. minutum** Lindl. Shallow soil of open rocky slopes, frequently on serpentine: Sausalito (the type locality of *E. Congdonii* Lévl., a synonym); Tiburon; Mount Tamalpais (Rock Spring, West Point, Rifle Camp, Berry Trail); Carson country.

A small-flowered form with petals less than 2 mm. long is var. *Biolettii* Greene, originally described from the south side of Mount Tamalpais. It has been seen in the Carson country as well as on Mount Tamalpais.

6. **E. paniculatum** Nutt. In typical or nearly typical form, this plant is rare in Marin County, having been seen only in low ground near the laguna in Chileno Valley. The following three are much more common:

6a. **E. paniculatum** Nutt. f. **adenocaulon** Hausskn. Hills and low fields that are wet in the spring: Manzanita; Frank Valley; Stinson Beach; Carson Ridge; Chileno Valley.

6b. **E. paniculatum** Nutt. f. **laevicaule** (Rydb.) St. John. Dry slopes and flats, becoming weedy around habitations: Mill Valley; Kentfield; Phoenix Lake, Mount Tamalpais; San Rafael; Ignacio.

6c. **E. paniculatum** Nutt. var. **jucundum** (Gray) Trel. Dry slope north of San Rafael. This tall airy plant with its large purplish rose flowers is rather attractive in a weedy sort of way.

5. Boisduvalia

a. Petals 6–12 mm. long; capsule septifragal; stems generally more than 3 dm.
    tall ....1. *B. densiflora*
a. Petals 1–4 mm. long; capsule loculicidal; stems generally less than 3 dm. tall. .b
b. Floral leaves oblong to ovate; petals 2–4 mm. long ....2. *B. glabella*
b. Floral leaves linear; petals 1–2 mm. long ....3. *B. stricta*

1. **B. densiflora** (Lindl.) Wats. Low places in the hills and valleys that are wet or marshy during the rainy season: Rodeo Lagoon and Ross to San Rafael, Carson Ridge, Olema Marshes, and Chileno Valley. In the commonest form in Marin County, f. *imbricata* (Greene) Munz, the floral leaves overlap and conceal the capsules.

2. **B. glabella** (Nutt.) Walp. var. **campestris** (Jeps.) Jeps. Dried beds of ponds, known in Marin County only from Lagunitas Meadows and Lake Lagunitas, *Sutliffe*.

3. **B. stricta** (Gray) Greene. Beds of shallow puddles on Carson Ridge, the only known station in Marin County.

6. Clarkia

a. Petals entire; stamens 8 ....1. *C. elegans*
a. Petals 3-lobed; stamens 4 ....2. *C. concinna*

1. **C. elegans** Dougl. Loose clayey soil and gravelly banks of hills and canyons, occasional: Mount Tamalpais, acc. Jepson; Fairfax; Gallinas Valley, *Leschke*.

It is from this native species that such an amazing series of horticultural forms has been developed for garden culture, chiefly by European hybridizers.

2. **C. concinna** (F. & M.) Greene. Shallow soil of steep rocky banks in partial shade: Northside Trail, Mount Tamalpais; Fairfax Hills; San Rafael Hills; Lagunitas.

7. Godetia

a. Inflorescence nodding in bud ....5. *G. lassenensis*
a. Inflorescence erect in bud ....b
b. Stigma lobes broadly deltoid, about 1 mm. long and broad; hypanthium 2–7 mm. long. Sepals reflexed in pairs or separately; capsule 8-ribbed when immature c
b. Stigma lobes elliptic to linear, 1.5–7 mm. long, about 0.5 mm. wide; hypanthium 4–12 mm. long ....f
c. Capsule slender, oblongish to linear, not noticeably enlarged at the center; flowers scattered, not becoming congested at the end of stems ....d
c. Capsule stout, noticeably enlarged at the center; flowers usually becoming crowded at the end of stems ....e
d. Leaves linear-lanceolate to linear-oblanceolate, acute ....1. *G. quadrivulnera*
d. Leaves oblanceolate to elliptic or obovate, obtuse
....1a. *G. quadrivulnera* var. *Davyi*
e. Leaves mostly more than 1 cm. wide; capsule pubescent or glabrate
....2. *G. purpurea*
e. Leaves mostly less than 1 cm. wide; capsule pubescent, often densely so
....................................2a. *G. purpurea* var. *parviflora*
f. Sepals reflexed in pairs or separately; capsule 8-ribbed when immature
....3. *G. viminea*
f. Sepals reflexed united; capsule 4-sulcate when immature ....4. *G. amoena*

1. **G. quadrivulnera** (Dougl.) Spach. Grassland and brushy slopes, occasional: Tiburon; Bootjack and Phoenix Lake, Mount Tamalpais; San Anselmo Canyon; San Rafael Hills.

1a. **G. quadrivulnera** (Dougl.) Spach var. **Davyi** Jeps. Coastal slopes or flats, commonly in sandy soil: Point Reyes Peninsula (Point Reyes, the type locality, Mount Vision, acc. Hitchcock, Point Reyes dunes); near Tomales.

2. **G. purpurea** (Curt.) G. Don. Rather rare on coastal hills: Sausalito; Tomales Point, acc. Hitchcock.

2a. **G. purpurea** (Curt.) G. Don var. **parviflora** (Wats.) Hitchc. Occasional on grassy or brushy slopes: Sausalito, acc. Hitchcock; Tiburon; Mill Valley; Frank Valley; Bootjack, Mount Tamalpais; Black Point.

3. **G. viminea** (Dougl.) Spach. Rare on grassy hillsides: Carson country, *Eastwood;* Olema, acc. Hitchcock.

4. **G. amoena** (Lehm.) G. Don. Widespread on brushy or open grassy slopes and variable. On hills immediately adjacent to the coast the flowers are unusually large and the color is a very attractive pink: Rodeo Lagoon; Sausalito; Point Reyes Peninsula (Point Reyes, McClure Beach). More generally distributed is the form with slightly smaller magenta-pink flowers: Sausalito, Angel Island, and Tiburon to Mount Tamalpais (Bootjack, Potrero Meadow), Fairfax Hills, and San Rafael Hills.

5. **G. lassenensis** Eastw. var. **concolor** (Jeps.) J. T. Howell. Occasional in

grassy or brushy places on dry open hills: between Mountain Theater and Potrero Meadow, Mount Tamalpais; Carson country; Fairfax Hills; San Rafael Hills.—*G. amoena* (Lehm.) G. Don var. *concolor* Jeps.

This species is not known farther south.

8. Oenothera

a. Petals 2–5 cm. long; lobes of stigmas 4, linear ....1. *O. Hookeri*
a. Petals less than 2 cm. long; stigmas entire, capitate ....b
b. Stems subterranean, the leaves clustered in a rosette; hypanthium pedicel-like, up to 10 cm. long ....5. *O. ovata*
b. Stems evident, elongate and leafy; hypanthium not drawn down into a slender pedicel-like tube c
c. Capsules terete; stems slender ....2. *O. contorta*
c. Capsules 4-sided; stems stout ....d
d. Pubescence of leaves hirsutulous; petals 2–4 mm. long ......3. *O. micrantha*
d. Pubescence of leaves appressed-silky; petals 5–9 mm. long. 4. *O. cheiranthifolia*

1. **O. Hookeri** T. & G. Evening Primrose. Occasional in sandy soil near the coast or in clay soil bordering marshy depressions: Stinson Beach; Olema Marshes; east of Aurora School.

2. **O. contorta** Dougl. var. **strigulosa** (F. & M.) Munz. Dunes and sandy flats along the ocean: Stinson Beach; Dillons Beach; Point Reyes dunes.

3. **O. micrantha** Hornem. Dunes near the radio station on Point Reyes Peninsula seem to be the northernmost known station for this species.

4. **O. cheiranthifolia** Hornem. Maritime dunes and beach sand just above high tide: Rodeo Lagoon; Stinson Beach; Dillons Beach; Point Reyes Peninsula (Shell Beach, dunes near radio station, lighthouse).—*O. spiralis* Hook.

5. **O. ovata** Nutt. Sun-cups. Widespread and sometimes locally numerous on open grassy hills and flats: Sausalito, Angel Island, and Tiburon to San Rafael, Ignacio, and Tomales.

9. Circaea. Enchanter's Nightshade

1. **C. pacifica** Aschers. & Magnus. In deep shade and moist soil: Camp Taylor; Lagunitas and Olema creeks, acc. Eastwood. Papermill Creek (i.e., Lagunitas Creek) is probably the type locality according to Jepson. This widely distributed but rare western American plant is not known south of Marin County in the Coast Ranges.

HALORAGACEAE. Water-milfoil Family

a. Submerged and aerial leaves alike and entire ....1. *Hippuris*
a. Submerged leaves with elongate, linear, or filiform divisions, the aerial similar or less divided to entire ....2. *Myriophyllum*

1. Hippuris. Mare's Tail

1. **H. vulgaris** L. Apparently this very widely distributed aquatic plant has not been recently collected in Marin County, the only known record being one by Bigelow in 1854 from "Tomales Bay."

2. **Myriophyllum.** MILFOIL

a. Flowers in the axils of bracts, the bracts shorter or but little longer than the
flowers ....1. *M. exalbescens*

a. Flowers in the axils of well-developed aerial leaves ....b

b. Submerged leaves divided into filiform segments, aerial leaves linear-oblong,
pinnately lobed to subentire; plants monoecious, the flowers unisexual or
perfect ....2. *M. hippuroides*

b. Submerged and aerial leaves alike, divided into linear segments; plants
dioecious ....3. *M. brasiliense*

1. **M. exalbescens** Fern. Rare in fresh-water ponds: Lagunitas, acc. Stacey; Camp
Taylor, acc. Behr. A sterile milfoil collected in Lily Lake is probably this
species.—*M. spicatum* L. var. *exalbescens* (Fern.) Jeps.

2. **M. hippuroides** Nutt. In Marin County known only from the pond south
of Olema, the southernmost station to be reported in the Coast Ranges.

3. **M. brasiliense** Camb. PARROT FEATHER. Pistillate plants are locally established
in marshy ground near Point Reyes Station. This native of South America has
been cultivated in aquaria for many years as *M. proserpinacoides* Gill.

ARALIACEAE. ARALIA FAMILY

a. Robust herb with large compound leaves ....1. *Aralia*

a. Woody trailing vine or climbing shrub with simple evergreen leaves
....2. *Hedera*

1. Aralia

1. **A. californica** Wats. ELK CLOVER. Occasional in wet soil around springs or
along streams, generally in deep shade: Mount Tamalpais (Blithedale Canyon,
Nora Trail, Steep Ravine, Bolinas Ridge); Lagunitas Canyon; Papermill Creek;
near Mud Lake and Inverness, Point Reyes Peninsula.

2. Hedera

1. **H. Helix** L. ENGLISH IVY. Widely cultivated and at times becoming naturalized around towns or in coastal hills: Sausalito; Belvedere; Cascade and
Blithedale canyons, Mill Valley; Ross; Ignacio; Olema country, acc. Menzies; Inverness. Native of Europe.

Ivy has two kinds of shoots: the juvenile with stems which cling closely to
ground, wall, or tree and bear angular cordate leaves, and the mature with stems
which are loose and bushy and bear round-ovate, truncate or broadly cuneate
leaves.

UMBELLIFERAE. PARSLEY FAMILY

a. Fruits smooth, glabrous or pubescent. Plants mostly perennial (*Ammi, Apium,
Bowlesia, Conium,* and *Pastinaca* annual or biennial) ....b

a. Fruit spiny, papillate, scaly, or bristly. Plants frequently annual or biennial
(*Eryngium, Osmorhiza,* and *Sanicula* perennial) ....t

b. Ribs of fruits not winged; fruits terete or somewhat flattened laterally c

b. Lateral ribs of fruits winged; fruit more or less flattened dorsally, usually
conspicuously so ....o

c. Leaves compound, frequently much dissected ....d
c. Leaves simple or blades reduced to the rachis ....m
d. Leaflets entire, linear or filiform ....e
d. Leaflets serrate to deeply lobed or divided, narrowly lanceolate to broadly
    ovate ....f
e. Stems stout, quite leafy; petals yellow ....12. *Foeniculum*
e. Stems slender, sparsely leafy; petals white or pink ....16. *Perideridia*
f. Umbels sessile or short-pedunculate. Plants of marshy places, biennial
    ....13. *Apium*
f. Umbels pedunculate, the peduncles usually longer than the rays ....g
g. Plants of perennially wet ground in or around springs, streams, or marshes;
    fruit or ribs of fruit corky-thickened ....h
g. Plants of moist or relatively dry ground; fruit or ribs not corky ....j
h. Ribs of fruit filiform, the broad intervals between the ribs corky; all leaves
    pinnate ....18. *Berula*
h. Ribs of fruit corky-thickened; generally most leaves 2- or 3-pinnate i
i. Leaflets lanceolate or ovate-lanceolate; rays more than 3 cm. long; styles short
    ....19.   *Cicuta*
i. Leaflets ovate; rays less than 3 cm. long; styles elongate ....20. *Oenanthe*
j. Involucre conspicuous, the bracts divided into filiform segments. Ribs filiform
    ....15. *Ammi*
j. Involucre inconspicuous or none ....k
k. Plants annual or biennial, generally 2–3 m. tall; stems red-spotted. Ribs blunt,
    prominent ....14. *Conium*
k. Plants perennial, generally about 1 m. tall or less; stems not red-spotted l
l. Petals yellow; ribs filiform ....5. *Tauschia*
l. Petals white or rose; ribs prominent, acute ....17. *Ligusticum*
m. Leaf blades reduced to terete jointed rachises. Plants glabrous, perennial with
    creeping rootstalks ....21. *Lilaeopsis*
m. Leaves with round or nearly round blades ....n
n. Plants of wet places, perennial, glabrous; stems or root-stocks creeping
    ....1. *Hydrocotyle*
n. Plants of moist places, annual, stellate-pubescent; stems trailing or ascending
    ....2.   *Bowlesia*
o. Ribs on the back of the fruit filiform, not winged or prominently acute ....p
o. Ribs on the back of the fruit winged or prominently acute ....r
p. Leaves with 3 large leaflets; petals white, unequal, the outer larger
    ....27. *Heracleum*
p. Leaves with more numerous smaller leaflets; petals equal ....q
q. Leaves 2- or 3-pinnate or finely dissected into many small segments
    ....25. *Lomatium*
q. Leaves pinnate ....26. *Pastinaca*
r. Dorsal and lateral wings about equally prominent; stems very short, the leaves
    and inflorescence resting on the ground ....22. *Cymopterus*
r. Dorsal wings much narrower than the lateral or the dorsal ribs merely acute;
    stems evident, 0.2–2 m. tall ....s
s. Leaflets pinnately lobed, parted, or divided, glabrous ....23. *Conioselinum*
s. Leaflets serrate, tomentose or hirsutulous beneath ....24. *Angelica*

t. Bracts and bractlets spiny-tipped; fruit scaly ....28. *Eryngium*
t. Bracts and bractlets not spiny-tipped; fruit not scaly ....u
u. Fruit (or fruit and beak) linear v
u. Fruit not linear ....w
v. Leaves twice ternate, the leaflets broad; fruit short-beaked ......6. *Osmorhiza*
v. Leaves dissected into small narrow segments; fruit long-beaked ....7. *Scandix*
w. Umbels sessile or subsessile x
w. Umbels pedunculate ....y
x. Stem and leaves glabrous; fruits broader than long, finely papillate
    ....4. *Apiastrum*
x. Stem and leaves scabrous-pubescent; fruits longer than broad, the outermost
    carpels with hooked bristles, the other carpels tuberculate ....8. *Torilis*
y. Flowers yellow or purple, sessile or nearly so, the umbellets subcapitate. Fruit
    without ribs, tuberculate, the tubercles generally tipped by a hooked bristle
    ....3. *Sanicula*
y. Flowers white or pinkish, generally pedicellate ....z
z. Involucre none or represented by 1 bract; ribs none, the fruit with scattered
    short hooked bristles ....11. *Anthriscus*
z. Involucre present, of conspicuous and leaf-like bracts; ribs present, bearing
    elongate hooked or barbed bristles ....A
A. Umbel irregular, the rays few and very unequal, much longer than the in-
    volucre ....9. *Caucalis*
A. Umbel regular, round, concave, the rays shorter than the involucre..10. *Daucus*

1. Hydrocotyle

a. Leaves peltate ....1. *H. verticillata*
a. Leaves round-reniform ....2. *H. ranunculoides*

1. **H. verticillata** Thunb. Wet or swampy places in low meadows or on borders of marshes, rather rare: Potrero Meadow, Mount Tamalpais; Olema Marshes; Dillons Beach; Ledum Swamp and Drakes Estero, Point Reyes Peninsula.

At Abbotts Lagoon on Point Reyes Peninsula, var. *triradiata* (A. Rich.) Fern. has been collected. In it the pedicels are 3–5 mm. long or more, whereas in the typical variant they are usually less than 2 mm.

2. **H. ranunculoides** L. f. On swampy ground or in shallow fresh water ponds, often becoming locally abundant: Rodeo Lagoon; near Stinson Beach; Olema Marshes; Dillons Beach, *Leschke;* Point Reyes Peninsula (Bolinas, Mud Lake, Abbotts Lagoon).

2. Bowlesia

1. **B. incana** R. & P. Rare on shaded rocky slopes in moist soil: Salmon Creek Canyon; Tomales; Dillons Beach.—*B. lobata* of North American references.

3. Sanicula. SNAKE-ROOT

a. Leaves palmately divided, the divisions sometimes pinnately divided ....b
a. Leaves 3 or 4 times pinnately divided or compound ....d
b. Bractlets showy, yellowish green, exceeding the flowers; hooked bristles mostly
    near top of fruit ....1. *S. arctopoides*

b. Bractlets not showy, generally shorter than the flowers; hooked bristles extending to base of fruit c
c. Flower heads bright yellow; fruit sessile .... 2. *S. laciniata*
c. Flower heads greenish yellow; fruit pedicellate .... 3. *S. crassicaulis*
d. Flowers purple, rarely yellow; fruit with hooked bristles .... 4. *S. bipinnatifida*
d. Flowers yellow; fruit with tubercles but without hooked bristles .5. *S. tuberosa*

1. **S. arctopoides** H. & A. FOOTSTEPS-OF-SPRING. Shallow or rocky soil of open hills and ridges near the coast: Sausalito, Tiburon, and Mount Tamalpais (acc. Stacey) to Tomales and Point Reyes Peninsula (Mud Lake, Point Reyes).

The bright yellow and yellow-green pad of leaves and flowers in fresh green grass is a familiar and welcome sight in early spring on coastal hills.

2. **S. laciniata** H. & A. Widespread and rather common in moist grassy places and along the edge of brush or woods, generally back from the coast: Sausalito (acc. Lovegrove) and Mount Tamalpais (Corte Madera Ridge, Throckmorton Ridge, Mountain Theater) to Carson country, San Rafael Hills, Big Rock Ridge, and Black Point.

The Marin County plants in which the lower leaves have narrow, pinnately divided lobes have been referred to var. *serpentina* (Elmer) Jeps.

3. **S. crassicaulis** Poepp. Common and widespread in shaded canyons and on moist wooded slopes: Sausalito, Angel Island, Tiburon, and Mount Tamalpais (Blithedale and Cascade canyons, Cataract Gulch) to San Rafael, Big Rock Ridge, Tomales, and Point Reyes Peninsula (Mud Lake, Bolinas).—*S. Menziesii* H. & A.

4. **S. bipinnatifida** Dougl. PURPLE SANICLE. Open grassy slopes and meadows and borders of the chaparral, rather common: Sausalito (acc. Lovegrove), Tiburon, Angel Island, and Mount Tamalpais (Bootjack, Rock Spring, Phoenix Lake) to San Rafael Hills, Big Rock Ridge, and Tomales. On Tiburon the usual purple-flowered plants are sometimes accompanied by the yellow-flowered form which has been called var. *flava* Jeps.

5. **S. tuberosa** Torr. Rocky slopes of open ridges, commonly on serpentine: Tiburon; Mount Tamalpais (Corte Madera Ridge, Fern Canyon, Laurel Dell); Carson Ridge; San Geronimo Ridge.

4. Apiastrum

1. **A. angustifolium** Nutt. Loose or gravelly soil of slopes in the chaparral, rare in Marin County: Mount Tamalpais, acc. Stacey; Carson Ridge; San Rafael Hills.

5. Tauschia

1. **T. Kelloggii** (Gray) Macbr. Occasional in moist soil in grassy places more or less shaded by trees or chaparral: Sausalito; Tiburon; Mount Tamalpais (Blithedale Canyon, Throckmorton Ridge, Rocky Ridge Fire Trail); Corte Madera, acc. Stacey; San Rafael Hills.—*Velaea Kelloggii* (Gray) C. & R.

Bolinas Bay is the type locality for this attractive, yellow-flowered parsley.

6. Osmorhiza. SWEET CICELY

1. **O. chilensis** H. & A. Rather common on wooded or brushy slopes: Sausalito, Tiburon and Mount Tamalpais (Blithedale Canyon, Cataract Gulch) to San Rafael Hills and Point Reyes Peninsula (Black Forest, Mud Lake).—*O. nuda* Torr.

7. Scandix

1. **S. Pecten-Veneris** L. SHEPHERD'S NEEDLE. Occasional but widespread in grassy places along roads, in meadows, or amid brush: Sausalito, Tiburon, and Mount Tamalpais (Blithedale Canyon, Phoenix Lake) to San Rafael, Bolinas, and Tomales. Native of the Old World.

8. Torilis. HEDGE PARSLEY

1. **T. nodosa** (L.) Gaertn. Occasional in grassy places around towns or on the edge of woods or brush: Sausalito; Tiburon; Muir Beach; Cascade Canyon and Lagunitas Meadows, Mount Tamalpais; San Anselmo Canyon; San Rafael. Native of the Mediterranean region.

9. Caucalis

1. **C. microcarpa** H. & A. Rather rare in grassy places in the chaparral or on maritime slopes: Sausalito; Tiburon; Fairfax Hills; Carson country; Dillons Beach; Point Reyes Lighthouse.

10. Daucus

a. Segments of involucral bracts elongate, linear; plant biennial, mostly 0.5–1 m. tall ....1. *D. Carota*
a. Segments of involucral bracts short, oblongish; plant annual, mostly less than 0.5 m. tall ....2. *D. pusillus*

1. **D. Carota** L. CARROT. Escaping from cultivation and occasionally established: Sausalito; Mill Valley; Stinson Beach, acc. Stacey; highway north of San Rafael; east of Tomales. Native in Europe and Asia.

2. **D. pusillus** Michx. RATTLESNAKE WEED. Widespread and common on grassy hills and meadows: Sausalito, Angel Island, Tiburon, and Mount Tamalpais (Lake Lagunitas, Corte Madera Ridge, Bootjack) to San Rafael Hills, Dillons Beach, and Point Reyes Peninsula (Bolinas, Mud Lake, Shell Beach, Point Reyes).

11. Anthriscus

1. **A. neglecta** Bois. & Reut. var. **Scandix** (Scop.) Hyl. BUR CHERVIL. In Marin County known only from Devils Gulch above Camp Taylor and from bluffs near Tomales; to be expected in waste ground or in wooded or brushy places about towns.—*A. vulgaris* (L.) Pers.

12. Foeniculum

1. **F. vulgare** Mill. SWEET FENNEL. Common and conspicuous roadside weed of summer and autumn: Sausalito, Angel Island, and Tiburon to Stinson Beach (acc. Stacey), San Rafael, and Point Reyes Station. Naturalized from the Mediterranean region.

13. Apium

1. **A. graveolens** L. CELERY. Escaping from cultivation and becoming naturalized along the edge of salt marshes: Bolinas Lagoon; near Chinese Camp; Black Point. Native of Eurasia.

14. Conium

1. **C. maculatum** L. POISON HEMLOCK. Rank and disagreeable weed, widespread and common along roads and trails: Sausalito, Stinson Beach, and Mount Tamalpais (Blithedale Canyon, Phoenix Lake) to San Rafael, Tomales, and Point Reyes Peninsula (Bolinas, Inverness). Native of the Old World.

In the Black Forest on Point Reyes Peninsula the plants were unusually rank with stems 12 feet tall. It was from the root of this very poisonous species that the death potion of Socrates was concocted.

15. Ammi

1. **A. majus** L. BISHOP'S WEED. In low ground along roads, rare in Marin County: Mill Valley; highway north of San Rafael. Native of the Old World.

16. Perideridia

a. Styles slender, elongate; stems from a fleshy tuberous root .....1. *P. Gairdneri*
a. Styles short, on a conical thickened base; stems from a cluster of coarse roots
....2. *P. Kelloggii*

1. **P. Gairdneri** (H. & A.) Mathias. SQUAW POTATO. Low meadows that are wet in the spring, sometimes locally common: Lagunitas Meadows, Mount Tamalpais; Tomales; Inverness, acc. Stacey.—*Carum Gairdneri* (H. & A.) Gray.

2. **P. Kelloggii** (Gray) Mathias. Rather common and widespread on grassy, brushy, or wooded hills, in clayey or rocky soil: Sausalito, Angel Island, Tiburon, and Mount Tamalpais (Phoenix Lake, Lagunitas Meadows, Rock Spring) to San Rafael Hills and Point Reyes Peninsula (Bolinas, McClure Beach road).—*Carum Kelloggii* Gray.

Unquestionably the collection made by Kellogg at Bolinas should be considered the type of this species named for him, even though his specimen is not the first cited.

17. Ligusticum. LOVAGE

1. **L. apiifolium** (Nutt.) Gray. Occasional on moist brushy canyonsides more or less shaded, or on more open slopes near the ocean: Sausalito, Tiburon, Mount Tamalpais (Potrero Meadow) to San Rafael Hills, Tomales, and Point Reyes Peninsula (Bolinas, Inverness Ridge, Point Reyes).—*L. apiodorum* (Gray) C. & R.

18. Berula

1. **B. erecta** (Huds.) Cov. Swampy meadow on east side of Bolinas Lagoon, the only station known in Marin County for a widely distributed plant ranging even to the Old World.

19. Cicuta. WATER HEMLOCK

a. Fruit not very oily, the intervals narrower than the ribs ....1. *C. Douglasii*
a. Fruit very oily, the intervals equaling the ribs or wider ......2. *C. Bolanderi*

1. **C. Douglasii** (DC.) C. & R. Occasional in wet ground around springs or in low marshy places near the coast: Sausalito, acc. Stacey; Stinson Beach; Azalea

Flat and Willow Meadow, Mount Tamalpais; Papermill Creek.—Including *C. californica* Gray, a form in which the leaves are largely 2-pinnate.

2. **C. Bolanderi** Wats. Swampy ground bordering the salt marshes: Olema Marshes; Drakes Estero.

20. Oenanthe

1. **O. sarmentosa** Presl. Common and widespread in wet springy places or in shallow water of ponds and marshes: Sausalito, Tiburon, and Mount Tamalpais (Potrero Meadow) to Tomales and Point Reyes Peninsula (Bolinas, Mud Lake, Ledum Swamp).

21. Lilaeopsis

1. **L. occidentalis** C. & R. Locally common on moist flats bordering coastal ponds, inconspicuous and perhaps often overlooked: Rodeo Lagoon; marshes below Mill Valley, acc. Eastwood; Abbotts Lagoon and Drakes Estero, Point Reyes Peninsula; Dillons Beach.—*L. lineata* (Michx.) Greene var. *occidentalis* (C. & R.) Jeps.

This plant, so readily recognized by its remarkable leaves, is not known south of the Golden Gate.

22. Cymopterus

1. **C. littoralis** Gray. In deep beach sand just above high tide: near Dillons Beach, the southernmost station for a rare plant known in California from less than a half dozen places.

23. Conioselinum

1. **C. Gmelinii** (C. & S.) C. & R. A form of this plant was described as *Selinum pacificum* Wats. from a Kellogg and Harford specimen said to have been collected in the Sausalito Hills. Although Jepson has stated that this locality was probably a mistake for one in Mendocino County, it seems quite likely that the locality was correct and that this very rare plant should still be watched for on maritime mesas and slopes along the Marin County coast.

24. Angelica

a. Leaves finely villous or scabrous beneath; flowers long-pedicellate
....1. *A. tomentosa*
a. Leaves closely tomentose beneath; flowers short-pedicellate ..2. *A. Hendersonii*

1. **A. tomentosa** Wats. Occasional on wooded or brushy canyonsides: Angel Island, Tiburon and Mount Tamalpais (Phoenix Lake) to San Rafael Hills, Salmon Creek Canyon, and Point Reyes Peninsula (Mud Lake).

2. **A. Hendersonii** C. & R. Coastal hills and maritime bluffs: Rodeo Lagoon; near Muir Beach; Dillons Beach; Point Reyes Peninsula (Bolinas, Inverness Ridge, Point Reyes).

25. Lomatium. HOG FENNEL

a. Stems mostly more than 3 dm. tall; leaflets large, 2–5 cm. long; fruits with
    narrow thickish wings ....1. *L. californicum*
a. Stems mostly less than 3 dm. tall; leaves finely dissected into many small segments ....b

b. Bractlets broadly obovate; fruits glabrous. Petals glabrous ....2. *L. utriculatum*
b. Bractlets oblong, lanceolate, or oblanceolate; fruits hairy or sometimes glabrous
....c
c. Petals glabrous; fruit narrowly oblong ....3. *L. macrocarpum*
c. Petals hairy; fruit broadly oblong to roundish ....4. *L. dasycarpum*

1. **L. californicum** (Nutt.) Math. & Const. Partial shade of wooded or brushy slopes: Lake Lagunitas, Mount Tamalpais, acc. Stacey; Fairfax Hills; San Rafael Hills; Big Rock Ridge; Salmon Creek Canyon.—*Leptotaenia californica* Nutt.

2. **L. utriculatum** (Nutt.) C. & R. Common and widespread on grassy or rocky slopes and flats and on the edge of brush: Sausalito, Tiburon, and Mount Tamalpais (Bootjack, Rock Spring) to the Carson country, Big Rock Ridge, and Black Point.

As with the other hog fennels of Marin County, this species presents peculiar forms in some localities that are not easy to place. In the Carson country is a form in which the body of the fruit is mostly above the middle. This would perhaps be included in Jepson's treatment of *L. Vaseyi* but the Marin plant seems different from the typical southern California plant. Tall leafy plants from Black Point are perhaps var. *anthemifolium* Jeps. Plants from Bootjack with habit and foliage of *L. utriculatum* but with pubescent fruits may represent a cross between this species and *L. dasycarpum*.

3. **L. macrocarpum** (Nutt.) C. & R. This widespread western North American species occurs in rather atypical form with short fruits on serpentine slopes near Bootjack. These plants are quite different in aspect and in fruit from variants of either *L. utriculatum* or *L. dasycarpum*. On Bald Mountain west of San Anselmo the plants are more typical.

4. **L. dasycarpum** (T. & G.) C. & R. Common and rather widely distributed on rocky slopes and ridges in grassland or chaparral: Tiburon; Corte Madera Ridge and West Point, Mount Tamalpais; Carson country; San Rafael Hills; Big Rock Ridge; Tomales.

A form is found on Tiburon with the entire fruit as hairy as in *L. tomentosum* of the interior of California, but the Marin County plant seems definitely referable to *L. dasycarpum* in that the body of the fruit extends quite to the bottom of the fruit. On serpentine areas on Carson Ridge a form with glabrous petals and subglabrous fruits is common. This may represent an undescribed variety or, more likely, it may have originated by hybridization between this species and *L. utriculatum*.

26. Pastinaca

1. **P. sativa** L. PARSNIP. Escaping from cultivation but probably not persisting, in Marin County known only from a Mill Valley record by Stacey.

27. Heracleum

1. **H. maximum** Bartr. COW PARSNIP. Common and showy on moist hillsides, often in brushy or openly wooded places near the coast: Sausalito, Angel Island, and Tiburon to Tomales and Point Reyes Peninsula (Bolinas, Inverness, Point Reyes).—*H. lanatum* Michx.

28. Eryngium

a. Bracts entire, the margin without spiny teeth ....1. *E. armatum*
a. Bracts with 2 or more spiny teeth ....b
b. Sepals 1.5 mm. long, about half as long as the elongate styles
....2. *E. aristulatum*
b. Sepals 3 mm. long, little shorter than the styles ....3. *E. oblanceolatum*

1. **E. armatum** (Wats.) C. & R. Widespread but usually not very common in low fields where water seeps or stands in the spring or on sandy flats near the ocean: Rodeo Lagoon; Tiburon; Baltimore Park; Hamilton Field; Chileno Valley; Olema Marshes; Bolinas; Point Reyes dunes.

2. **E. aristulatum** Jeps. Low ground where water stands in the spring: Hidden Lake, Mount Tamalpais; east of Aurora School.

3. **E. oblanceolatum** C. & R. Summer-dried bed of a vernal pool at Burdell Station is the only place in Marin County where this species has been found.

CORNACEAE. Dogwood Family

1. Cornus. Dogwood

a. Flowers in heads subtended by several large petal-like bracts; trees
....1. *C. Nuttallii*
a. Flowers in loose clusters, without petal-like bracts; shrubs ....b
b. Leaves green above, pale and thinly hairy below ....2. *C. californica*
b. Leaves nearly the same above and below, very sparsely appressed-hairy
....3. *C. glabrata*

1. **C. Nuttallii** Aud. This, one of western America's most beautiful flowering trees, is at present known in Marin County from only a single individual on Mount Tamalpais near Laurel Dell. Although Miss Eastwood has expressed the belief that the species is not native in Marin County, an earlier record by Behr reported it as "not common" on Bolinas Ridge. The Nuttall dogwood might well be indigenous in Marin County since it occurs southward through the Coast Ranges to the Santa Cruz and Santa Lucia mountains.

2. **C. californica** C. A. Mey. CREEK DOGWOOD. Widespread and rather common in wet ground along streams or around marshes: Muir Beach, Mill Valley, and Mount Tamalpais (Willow Meadow, Camp Hogan) to Chileno Valley and Inverness.

3. **C. glabrata** Benth. In Marin County known only from the moist border of the laguna in Chileno Valley.

GARRYACEAE. Silk Tassel Family

1. Garrya

a. Leaves with undulate and somewhat revolute margins, tomentose beneath
....1. *G. elliptica*
a. Leaves with plane or nearly plane margins, glabrous or nearly so
....2. *G. Fremontii*

1. **G. elliptica** Dougl. SILK TASSEL BUSH; QUININE BUSH. Occasional on rocky slopes in the chaparral or on coastal bluffs: Sausalito; Mount Tamalpais (Corte

Madera Ridge, West Point, Phoenix Lake); Carson country; San Rafael Hills; Big Rock Ridge; Tomales; Shell Beach, Point Reyes Peninsula.

The graceful catkins of the staminate plants make the silk tassel bush one of the most beautiful shrubs in the chaparral. The flowers are among the earliest, opening in December and January, harbingers of spring in the midst of winter. The leaves have a bitter principle and were used by early-day settlers as a substitute for quinine.

2. **G. Fremontii** Torr. Rare in the chaparral along the summit ridge of Mount Tamalpais: West Peak; Northside Trail; Fern Canyon. The Tamalpais plant belongs to a variant found on Coast Range peaks of central California which was described as G. *rigida* Eastw., Mount Tamalpais, the type locality.

ERICACEAE. Heather Family

a. Shrubs or trees with numerous broad green leaves ....b
a. Herbs, generally without green leaves, the leaves reduced and scalelike ....h
b. Bark of stems or younger branches terra-cotta, reddish, or chocolate in color, smooth (the bark of *Arctostaphylos Uva-ursi,* a prostrate or decumbent shrub, generally rough). Fruit a dry or juicy berry c
b. Bark rough, grayish or brownish. Stems erect (except in the maritime form of *Gaultheria*) ....e
c. Trees with large leaves more than 5 cm. long; fruit a juicy berry ..1. *Arbutus*
c. Erect or prostrate shrubs with leaves generally less than 5 cm. long; fruit a berry dry at maturity ....d
d. Corolla 5-lobed; stamens 10; fruit not breaking to pieces on the plant
   ....2. *Arctostaphylos*
d. Corolla 4-lobed; stamens 8; fruit breaking to pieces on the plant
   ....3. *Schizococcus*
e. Leaves entire; fruit a capsule ....f
e. Leaves toothed; fruit a fleshy berry ....g
f. Leaves resinous-dotted beneath; corolla whitish, divided to the base .4. *Ledum*
f. Leaves not resinous-dotted; corolla white or purplish pink, campanulate, 5-lobed ....5. *Rhododendron*
g. Leaves more than 4 cm. long; ovary superior ....6. *Gaultheria*
g. Leaves less than 4 cm. long; ovary inferior ....7. *Vaccinium*
h. Stems slender, scapelike, bractless or nearly so, the leaves basal; petals distinct, rounded ....8. *Pyrola*
h. Stems fleshy, leafy with scalelike bracts; petals or corolla lobes acute i
i. Sepals 4 or 5; corolla divided to the base or nearly so, the petals 4 or 5
   ....9. *Pityopus*
i. Sepals 2 or 4; corolla campanulate, 4–6-lobed ....10. *Newberrya*

1. Arbutus

1. **A. Menziesii** Pursh. Madroño (frontis.). Common forest tree of slopes and valley flats, usually large and handsome, occasionally becoming stunted and dwarfed on the edge of the chaparral: Sausalito, Angel Island, Tiburon, and Mount Tamalpais (Blithedale and Cascade canyons, Rock Spring, Lake Lagunitas) to San Rafael Hills, Big Rock Ridge, and Point Reyes Peninsula (Inverness Ridge).

The madroño, as Jepson has written, is "a tree than which none other in the western woods is more marked by sylvan beauty . . ."; and, it might be added, by flowers and fruits beyond compare—the former like sculptured ivory urns, the latter like etched carnelian globes.

2. Arctostaphylos. MANZANITA

a. Bracts of the inflorescence small, appressed, not leaflike in texture; ovary generally glabrous; sepals appressed to the base of the corolla . . . . .b

a. Bracts, at least near the base of the inflorescence, large, spreading, leaflike in texture; ovary hairy; margins of sepals reflexed from the base of the corolla . . . . .f

b. Stems prostrate, creeping and rooting, the bark usually rough, not red; leaves widest above the middle, the upper and lower sides unlike . .1. *A. Uva-ursi*

b. Stems erect (or frequently spreading and decumbent in *A. montana*), the bark smooth and red; leaves widest at the middle or below, the upper and lower sides alike c

c. Leaves small, usually less than 2.5 cm. long, mostly pungently acute; young stems and inflorescence white-hairy . . . . .2. *A. montana*

c. Leaves larger, usually more than 2.5 cm. long, not pungently acute, mostly obtuse; young stems and inflorescence glabrous or finely hairy . . . . .d

d. Leaves pallid; pedicels glandular-hairy; fruit viscid-glandular . . . .5. *A. viscida*

d. Leaves green; pedicels glabrous; fruit not glandular .e

e. Branches of the inflorescence mostly glabrous and very slender; corolla about 5 mm. long; low shrub, mostly 2 m. tall or less . . . . .3. *A. Stanfordiana*

e. Branches of the inflorescence finely puberulent, stouter; corolla generally more than 5 mm. long; tall shrub, mostly more than 2 m. tall . . . .4. *A. Manzanita*

f. Leaves hairy, pale and hoary, nearly alike on both surfaces . .6. *A. canescens*

f. Leaves glandular-hairy or nonglandular and glabrate, green or glaucous but not hoary, upper and lower sides of leaves different . . . . .g

g. Leaves not glandular-hairy; branchlets not bristly-hairy, the hairs more or less appressed and crinkly or none. Stems forming woody platforms at the surface of the ground . . . . .9. *A. Cushingiana*

g. Leaves glandular-hairy and slightly scabrous to the touch; branchlets more or less bristly and glandular, the hairs spreading . . . . .h

h. Stems not forming woody swellings or platforms at the surface of the ground; leaves frequently ovate-lanceolate . . . . .7. *A. virgata*

h. Stems forming woody swellings and platforms; leaves generally broader, ovate . . . . .8. *A. glandulosa*

1. **A. Uva-ursi** (L.) Spreng. KINNIKINNICK. Occasional on exposed or open maritime hills and slopes on Point Reyes Peninsula: near the lighthouse; McClure Beach road. These are the southernmost stations in California.

In typical *A. Uva-ursi* the stems are prostrate and the pedicels and ovaries are glabrous. When the areas occupied by this species are adjacent to the bishop pine forests where *A. virgata* abounds, low plants exhibiting somewhat divergent characters are sometimes found which may have originated through the hybridization of these two species. In such plants the stems are trailing and assurgent and the pedicels and ovaries are thinly hairy. These suspected hybrids have been reported from Marin County as *A. media* Greene but that name should be restricted to a similar plant that is probably derived from the hybridization of *A. Uva-ursi* and the north-coast relative of *A. virgata, A. columbiana* Piper.

2. **A. montana** Eastw. TAMALPAIS MANZANITA (plate 14). Common on rocky flats and slopes of serpentine, occasional on other kinds of rocks: Mount Tamalpais (West Point, Bootjack, Rifle Camp, Lagunitas Meadows); Carson Ridge; San Geronimo Ridge; Inverness Ridge, *Brydon*.

Generally, as the plant occurs on exposed open slopes, it has a low, broad, bushy, or even matlike habit, but in more protected places it becomes erect and tall and sometimes arborescent. On Mount Tamalpais this is the last species of manzanita to bloom, the plants being attractively dotted with tight little flower clusters in March and April. Although the Tamalpais manzanita seems to be nearly or quite restricted to Marin County, it is closely related to *A. pungens* H.B.K., a widely distributed species that ranges far south into Mexico.

3. **A. Stanfordiana** Parry. Although the Stanford manzanita is generally regarded as restricted to the Coast Ranges north and east of Marin County, plants that are unquestionably of this trim distinctive species have been found on Point Reyes Peninsula north of Shell Beach by Gregory Lyon. The nature of its occurrence there in open parts of the bishop pine forest (a habitat quite unlike the one in which it grows in the Sonoma and Napa mountains) would seem to preclude the suggestion that it is not indigenous.

4. **A. Manzanita** Parry. Occurring generally as scattered shrubs on grassy or openly wooded hills, rarely as a component of the chaparral or dense brush: hills between Kentfield and San Rafael, acc. M. G. Smith; San Rafael Hills; Big Rock Ridge; Black Point. The species does not grow south of Marin County in the Coast Ranges.

In Marin County *A. Manzanita* is usually a shrub in size, though on Big Rock Ridge some individuals are large and arborescent. This is one of the earliest species to bloom; plants have been seen in flower before the end of November, but usually the height of bloom is about mid-January.

5. **A. viscida** Parry. In Marin County known only from the north side of Black Canyon in the San Rafael Hills, the southernmost station in the Coast Ranges.

6. **A. canescens** Eastw. HOARY MANZANITA. Mostly restricted to steep and otherwise barren sandstone or shale slopes and ridges, rarely in dense chaparral, in Marin County occurring only on Mount Tamalpais, the type locality: East Peak; Throckmorton Ridge; Rock Spring and Matt Davis trails.

When, in late December and early January, the low sprawling bushes are laden with clusters of rosy waxen flowers, this species is generally conceded to be one of the most lovely in this genus of handsome flowering shrubs. On Mount Tamalpais, where the sterile barrens dwarf the plants to pygmy proportions, there is developed that picturesque type of habit which one associates with Japanese horticultural effects and it adds much to the attractiveness of the beautiful corollas.

7. **A. virgata** Eastw. MARIN MANZANITA. Rocky brushy slopes on the forest borders of bishop pine and redwood: head of Muir Woods, the type locality, and Sierra Trail, Mount Tamalpais; Bolinas Ridge; Mud Lake and Inverness Ridge,

Point Reyes Peninsula.—*A. glandulosa* Eastw. var. *virgata* (Eastw.) Jeps.; *A. columbiana* Piper var. *virgata* (Eastw.) McMinn.

Although this species is closely related to certain other manzanitas that grow along the coast as far north as Vancouver Island, it is marked by distinctive characters that have not been found in plants outside of Marin County. It is another early-blooming species, some individuals above Muir Woods frequently being in attractive flower before the end of the year.

8. **A. glandulosa** Eastw. Locally common in the chaparral on rocky slopes and ridges: Mount Tamalpais, the type locality (rare to the east of West Point, common to the west, Rock Spring to Potrero Meadows); Bolinas Ridge; Carson Ridge; Big Rock Ridge (only one plant seen).

The stumplike root crowns which are characteristic of this and the following species are gradually enlarged as the plants grow normally, but when regeneration of the aerial stems from this stump occurs after fires, lateral growth is much more rapid. Nevertheless, the larger of these crowns must be very ancient, and, as Miss Eastwood has observed, they may be as old as the giant redwoods themselves.

9. **A. Cushingiana** Eastw. Widespread and abundant on rocky hills and mountains, a characteristic member of the chaparral: Angel Island; southwest of Almonte; Mount Tamalpais (Corte Madera Ridge, Azalea Flat, above Muir Woods, the type locality, Berry Trail, Phoenix Lake); Carson Ridge; San Rafael Hills; Big Rock Ridge; Inverness Ridge.

The Cushing manzanita is not only the most common species in Marin County but it is also the most variable. Field observation would seem to indicate that some of the variation is due to hybridization with related species, especially on Mount Tamalpais. Thus about East Peak where *A. canescens* is particularly abundant, the common form of *A. Cushingiana* is densely white-hairy, so much so in some individuals that it is necessary to establish its identity by means of its root crown. To the westward and elsewhere, *A. Cushingiana* merges with *A. glandulosa* and this intergradation is perhaps also due to hybridization, for, in typical condition, both these species are marked by distinctive characters. Elsewhere in Marin County beyond the contact zones between these three species, *A. Cushiangiana* is quite uniform in appearance and in vesture, except locally on barren granitic hogbacks of Inverness Ridge. There, in exposed rocky places where the bishop pine is stunted to a tenth its normal size, the Cushing manzanita has developed a matlike habit and the prostrate stems bear leaves much smaller than usual. This plant, which has been called the huckleberry manzanita because of its small leaves, is *A. Cushingiana* Eastw. f. *repens* J. T. Howell, a distinctive ecologic form, but everywhere in the vicinity it is entirely confluent with the general population of the species in typical form. Not far away, a lone glandular individual with an unmistakable rootcrown was interpreted as a hybrid derivative between *A. virgata* and *A. Cushingiana*. Certainly in Marin County, variation in the crown-forming manzanitas centers in *A. Cushingiana*.

3. Schizococcus

1. **S. sensitivus** (Jeps.) Eastw. SHATTERBERRY. Rocky ridges and flats of Mount Tamalpais, the type locality, often locally abundant in the chaparral: Blithedale

Canyon; Throckmorton Ridge; Azalea Flat; Bolinas Ridge.—*Arctostaphylos sensitiva* Jeps.

From early winter to spring, small, dainty flowers adorn the trim leafy shrubs. After the corollas fall the pedicels curve into an upright position and at their tips develop the peculiar little fruits that are so unlike the plump *manzanitas* of the related genus *Arctostaphylos*.

4. Ledum

1. **L. glandulosum** Nutt. LABRADOR TEA. Springy slopes and coastal swales on Point Reyes Peninsula: Inverness; Shell Beach; Ledum Swamp.

5. Rhododendron

a. Leaves evergreen; corolla rose ....1. *R. macrophyllum*
a. Leaves deciduous; corolla mostly white, sometimes pink-tinged
....2. *R. occidentale*

1. **R. macrophyllum** G. Don. WESTERN RHODODENDRON. Rare in the chaparral on Mount Tamalpais: Pipeline Trail; head of Blithedale Canyon; north slope of Corte Madera Ridge.—*R. californicum* Hook.
2. **R. occidentale** (T. & G.) Gray. WESTERN AZALEA. Common around springs, along streams, and on marshy flats: Rodeo Lagoon; Muir Woods; Mount Tamalpais (Cascade Canyon, Azalea Flat, Bootjack, Potrero Meadow); Lagunitas Canyon.

The beauty of the western azalea is not restricted to what is seen in graceful shrubbery or attractive blossoms but when in flower its beauty pervades the mountain or canyon with a most delicious fragrance. May and June in Bootjack Canyon or Potrero Meadow are times to smell as well as to look. Though late spring is the peak season for azalea blooms, there are bushes on moist slopes near Rattlesnake Camp where flowers can be found almost every month of the year. And even in December the flowers are fragrant.

6. Gaultheria

1. **G. Shallon** Pursh. SALAL (plate 11). Wooded and brushy slopes near the coast or on exposed maritime headlands: Sausalito; Muir Woods; Cascade Canyon, Mount Tamalpais; San Geronimo Ridge; Ledum Swamp and near the lighthouse, Point Reyes Peninsula.

7. Vaccinium

1. **V. ovatum** Pursh. HUCKLEBERRY. Rather common and sometimes locally abundant on moist shaded slopes in the southern part of the county and on Point Reyes Peninsula: Sausalito; Mount Tamalpais (Cascade Canyon, Azalea Flat, Willow Meadow, Phoenix Lake); Bolinas Ridge; Lagunitas Canyon; San Geronimo Ridge; San Rafael Hills; Point Reyes Peninsula (Mud Lake, Shell Beach, Point Reyes).

The berries are of two kinds, a rarer form with a bloom and one more common without a bloom. Both are delicious but there is a discernible difference in flavor even when the two kinds are on bushes growing side by side, as is often the case. The form with the bloom is var. *saporosum* Jeps.

8. Pyrola

1. **P. picta** Smith f. **aphylla** (Smith) Camp. Rare and local in deep soil of woods: Mount Tamalpais (Bootjack Canyon, Rock Spring, acc. Stacey, Cataract Gulch, Swede George Gulch, acc. Leschke); Lagunitas Canyon.—*P. aphylla* Smith.

Usually this plant, which is almost a complete saprophyte without green leaves, occurs with leafy plants which are typical *P. picta*. On Mount Tamalpais all of the plants are practically without green leaves, only one plant with a single small green leaf having been seen from the county.

9. Pityopus

1. **P. californicus** (Eastw.) Copeland f. This is perhaps the rarest plant in Marin County where it is known only from the small colony of plants discovered by Alice Eastwood in May, 1901, in Little Carson Canyon, the type locality. Also it is one of California's rarest plants, being known outside of Marin County only from Humboldt and Fresno counties.

10. Newberrya

1. **N. congesta** Torr. This species is known in Marin County only from plants discovered by Dr. and Mrs. Paul Wilson on the east side of Mount Tamalpais on the Morningside Trail above Kentfield. Although in Marin County it is as rare as *Pityopus californicus,* beyond the county it is known from a number of stations that range from Monterey County in California north to Washington.

PRIMULACEAE. PRIMROSE FAMILY

a. Leaves about equally distributed along the stems, the flowers solitary in the axils. Corolla, when present, less than 1 cm. long ....b
a. Leaves clustered at the base or top of scapelike stems; flowers in umbels. Perennials ....d
b. Leaves alternate. Inconspicuous annuals ....1. *Centunculus*
b. Leaves opposite c
c. Annual herb of hills and valleys; flowers long-pedicellate with showy corolla ....2. *Anagallis*
c. Succulent perennial herb of coastal or saline habitats; flowers nearly sessile, apetalous ....3. *Glaux*
d. Foliage leaves whorled at the top of the stem; corolla rotate, less than 1 cm. long ....4. *Trientalis*
d. Foliage leaves basal; corolla reflexed, more than 1 cm. long ...5. *Dodecatheon*

1. Centunculus

1. **C. minimus** L. CHAFFWEED. Grassy slopes or flats that are wet in the spring: Tiburon; Mount Tamalpais (Rock Spring, Potrero and Lagunitas meadows); Baltimore Park; Carson Ridge; San Rafael; near Chinese Camp; Point Reyes Peninsula (Bolinas, Ledum Swamp, Drakes Estero).

2. Anagallis

1. **A. arvensis** L. PIMPERNEL. Common and widespread about towns and in hills and meadows: Sausalito, Angel Island, Tiburon, and Mount Tamalpais

(Azalea Flat, Blithedale and Cascade canyons) to San Rafael, Tomales, and Point Reyes Peninsula (Bolinas, Inverness, McClure Beach).

Usually the corollas are coral or reddish but on hills overlooking Drakes Bay f. *azurea* Hyl. (var. *coerulea* of Jepson's Manual) with deep blue corollas is locally common. Both forms are indigenous to the Old World.

3. Glaux

1. **G. maritima** L. SEA-MILKWORT. Rare among maritime dunes or in salt marshes: Burdell; Tomales Bay, acc. Stacey; Point Reyes, acc. Jepson.

4. Trientalis

1. **T. latifolia** Hook. STAR-FLOWER. Occasional in deep soil of moist shaded slopes: Sausalito, acc. Lovegrove; Mount Tamalpais (Steep Ravine, Fern Canyon, Willow Meadow, Phoenix Lake); Lagunitas Canyon; Carson Ridge; Mud Lake, Point Reyes Peninsula.—*T. europaea* L. var. *latifolia* (Hook.) Torr.

5. Dodecatheon

1. **D. Hendersonii** Gray. SHOOTING STAR. Common and widespread on moist grassy flats and slopes and in brush or open woods: Sausalito, Tiburon, and Mount Tamalpais (Corte Madera Ridge, Bootjack, Cataract Gulch) to San Rafael Hills and Tomales.

In Marin County, the shooting stars are early bloomers and, like rosy-fingered Aurora who precedes dazzling Phoebus, their bright and cheery flowers of late winter are harbingers of the floral flood that comes with spring.

PLUMBAGINACEAE. THRIFT FAMILY

a. Leaves linear; inflorescence a condensed head-like panicle ....1. *Armeria*
a. Leaves oblanceolate to oblong- or elliptic-obovate; inflorescence an openly
    branched panicle ....2. *Limonium*

1. Armeria

1. **A. maritima** (Mill.) Willd. var. **californica** (Boiss.) Lawr. SEA-PINK; THRIFT. Rocky slopes and sandy mesas near the ocean, in Marin County known only on Point Reyes Peninsula: head of Drakes Estero and near the lighthouse.—*Statice arctica* (Cham.) Blake var. *californica* (Boiss.) Blake.

2. Limonium

1. **L. commune** S. F. Gray var. **californicum** (Boiss.) Greene. STATICE; SEA-LAVENDER; MARSH ROSEMARY. Common in salt marshes bordering the bay or ocean: Sausalito; Tiburon; Escalle; Black Point; Stinson Beach; Inverness; Drakes Estero.

OLEACEAE. OLIVE FAMILY

1. Fraxinus

1. **F. oregona** Nutt. OREGON ASH. Small or large trees of stream banks or low moist valley lands: Phoenix Lake and Lagunitas Meadows, Mount Tamalpais;

Ross; San Anselmo and Lagunitas canyons; Papermill Creek; Olema Marshes; Gallinas Valley, acc. Leschke; San Antonio Creek.

GENTIANACEAE. GENTIAN FAMILY

a. Corolla blue, 3–4 cm. long; plants perennial ....1. *Gentiana*
a. Corolla pink or yellow, rarely whitish, 2 cm. long or less; plants annual ....b
b. Corolla yellow; calyx campanulate, 4-toothed at top ....2. *Microcala*
b. Corolla pink, rarely white; calyx divided nearly to the base into 4 or 5 linear
   divisions ....3. *Centaurium*

1. Gentiana. GENTIAN

1. **G. oregana** Engelm. Rare and local on grassy or brushy slopes, not blooming until midsummer: Willow Meadow and Hidden Lake Fire Trail, Mount Tamalpais; Carson Ridge; McClure Beach road, Point Reyes Peninsula.

This attractive coastal gentian is not known south of Marin County. A Bolander collection from Mount Tamalpais served Gray as the type of *G. affinis* Griseb. var. *ovata*.

2. Microcala

1. **M. quadrangularis** (Lamk.) Griseb. Clay soil of slopes and flats where water seeps or stands in the spring: Tiburon; Lagunitas Meadows, Mount Tamalpais; Carson Ridge; Baltimore Park; San Rafael, acc. Stacey; Black Point; Tomales; Ledum Swamp and Drakes Estero, Point Reyes Peninsula.

3. Centaurium

a. Pedicels of lower flowers half as long as the calyx or longer ..1. *C. Muhlenbergii*
a. Pedicels very short or none ....b
b. Style scarcely branched, the 2-lobed stigma less than 1 mm. long; seeds
   0.5–0.75 mm. long ....2. *C. trichanthum*
b. Style branched, the branches and stigmas about 1 mm. long; seeds 0.25–0.33
   mm. long ....3. *C. floribundum*

1. **C. Muhlenbergii** (Griseb.) Wight. Occasional near the coast on slopes or flats wet in the spring: Angel Island; Tiburon; Lagunitas Meadows and Kent Trail, Mount Tamalpais; Black Point; Inverness and near the radio station, Point Reyes Peninsula.—*C. exaltatum* (Griseb.) Wight var. *Davyi* Jeps.

2. **C. trichanthum** (Griseb.) Robins. In typical form, *C. trichanthum* is known in Marin County only from low ground bordering the salt marsh near Burdell Station. A small-flowered variant is found in hills to the south on open rocky slopes, generally in serpentine areas: Tiburon; Mount Tamalpais (Bootjack, Laurel Dell Trail, Barths Retreat); Carson Ridge (some albino).

3. **C. floribundum** (Benth.) Robins. Clay soil of open slopes, rare in Marin County: along the railroad north of San Rafael; Tamalpais Valley.

APOCYNACEAE. DOGBANE FAMILY

a. Flowers axillary, solitary; corolla large, lavender-blue ....1. *Vinca*
a. Flowers in small terminal clusters; corolla small, 3–4 mm. long, whitish
   ....2. *Apocynum*

1. Vinca

1. **V. major** L. PERIWINKLE. Escaping from cultivation and becoming locally naturalized on moist shaded slopes: Sausalito (acc. Lovegrove), Mill Valley, and Mount Tamalpais (Lagunitas Meadows) to San Rafael, Ignacio, Inverness, and Tomales.

2. Apocynum. DOGBANE

1. **A. cannabinum** L. var. **glaberrimum** A. DC. A rare plant in Marin County, known only from wooded slopes in the Fairfax Hills.

The common name, Indian hemp, often given to this species refers to the fine strong fiber of the stems used by the Indians to make cords.

ASCLEPIADACEAE. MILKWEED FAMILY

1. Asclepias. MILKWEED

a. Leaves linear-lanceolate, glabrous; hoods short, broadly obtuse
....1. *A. fascicularis*
a. Leaves ovate-lanceolate to ovate, tomentose; hoods produced into a long, lanceolate tip ....2. *A. speciosa*

1. **A. fascicularis** Decne. Rare in clay soil of meadows and fields: Willow Meadow, Mount Tamalpais; north of San Rafael.—*A. mexicana* of California references.

2. **A. speciosa** Torr. In Marin county confined to rocky or grassy flats along streams on the north side of Mount Tamalpais: east of Rock Spring; Lagunitas Meadows. Not known farther south in the Coast Ranges.

The flowers exhale a heavy fragrance of honey which is very alluring to insects. Bees and various flies may get their legs inextricably caught in the complicated hood structure of the flowers and not infrequently can be seen dangling from their attractive death traps.

CONVOLVULACEAE. MORNING-GLORY FAMILY

a. Plants yellowish or golden throughout, without green leaves, parasitic; styles 2, distinct ....4. *Cuscuta*
a. Plants with green leaves, not parasitic; styles 1 or 2 ....b
b. Flowers large, 1.5–8 cm. long; style 1, simple or 2-parted at the apex; stems erect or trailing, or frequently twining ....3. *Convolvulus*
b. Flowers small, less than 1 cm. long; styles 2, distinct or nearly so; stems creeping or erect, not twining c
c. Stems creeping; leaves roundish-reniform, petiolate; fruit deeply 2-lobed
....1. *Dichondra*
c. Stems erect or diffuse; leaves lanceolate to elliptic or ovate, subsessile; fruit not 2-lobed ....2. *Cressa*

1. Dichondra

1. **D. repens** Forst. Occasional on brushy or grassy hills near the coast: Tiburon; Sausalito; Elk Valley; Frank Valley; Stinson Beach; Kentfield, *Eastwood;* south of Olema. Appearing indigenous but perhaps introduced from the American tropics.

2. Cressa

1. **C. truxillensis** H.B.K. Rare in low ground bordering salt marshes: Corte Madera; San Antonio.—*C. cretica* of California references; *C. cretica* L. var. *truxillensis* (H.B.K.) Choisy.

3. Convolvulus. MORNING-GLORY

a. Flowers sessile in a pair of large sepal-like bracts that are closely appressed to the calyx . . . . .b

a. Flowers pedicellate, the bracts smaller and narrower than the sepals and either close to the base of the calyx (*i.e.*, the pedicels about 1–3 mm. long) or more distant . . . . .e

b. Stems and leaves glabrous or nearly so, the stems short and trailing or elongate and twining; bracts broadly ovate to round c

b. Stems and leaves thinly to densely pubescent with spreading hairs, the stems usually very short, not twining; bracts oblong to ovate . . . . .d

c. Leaves broadly reniform, obtuse or retuse, fleshy; stems trailing; corolla pink . . . . .1. *C. Soldanella*

c. Leaves ovate or triangular, acute, herbaceous; stems twining; corolla white . . . . .2. *C. sepium*

d. Stems and leaves densely white-velvety-pubescent . . . . .3. *C. malacophyllus*

d. Stems and leaves greenish, villous-pubescent, sometimes sparsely so . . . . .4. *C. subacaulis*

e. Pedicels very short, 1–2 mm. long, the bracts loosely subtending the calyx. Stems and leaves finely pubescent; corolla usually more than 3 cm. long . . . . .5. *Ċ. polymorphus*

e. Pedicels longer, the bracts more or less distant from the base of the calyx . . . .f

f. Stems and leaves glabrous; leaves mostly acute; corolla more than 3 cm. long . . . . .6. *C. occidentalis*

f. Stems and leaves pubescent; leaves mostly obtuse; corolla less than 3 cm. long . . . . .7. *C. arvensis*

1. **C. Soldanella** L. BEACH MORNING-GLORY. Deep sand of dunes along the ocean: Rodeo Lagoon; Stinson Beach; near the radio station, Point Reyes Peninsula.

2. **C. sepium** L. A rampant climber in low wet ground at Stinson Beach. Introduced from Europe.

3. **C. malacophyllus** Greene. Locally common on open rocky slopes of serpentine areas: Bootjack and Rifle Camp, Mount Tamalpais; Carson Ridge.—*C. villosus* (Kell.) Gray.

4. **C. subacaulis** (H. & A.) Greene. Rather common on open grassy hills: Angel Island; Tiburon; Bootjack and Lagunitas Meadows, Mount Tamalpais; Greenbrae and Fairfax hills; Hamilton Field; Lucas Valley; Chileno Valley.

5. **C. polymorphus** Greene. Occasional in grassland of hills near the coast: Sausalito; Angel Island; Muir Beach; pond near Olema.

These are the southernmost stations recorded for this morning-glory which ranges northward to Oregon and eastward to the northern Sierra Nevada. The bracts are not as far removed from the calyx as in some specimens but in all the Marin plants the abbreviated pedicel is usually quite apparent. Our plants superficially resemble trailing forms of *C. subacaulis* but the pubescence of the two seems to be essentially different (subappressed in the present species but divaricately spreading in *C. subacaulis*) and the bracts differ in general appearance as well as in insertion. At Sausalito an apparent hybrid between *C. polymorphus* and *C. occidentalis* var. *purpuratus* has been found.

6. **C. occidentalis** Gray. In its several forms this attractive and widespread morning-glory is the most common, as well as the most variable, in Marin County. The typical form, with broadly rounded puberulent leaves and 2-flowered peduncles, was originally collected "near San Francisco" and should be watched for. In the southern part of the county a beautiful variety with white or pinkish flowers that become purplish-rose in age grows along the coast, trailing widely over the ground or twining on shrubs. This plant, originally described from the San Francisco area and Marin County, is *C. occidentalis* var. *purpuratus* (Greene) J. T. Howell, and is abundant at Sausalito, Angel Island, Tiburon, northward to Bolinas and Inverness on Point Reyes Peninsula.—*C. purpuratus* (Greene) Greene; *C. luteolus* Gray var. *purpuratus* (Greene) Jeps.

In the northwestern part of the county another member of this complex is found, *C. occidentalis* var. *saxicola* (Eastw.) J. T. Howell, a plant with smaller, round-ovate leaves and frequently lobed bracts, known only from southern Sonoma County and adjacent Marin County (McClure Beach, Point Reyes Peninsula, and north of Dillons Beach). Near the Point Reyes Lighthouse, plants are usually like var. *purpuratus* but intermediates varying towards var. *saxicola* may also be found.—*C. luteolus* Gray var. *saxicola* (Eastw.) Jeps.; *C. purpuratus* (Greene) Greene var. *saxicola* (Eastw.) Jeps.

Inland, away from the coast, the western morning glory is common and widespread on wooded or brushy hills or rocky bluffs. It is variable in many ways: the shape and texture of the leaves, the size and shading of the corollas, the shape and position of the bracts, and the number of flowers on the peduncles. This is var. *solanensis* (Jeps.) J. T. Howell and is the plant that was formerly known as *C. luteolus* Gray. In Marin County it is found from Sausalito, Elk Valley, and Mount Tamalpais (Cascade Canyon, West Peak) north to Carson Ridge, San Rafael Hills, and Tomales. A form of it from Mount Tamalpais with cuspidate-acuminate sepals was described as *C. illecebrosus* House. After the fire of 1945, the western morning-glory was a common fireweed and over hundreds of acres of the Carson country the trailing twining stems formed a veritable tangle.

7. **C. arvensis** L. Orchard Morning-glory; Field Bindweed. Widespread weed of roadside and of cultivated ground: Sausalito, Angel Island, Tiburon, and Mount Tamalpais (Blithedale Canyon, West Point, Lagunitas Meadows) to Fairfax, San Rafael, and Point Reyes Peninsula (Bolinas, Inverness). Introduced from Europe.

4. **Cuscuta.** Dodder

a. Calyx lobes obtuse; capsule prominently exserted beyond the corolla
....1. *C. campestris*
a. Calyx lobes acute; capsule enclosed by the corolla ....b
b. Corolla tube cylindric, usually conspicuously longer than the calyx. Fringed scales present in corolla tube ....2. *C. subinclusa*
b. Corolla tube campanulate or bowlshaped c
c. Calyx equaling or longer than the corolla tube; corolla lobes ovate, abruptly apiculate; scales present in the corolla tube ....3. *C. salina*

c. Calyx generally shorter than the corolla tube; corolla lobes lanceolate-
acuminate; scales lacking ....4. *C. californica*

1. **C. campestris** Yuncker. In Marin County known only from low ground bordering the laguna in Chileno Valley where it is parasitic on *Anthemis, Lactuca.
Senecio,* and *Xanthium.—C. arvensis* Beyr.

2. **C. subinclusa** D. & H. Occurring abundantly on *Grindelia* in the salt marshes
and also widespread in meadows and on dry slopes in the chaparral where it is
parasitic on *Rosa, Pickeringia, Thermopsis, Rhus, Arctostaphylos, Eriodictyon,*
and *Artemisia:* Mount Tamalpais (West Peak, Potrero Meadow, Camp Handy);
Carson Ridge; salt marshes at San Rafael, Santa Venetia, and Burdell.

3. **C. salina** Engelm. var. **major** Yuncker. Widespread and often common in the
salt marshes where it occurs chiefly on *Salicornia* but also on *Triglochin,
Frankenia,* and *Jaumea:* Sausalito; Almonte; Burdell; Inverness; Drakes Estero.

4. **C. californica** H. & A. Dry rocky slopes and flats, rather common but in
Marin County not widely distributed: Tiburon; Mount Tamalpais (Cascade
and Fern canyons, Bootjack, Barths Retreat); Carson Ridge.

This dodder seems to thrive equally well on a great variety of hosts which
include both annual and perennial herbs as well as hardwood shrubs of the
chaparral. The following genera have been noted as hosts in Marin County:
*Gastridium, Allium, Chlorogalum, Eriogonum, Silene, Thelypodium, Adenostoma,
Photinia, Lotus, Lupinus, Pickeringia, Ceanothus, Sanicula, Convolvulus, Navarretia, Eriodictyon, Monardella, Galium, Achillea, Aplopappus, Eriophyllum,* and
Gnaphalium.

POLEMONIACEAE. Gilia Family

a. Leaves opposite, at least below the inflorescence ....b
a. Leaves alternate c
b. Leaves entire ....3. *Phlox*
b. Leaves palmately divided ....5. *Linanthus*
c. Corolla funnelform, 1.5–2.5 cm. across, the tube short ....1. *Polemonium*
c. Corolla tubular or salverform, 1.5 cm. across or less ....d
d. Calyx lobes pungently spiny ....6. *Navarretia*
d. Calyx lobes not spiny ....e
e. Fruiting calyx developing a reflexed lobe in the sinus ....2. *Collomia*
e. Fruiting calyx without a reflexed lobe in the sinus ....4. *Gilia*

1. Polemonium

1. **P. carneum** Gray. Rare on grassy and brushy slopes: Sausalito, *Kellogg;*
Point Bonita and Angel Island, acc. Stacey; perhaps also Salmon Creek.

2. Collomia

a. Leaves irregularly pinnately lobed or divided; corolla limb bright rose
....1. *C. heterophylla*
a. Leaves entire; corolla limb yellowish to salmon-color ....2. *C. grandiflora*

1. **C. heterophylla** Hook. Common in loose soil in woodland and brush, becom-

ing abundant after fires: Mount Tamalpais (Corte Madera Ridge, Fern Canyon, Berry Trail); Carson country; Inverness Ridge, Point Reyes Peninsula.

2. **C. grandiflora** Dougl. The only Marin County record of this widespread western American annual is one reported from Lagunitas by Stacey.

3. Phlox

1. **P. gracilis** (Hook.) Greene. Widespread and rather common in loose clayey or gravelly soils in brush or grassland: Sausalito (acc. Lovegrove), Tiburon, and Mount Tamalpais (Bolinas Ridge) to San Rafael Hills, Tomales, and Point Reyes Peninsula (lighthouse and dunes near the radio station).—*Gilia gracilis* Hook.

4. Gilia

a. Inflorescence capitate, pedicels if present less than 1 mm. long, heads generally many-flowered ....b
a. Inflorescence loosely to subcapitately paniculate, pedicels generally elongate, occasionally very short and only about 1 mm. long, panicles mostly few-flowered c
b. Corolla lobes narrowly oblong, 3–4 times longer than wide, light blue ....1. *G. capitata*
b. Corolla lobes elliptic to obovate, 1–2 times longer than wide, deep blue ....2. *G. Chamissonis*
c. Corolla throat tubular-campanulate, slightly ampliate upwards, bluish-purple spotted with blackish-purple; seeds nearly smooth ....3. *G. millefoliata*
c. Corolla throat turbinate-campanulate, strongly ampliate upwards, whitish or pale lavender; seeds conspicuously rugulose ....4. *G. multicaulis*

1. **G. capitata** Dougl. Gravelly soil of steep rocky hills, not very common: Fish Grade, Mount Tamalpais; Fairfax Hills; San Rafael Hills; Carson country; between Bolinas and Olema; Tomales, *Leschke.*

2. **G. Chamissonis** Greene. Sandy flats and dunes, Point Reyes Peninsula. The Marin County plant was described by Jepson as *G. capitata* var. *regina.*

3. **G. millefoliata** F. & M. In typical form, this is a plant of maritime dunes and sandy flats characterized by the strongly accrescent fruiting calyx. It is not known south of Point Reyes Peninsula (Abbotts Lagoon).

Gilia inconspicua (Smith) Sweet is the name Jepson has assumed for our coastal plant. Because there is still some doubt about the identity of *G. inconspicua*, *G. millefoliata* is retained here.

Back from the coast, a closely related form in which the fruiting calyx is not so much enlarged has been found at the following stations: between Summit Meadow and Rock Spring, Mount Tamalpais; Larkspur, *Eastwood;* Drakes Estero, Point Reyes Peninsula. This is perhaps the plant that is treated as *G. multicaulis* var. *clivorum* Jeps., a plant Jepson would restrict to the South Coast Ranges.

4. **G. multicaulis** Benth. Widespread and sometimes locally common on grassy hills, chaparral slopes, or rocky slides: Sausalito, Tiburon, and Mount Tamalpais (West Point, Rocky Ridge Fire Trail) to the Carson country and San Rafael Hills.

An occasional form in which the pedicels are longer and more slender and the flowers less congested is var. *tenera* Gray, occurring on Mount Tamalpais (Blithedale Canyon, Nora Trail) and in the San Rafael Hills. This may be the plant that is listed by Stacey from Mount Tamalpais as *G. gilioides* (Benth.) Greene, a species widespread in California but not known to occur in Marin County.

5. Linanthus

a. Flowers on elongate slender pedicels, not congested in headlike clusters ....b
a. Flowers on very short pedicels, congested in headlike clusters c
b. Corolla generally less than 5 mm. long, equaling or shorter than the calyx ....1. *L. pygmaeus*
b. Corolla generally more than 10 mm. long, conspicuously longer than the calyx ....2. *L. liniflorus*
c. Corolla funnelform, the widened throat about equal to the slender tube ....3. *L. grandiflorus*
c. Corolla salverform, the throat very much shorter than the elongate tube ....d
d. Corolla golden-yellow; stamens well exserted beyond the corolla throat, about ½-¾ the length of the lobes ....4. *L. acicularis*
d. Corolla white or yellowish-white to lavender and deep rose; stamens about equaling the corolla throat or a little exserted ....e
e. Calyx puberulent outside to the base, the scarious interval extending about half way to the base below the sinus ....5. *L. parviflorus*
e. Calyx glabrous except for the coarsely ciliate lobes, the scarious interval extending more than half way to the base below the sinus ..6. *L. androsaceus*

1. **L. pygmaeus** (Brand) J. T. Howell. Gravelly soil of steep canyonside in chaparral, San Rafael Hills, the only station known in Marin County.—*L. pusillus* of North American references.

2. **L. liniflorus** (Benth.) Greene. Known in Marin County only at Sausalito where it was collected by Alice Eastwood in 1894. This attractive plant is not known farther north in the Coast Ranges.

3. **L. grandiflorus** (Benth.) Greene. Sandy flats back of the dunes, Point Reyes Peninsula; San Rafael, acc. Stacey. This species is not known farther north.

4. **L. acicularis** Greene. Grassy slopes and flats in clay soil: Barths Retreat, Mount Tamalpais, acc. Sutliffe; Big Carson Canyon; Fairfax Hills; San Rafael Hills; Big Rock Ridge; between Ignacio and Novato.

Linanthus acicularis has been included by Jepson in *L. bicolor* (Nutt.) Greene but it differs not only in the golden color of the corolla and the more needlelike leaf segments but also in the character of the style. In *L. acicularis* the style is not at all or scarcely visible but the elongate branches exceed the anthers while in *L. bicolor* the shortly 3-cleft style is visibly exserted from the corolla tube.

5. **L. parviflorus** (Benth.) Greene. Grassy or brushy slopes and flats of hills and meadows, frequently common and very attractive: Sausalito, acc. Stacey; Mount Tamalpais (Bootjack, acc. L. S. Rose, Laurel Dell, Potrero Meadow, Lagunitas Meadows); Carson country; Big Rock Ridge, *Robbins;* Inverness Ridge.

Along the coast is an especially beautiful large-flowered form, var. *rosaceus* (Hook. f.) Jeps., with corollas of a bright rose and numerous paler shades that vary towards white. This form is found on sandy slopes and downs of Point Reyes Peninsula near the lighthouse and at Abbotts Lagoon.

6. **L. androsaceus** (Benth.) Greene. Grassy and brushy places in shallow soil, frequently on steep rocky canyonsides more or less shaded: Sausalito, Tiburon, and Stinson Beach (acc. Stacey) to the Carson country, San Rafael Hills, Big Rock Ridge (*Robbins*), and Point Reyes Peninsula (road to the lighthouse).

This species, another of the attractive spring flowers of our hill country, is closely related to *L. parviflorus*. The two, however, may be distinguished not only by the evident calyx characters indicated above but also by small differences in the corollas which seem to be responsible for slightly different flower-shapes. The anthers are much more apparent in *L. parviflorus,* perhaps because in that species the sinuses between the lobes extend almost to the top of the tube while in *L. androsaceus* the throat is longer and more funnelform.

6. Navarretia

a. Inflorescence bracts and calyx lobes white-hairy, not glandular; corolla white
. . . . .1. *N. intertexta*
a. Inflorescence bracts and calyx lobes more or less viscid-glandular, not white-hairy; corolla whitish to violet or blue . . . . .b
b. Rachis of leaves and involucral bracts conspicuously elongate, generally linear; corolla lobes 3–4 mm. long; anthers oblong . . . . .2. *N. viscidula*
b. Rachis of leaves and involucral bracts not linear-elongate; corolla lobes 1–3 mm. long; anthers round to elliptic c
c. Stamens included in the corolla tube or barely reaching its mouth . . . . .d
c. Stamens equaling the mouth of the corolla tube or well exserted . . . . .e
d. Herbage strongly mephitic; corolla mostly about 10 mm. long, the lobes 2–3 mm. long . . . . .3. *N. squarrosa*
d. Herbage odorous but not strongly mephitic; corolla mostly about 5 mm. long, the lobes 1–2 mm. long . . . . .4. *N. mellita*
e. Corolla longer than the calyx and showy, the tube much widened upward; stamens conspicuously exserted, the anthers elliptic, nearly 1 mm. long
. . . . .5. *N. heterodoxa*
e. Corolla equaling or somewhat shorter than the calyx and not showy, the tube not much widened upward; stamens not exserted or very slightly, the anthers round, less than 0.5 mm. long . . . . .6. *N. rosulata*

1. **N. intertexta** (Benth.) Hook. Low places in clay soil where water has stood, in Marin County restricted to meadows on the north side of Mount Tamalpais: Potrero Meadow; Willow Meadow; Lagunitas Meadows. These are the southernmost stations in the Coast Ranges.

2. **N. viscidula** Benth. Open grassy slopes or meadows in clay soil: Lagunitas Meadows, Mount Tamalpais; Ross Valley, acc. Jepson; Fairfax Hills; San Rafael Hills; Big Rock Ridge.

3. **N. squarrosa** (Esch.) H. & A. Skunkweed. Common and widespread in clay soil of hills and flats, becoming ruderal along roads and in fields: Sausalito, Angel Island, and Mount Tamalpais (Throckmorton Ridge) to San Rafael Hills, Novato, and Point Reyes Peninsula (Bolinas, Mud Lake, Inverness).

4. **N. mellita** (Greene) Greene. Sometimes locally abundant on clayey or gravelly hills, generally on the edge of brush: Mill Valley, *Eastwood;* Mount Tamalpais (Throckmorton Ridge, near Bootjack, Phoenix Lake, Bolinas Ridge, *Eastwood*); Carson Ridge.

A slender shade form of this species was described from the West Point Trail, Mount Tamalpais, as *N. Eastwoodae* Brand.

5. **N. heterodoxa** (Greene) Greene. Occasional on grassy and brushy slopes in shallow rocky, clayey soil: near Camp Hogan, Mount Tamalpais; San Anselmo Canyon; Carson Ridge; Fairfax, *Eastwood*.

6. **N. rosulata** Brand. Occasional in shallow rocky soil in areas of serpentine: Mount Tamalpais (Steep Ravine, acc. Sutliffe, Simmonds Trail, Potrero Meadow, *Eastwood*, Camp Handy); Carson Ridge; San Anselmo, the type locality.

This distinctive plant has been treated as a variety of *N. heterodoxa* by Jepson but the two are well differentiated by the morphological differences indicated in the key and they also seem to be separated physiologically or ecologically as well. In Marin County, *N. heterodoxa* has been reported only from sandstone or shale or clayey soil derived from those rocks, while *N. rosulata* has been found chiefly, if not exclusively, in areas of serpentine. *Navarretia rosulata* is known only from Marin and San Mateo counties.

HYDROPHYLLACEAE. Waterleaf Family

a. Shrubs with erect woody stems; styles 2, distinct ....6. *Eriodictyon*
a. Annual and perennial herbs, rarely forming a woody caudex at the ground; style 1, simple or 2-cleft ....b
b. Inflorescence scorpioid ....4. *Phacelia*
b. Inflorescence not scorpioid c
c. Ovary nearly 2-celled, the placentae about meeting in the center; style entire; capsule elliptic to oblong; perennial herbs with conspicuous tubers at the base of the stem ....5. *Romanzoffia*
c. Ovary 1-celled; style more or less 2-parted; capsule globose; annual herbs ..d
d. Calyx without small bracts in the sinuses ....3. *Eucrypta*
d. Calyx with small bracts or folds in the sinuses between the calyx lobes ......e
e. Petioles broadly winged, auriculate-clasping at the base; seeds reticulate, without a cucullus ....1. *Ellisia*
e. Petioles very narrowly or not at all winged; seeds smooth, pitted or tuberculate, cucullate ....2. *Nemophila*

1. Ellisia

1. **E. aurita** (Lindl.) Jeps. Climbing Nemophila. Although the climbing nemophila occurs in Marin County only on Angel Island, it is found northward in the inner North Coast Ranges as far as Lake County and several records have been reported by Constance from Contra Costa County across San Francisco Bay.— *Nemophila aurita* Lindl.

2. Nemophila

a. Corolla more than 1 cm. broad; style 3–5 mm. long ....1. *N. Menziesii*
a. Corolla less than 1 cm. broad; style 1–2.5 mm. long ....b
b. Corolla basinshaped, subrotate; style 2–2.5 mm. long ......2. *N. heterophylla*
b. Corolla campanulate; style 1–1.5 mm. long ....3. *N. parviflora*

1. **N. Menziesii** H. & A. Baby-blue-eyes. Widespread and frequently common on

grassy or brushy slopes in sun or partial shade: Sausalito, Mill Valley, and Mount Tamalpais (Summit Meadow, Laurel Dell, Cataract Gulch, Bolinas Ridge) to San Rafael Hills, Point Reyes Lighthouse, and Tomales.

Certainly this is one of the most beautiful and best-beloved wildflowers of the spring, a high favorite with everyone. Most of the plants in Marin County belong to the type in which the corolla is a light blue marked with deeper blue dots and lines, the variant named as *N. liniflora* F. & M.

In low wet fields and on springy slopes in the hills a quite different form with white corolla dotted with black or dark blue is sometimes locally abundant. This type is *N. Menziesii* var. *atomaria* (F. & M.) Chandler (*N. atomaria* F. & M.) and has been noted from Sausalito, Tiburon (acc. Orr), Mill Valley (*Carruth*), Lake Lagunitas (*Eastwood*), Novato (acc. L. S. Rose), and Dillons Beach.

2. **N. heterophylla** F. & M. Moist shaded slopes of hills and canyons, rather common: Sausalito, Angel Island, and Mount Tamalpais (Cataract Gulch) to San Rafael Hills, Big Rock Ridge, and Tomales. A detailed study of many plants from various parts of California has disclosed much variation in the group treated here as a simple species. The Marin County plant of this complex was described by Miss Eastwood as *N. nemorensis,* the type of which came from Fairfax.

3. **N. parviflora** Dougl. Like the preceding, this name also represents a widespread complex and in many instances distinctive variants have been named, but unlike *N. heterophylla, N. parviflora* is rather rare in Marin County. The only stations known for it are Mount Tamalpais (the type locality of *N. micrantha* Eastw.), Cascade Canyon above Mill Valley, Ross (acc. Constance), and Camp Taylor.

3. Eucrypta

1. **E. chrysanthemifolia** (Benth.) Greene. As with *Ellisia aurita,* this species is known in Marin County only from Angel Island (acc. Constance). The *Ellisia,* however, is found farther north of the San Francisco Bay region but the present plant is not known to occur north of Marin and Contra Costa counties.—*Ellisia chrysanthemifolia* Benth.

4. Phacelia

a. Plants perennial. Ovules 2 to each placenta; corolla bowlshaped or campanulate; leaves pinnately divided, the divisions entire ....b
a. Plants annual ....d
b. Shorter, nonbristly hairs of the inflorescence viscid-glandular or capitate-glandular; corolla 4–5 mm. long, sordid-buff or brownish-tinged
....3. *P. nemoralis*
b. Shorter, nonbristly hairs of the inflorescence not glandular; corolla mostly 5–7 mm. long c
c. Basal leaves generally with only 1 or 2 pairs of divisions below the terminal segment; corolla violet, the lobes thin; calyx segments oblong-lanceolate in fruit ....1. *P. californica*
c. Basal leaves generally with 3–5 pairs of divisions below the terminal segment; corolla ivory-white becoming sordid, the lobes thicker and papillate within;

calyx segments lanceolate to ovate in fruit ....2. *P. imbricata*
d. Leaves much divided, 2 or 3 times pinnatifid-dissected. Ovules 2 to each
placenta; corolla limb broad and spreading, whitish or lavender-tinged
....4. *P. distans*
d. Leaves entire, or if pinnately divided, the divisions entire or subentire ......e
e. Stems bristly-hirsute; corolla sordid-buff or brownish-tinged; stamens exserted from the corolla; ovules 2 to each placenta ....5. *P. malvaefolia*
e. Stems not bristly-hirsute; corolla blue, violet, or rose; stamens included, not
exceeding the corolla limb; ovules more than 2 to each placenta ....f
f. Leaves glandular and odorous, crenulate or serrate; corolla tubular-
campanulate, the limb rose, the tube yellow; style cleft $\frac{1}{6}$–$\frac{1}{4}$
....8. *P. suaveolens*
f. Leaves not glandular or odorous, entire or the divisions entire; corolla violet
or blue, bowlshaped or campanulate; style cleft $\frac{1}{3}$–$\frac{1}{2}$ ....g
g. Pedicels of lowest flowers shorter than the fruiting calyx; calyx lobes elongate
in fruit; capsule attenuate-acute ....6. *P. divaricata*
g. Pedicels of lowest flowers equaling or longer than the fruiting calyx; calyx
lobes broadly obovate in fruit; capsule obtuse ....7. *P. insularis*

1. **P. californica** Cham. Common and widespread on gravelly slopes and
rocky outcrops of ocean bluffs, hills, and steep canyonsides: Sausalito, Angel
Island, Tiburon, and Mount Tamalpais (Rattlesnake Camp, Rock Spring, Willow
Meadow, Alpine Dam) to San Rafael Hills, Big Rock Ridge, Dillons Beach, and
Point Reyes. The Angel Island plant was named *P. magellanica* (Lamk.) Cov.
f. *Jepsonii* Brand.

2. **P. imbricata** Greene. Occasional on rocks in hills generally away from the
coast: Bootjack, Mount Tamalpais; Fairfax Hills and Carson country; San
Rafael Hills; Lucas Valley.—*P. californica* Cham. var. *imbricata* (Greene) Jeps.

3. **P. nemoralis** Greene. Rather rare in partial shade on moist brushy or
wooded canyonsides: Sausalito; Angel Island; Fairfax Hills; San Rafael Hills;
Salmon Creek Canyon.

4. **P. distans** Benth. Common on sandy or rocky flats and slopes near the
ocean or rather rare around rocks in hills of the interior: Tiburon; Blithedale
Canyon and Bootjack, Mount Tamalpais; Stinson Beach; Carson country; San
Rafael; Dillons Beach; Point Reyes.

5. **P. malvaefolia** Cham. Rather common on moist partly shaded slopes near
the coast: Sausalito, Angel Island, and Tiburon to Stinson Beach, Tomales, and
Point Reyes Peninsula (Bolinas, Mud Lake, Point Reyes, McClure Beach).

6. **P. divaricata** (Benth.) Gray. Occasional on open slopes in rocky or gravelly
soil, becoming abundant and showy on burns in the chaparral: Sausalito;
Tiburon; Mount Tamalpais (West Point, Rock Spring); Carson country; Tomales.

7. **P. insularis** Munz var. **continentis** J. T. Howell. In Marin County occurring
on the exposed headland of Point Reyes; otherwise reported only from the coast
of Mendocino County. The species is found on the islands of Santa Rosa and
San Miguel in southern California.

8. **P. suaveolens** Greene. Usually very rare in gravelly soil in the chaparral but
appearing in vast numbers in the ashes of a burn: Mount Tamalpais (Potrero
Meadow, Artura Trail, Rocky Ridge Fire Trail); Carson Ridge.

5. Romanzoffia

1. **R. californica** Greene. MIST MAIDEN. Occasional on moist and shaded rocky slopes and bluffs: Mount Tamalpais (Cataract Gulch, Lake Lagunitas, above Phoenix Lake, Baltimore Canyon); Ribbon Falls, San Anselmo Canyon; Lagunitas, acc. Stacey; Big Rock Ridge, *Robbins;* Tomales.—*R. sitchensis* of coastal California references.

6. Eriodictyon

1. **E. californicum** (H. & A.) Greene. YERBA SANTA. In brush or chaparral on exposed rocky slopes, rarely in pine woods: Sausalito, acc. Lovegrove; Mount Tamalpais (Corte Madera Ridge, Cascade Canyon); Carson Ridge; San Geronimo Ridge; among pines, Inverness Ridge.

Following fires in the chaparral, *E. californicum* reproduces by seeds and regenerates itself by vigorous shoots from roots. These plants and shoots bloom by the second season after a fire.

BORAGINACEAE. BORAGE FAMILY

a. Corollas white (sometimes yellow or bluish in the center) ....b
a. Corollas yellow or blue (or rarely white in *Cynoglossum*) ....e
b. Leaves somewhat fleshy, glaucous; ovary not lobed ....1. *Heliotropium*
b. Leaves herbaceous, not glaucous; ovary deeply divided into 4 parts, maturing
    to form 1–4 nutlets c
c. Ventral (inner) face of nutlets with a fine longitudinal line or groove that is
    forked at the base ....2. *Cryptantha*
c. Ventral face of nutlets without a longitudinal groove but bearing an attach-
    ment scar at or below the middle with a more or less definite elevated keel
    or ridge above the scar ....d
d. Leaves alternate, the lowest frequently forming a rosette; attachment scar of
    nutlets near middle of ventral face ....3. *Plagiobothrys*
d. Leaves opposite near the base, sometimes alternate above, the lowest leaves
    not forming a rosette; attachment scar of nutlets below the middle near the
    base ....4. *Allocarya*
e. Corollas yellow ....5. *Amsinckia*
e. Corollas blue (sometimes white in *Cynoglossum*) ....f
f. Stamens much longer than the corolla; plant shrubby ....8. *Echium*
f. Stamens included in the tube of the corolla; plants herbaceous ....g
g. Corolla about 1 cm. or more broad, bearing conspicuous white central crests;
    nutlets large, roughened with barbed prickles ....6. *Cynoglossum*
g. Corolla about 0.5 cm. broad, the crests less conspicuous; nutlets small, smooth
    ....7. *Myosotis*

1. Heliotropium. HELIOTROPE

1. **H. curassavicum** L. var. **oculatum** (Hel.) Jtn. Occasional on gravelly slopes, sandy beaches, and dessicated strands: Alpine Lake, Mount Tamalpais; San Rafael; near Burdell Station; Rodeo Lagoon; dunes near the radio station and McClure Beach, Point Reyes Peninsula.

2. **Cryptantha.** Nievitas; White Forget-me-not

a. Nutlets, at least some of them, rough .... b
a. Nutlets smooth c
b. Calyx in fruit 2–4 mm. long; nutlets all alike, more than 1 mm. long
    .... 1. *C. muricata*
b. Calyx in fruit 1–2 mm. long; nutlets less than 1 mm. long, 3 nutlets papillate
    and 1 nutlet smooth or nearly so .... 2. *C. micromeres*
c. Nutlets generally 4 .... d
c. Nutlets 1 or 2 .... f
d. Inflorescence bearing bractlike leaves throughout; stems prostrate
    .... 3. *C. leiocarpa*
d. Inflorescence without bractlike leaves or these only near the base; stems erect
    or diffuse .... e
e. Nutlets broadly triangular-ovate .... 4. *C. Torreyana*
e. Nutlets much narrower, ovate-lanceolate .... 5. *C. hispidissima*
f. Hairs on calyx lobes straight .... 6. *C. microstachys*
f. Hairs on calyx lobes strongly recurved .... 7. *C. flaccida*

1. **C. muricata** (H. & A.) Nels. & Macbr. var. **Jonesii** (Gray) Jtn. Rare in gravelly soil in the chaparral on Mount Tamalpais: east side on Wheeler Trail; above West Point on West Peak Fire Trail.

2. **C. micromeres** (Gray) Greene. Rare in loose soil in brushy places, becoming locally common in chaparral burns: Pipeline Trail above Mill Valley; Mount Tamalpais, *Eastwood;* Carson country; Shell Beach, Point Reyes Peninsula.

This small-flowered, small-fruited species has not been reported north of Marin County.

3. **C. leiocarpa** (F. & M.) Greene. Locally common on beaches and sandy flats along the coast: Stinson Beach; Dillons Beach; dunes near the radio station and McClure Beach, Point Reyes Peninsula.

4. **C. Torreyana** (Gray) Greene. This widespread western American species is represented in Marin County by two varieties. Var. *calistogae* Jtn., with fruiting calyx-lobes 3.5–5 mm. long and style equaling or a little exceeding the nutlets, has been found in the Carson country, near Tomales, and at Shell Beach, Point Reyes Peninsula. Var. *pumila* (Hel.) Jtn., with fruiting calyx lobes only 2–3.5 mm. long and style shorter than the nutlets, is not uncommon on the edge of brush and in the chaparral in gravelly soil on Mount Tamalpais (Corte Madera Ridge, Bootjack) and in the Carson country. Mount Tamalpais is the type locality of *C. pumila* Hel.

5. **C. hispidissima** Greene. Gravelly soil of an opening in the pine forest on Inverness Ridge, Point Reyes Peninsula. This is the most northern station that has been reported for the plant.

6. **C. microstachys** (Greene) Greene. In Marin County this species has been found only in brush on the north side of Black Canyon in the San Rafael Hills.

7. **C. flaccida** (Dougl.) Greene. Occasional in shallow rocky soil of sunny southern slopes: Tiburon; near Mountain Theater, Mount Tamalpais, *L. S. Rose;* Fairfax Hills; Carson country; San Rafael Hills.

3. Plagiobothrys

a. Upper part of fruiting calyx deciduous, the lower part persistent; corolla
   generally more than 4 mm. broad ....1. *P. nothofulvus*
a. Upper part of fruiting calyx not deciduous; corolla generally less than 4
   mm. broad ....2. *P. tenellus*

1. **P. nothofulvus** (Gray) Gray. POPCORN FLOWER. Widely distributed and
sometimes locally common in the spring on open grassy slopes: Sausalito (acc.
Lovegrove) and Mount Tamalpais (Berry Trail, Bolinas Ridge) to the Carson
country, San Rafael Hills, and Point Reyes Station (acc. Stacey).
2. **P. tenellus** (Nutt.) Gray. Rare in open places in gravelly soil: Laurel Dell,
Mount Tamalpais; San Rafael Hills; Big Rock Ridge.

4. Allocarya

a. Scar of nutlet linear ....1. *A. undulata*
a. Scar of nutlet ovate or narrowly cuneate ....b
b. Keel on ventral side of nutlets lying in an elongate hollow, the keel thus
   separated from the wrinkles on the ventral face of nutlets ..2. *A. californica*
b. Keel on ventral side of nutlets not lying in an elongate hollow and more or
   less united with the wrinkles on the ventral face ....3. *A. bracteata*

1. **A. undulata** Piper. Rare on muddy flats and strands as water recedes in
ponds and lakes: Lake Lagunitas, Mount Tamalpais; the laguna, Chileno Valley.
2. **A. californica** (F. & M.) Greene. Near the coast on moist or swampy slopes
and flats where water seeps in the spring: Tomales; Point Reyes Peninsula (Mud
Lake, Ledum Swamp, Point Reyes).
3. **A. bracteata** Howell. Moist flats and beds of vernal rain pools, frequently
abundant: Lagunitas Meadows, Mount Tamalpais; Greenbrae; San Rafael; Black
Point; Chileno Valley.

5. Amsinckia. FIDDLENECK

a. Some calyx lobes partly united; nutlets about 2 mm. long ....3. *A. spectabilis*
a. All calyx lobes distinct to the base; nutlets about 3 mm. long ....b
b. Corolla limb well expanded, mostly 4–10 mm. wide; plants generally greenish
   ....1. *A. intermedia*
b. Corolla limb very small, mostly about 2 mm. wide; plants grayish .2. *A. Helleri*

1. **A. intermedia** F. & M. Along roads, on the edge of brush, and in grassy
fields, widespread but usually not very common locally: Tiburon and Mount
Tamalpais (above Cascade Canyon, Summit Meadow, Bolinas Ridge) to San
Rafael Hills, Big Rock Ridge, Chileno Valley, Tomales, and Inverness.—*A.
Douglasiana* of Jepson's Manual.
Suksdorf described the form of this species from Tennessee Valley as *A.
pullata.*
2. **A. Helleri** Brand. Rare on grassy flats in clay soil: Deer Park; San Anselmo
Canyon; Black Point.
3. **A. spectabilis** F. & M. Dunes, coastal slopes, and mesas, usually abundant:

Rodeo Lagoon; Dillons Beach; Point Reyes Peninsula (Point Reyes, near the radio station, McClure Beach).—*A. intermedia* of Jepson's Manual.

The form of *A. spectabilis* at Point Reyes was described as *A. truncata* Suksdorf.

6. Cynoglossum. HOUND'S TONGUE

1. **C. grande** Dougl. Common and widespread on moist wooded or brushy hillsides, generally in partial shade: Sausalito, Angel Island, Tiburon, and Mount Tamalpais (Blithedale Canyon, Nora Trail, Phoenix Lake) to the Carson country, San Rafael Hills, and Tomales. Rarely the corollas are pure white, as in specimens found by Berta Kessel on Tiburon Peninsula.

7. Myosotis. FORGET-ME-NOT

1. **M. sylvatica** (Ehrh.) Hoffm. Abundantly and attractively naturalized in moist shaded places near towns, especially common under redwoods: Sausalito; Belvedere; Mill Valley; Ross; San Rafael; Inverness. A native of Europe.

8. Echium

1. **E. fastuosum** Ait. Locally but luxuriantly naturalized on Angel Island on rocky bluffs overlooking San Francisco Bay. A native of the Canary Islands where also it inhabits coastal bluffs.

VERBENACEAE. VERBENA FAMILY

a. Calyx 2-lobed; stems creeping and rooting, forming a turf; flowers in dense roundish or oblong heads .... 1. *Lippia*
a. Calyx 5-toothed; stems erect or spreading, not forming a turf; flowers separated or approximate in oblong or linear spikes .... 2. *Verbena*

1. Lippia. MAT GRASS

1. **L. nodiflora** (L.) Michx. Escaping from cultivation and becoming established near habitations, along roads, and on the desiccated strands of reservoirs: Mill Valley; Mount Tamalpais (Phoenix Lake and Lake Lagunitas); Fairfax; between Chinese Camp and McNears Landing. Introduced from South America. According to H. N. Moldenke, the Marin County plant would be called *Phyla nodiflora* (L.) Greene var. *rosea* (D. Don) Moldenke.

2. Verbena. VERVAIN

a. Upper leaf surface hirsutulous-pubescent, not scabrous; inner face of nutlets minutely pubescent, without whitish papillae .... 1. *V. lasiostachys*
a. Upper leaf surface scabrous-pubescent; inner face of nutlets bearing numerous whitish processes .... b
b. Leaves ovate; inflorescence glandular-pubescent; processes on inner face of nutlets elongate-papillate .... 2. *V. robusta*
b. Leaves lanceolate; inflorescence not glandular; processes on inner face of nutlets tuberculate .... 3. *V. bonariensis*

1. **V. lasiostachys** Link. Rather rare in clay soil in open grassy places or on the

edge of brush: Mount Tamalpais (Lake Lagunitas, acc. Stacey, Hidden Lake Fire Trail, *Leschke*); Salmon Creek Canyon.—*V. prostrata* R. Br.

2. **V. robusta** Greene. Occasional about seepages or along the course of ephemeral vernal brooks: Angel Island; Tiburon; between Muir and Stinson beaches; Lagunitas Meadows, Mount Tamalpais; Lily Lake; Carson Ridge; San Rafael Hills.

3. **V. bonariensis** L. Locally abundant in low ground between Mill Valley and the Almonte Marshes, the only known station in Marin County. Introduced from South America.

LABIATAE. Mint Family

a. Calyx strongly 2-lipped ....b
a. Calyx regular to slightly irregular, the lobes equal or nearly so ....e
b. Fertile stamens 2 ....7. *Salvia*
b. Fertile stamens 4 c
c. Calyx lips entire, the upper hoodlike ....2. *Scutellaria*
c. Calyx lips lobed or toothed, not hoodlike ....d
d. Flowers in dense terminal headlike cluster ....4. *Prunella*
d. Flowers in axillary clusters ....8. *Melissa*
e. Calyx teeth 10, tipped by hooked spines ....3. *Marrubium*
e. Calyx teeth 5, not hooked at tip ....f
f. Stamens long-exserted; nutlets united ....1. *Trichostema*
f. Stamens included in the corolla tube or slightly exserted; nutlets distinct or
    nearly so ....g
g. Corolla irregular, strongly 2-lipped ....h
g. Corolla regular or nearly so, the lobes about equal ....k
h. Flowers solitary in the axils of leaves or bracts; shrubs or subshrubs i
h. Flowers several in clusters forming whorls; herbs ....j
i. Shrub with erect stems; corolla about 2 cm. long ....9. *Lepechinia*
i. Subshrub with trailing slender stems; corolla less than 1 cm. long ..10. *Satureja*
j. Annual; calyx lobes not spine-tipped ....5. *Lamium*
j. Perennial; calyx lobes shortly spine-tipped ....6. *Stachys*
k. Flowers in whorls, axillary or spikelike ....11. *Mentha*
k. Flowers in terminal heads ....12. *Monardella*

1. **Trichostema.** Blue Curls

1. **T. lanceolatum** Benth. Vinegar Weed. Rare on dry open hillsides, blooming in summer and autumn: San Rafael Hills; Hamilton Field.

2. **Scutellaria.** Skullcap

a. Corolla violet-purple; rootstocks slender, ending in small tubers ..1. *S. tuberosa*
a. Corolla white; rootstocks without tubers ....2. *S. californica*

1. **S. tuberosa** Benth. Occasional in grassy places or on the edge of brush: Bill Williams Gulch, Mount Tamalpais, acc. Sutliffe; Ross Valley, acc. Jepson; San Anselmo Canyon; San Rafael Hills.

2. **S. californica** (Gray) Gray. Gravelly soil of hillsides: Mount Tamalpais above Bootjack; San Rafael, acc. Stacey.

3. Marrubium

1. **M. vulgare** L. HOREHOUND. Widespread weed, common in waste ground along roads or near habitations: Mill Valley, acc. Stacey; East Peak and West Point, Mount Tamalpais; San Rafael; Ignacio; Bolinas and Mud Lake, Point Reyes Peninsula. Introduced from Europe.

4. Prunella. SELFHEAL

1. **P. vulgaris** L. This widespread herb is represented in Marin County by two distinct varieties. Var. *atropurpurea* Fern., with large dark purple corollas, is the handsome and prevalent form on moist coastal slopes that are either open and grassy or brushy from Sausalito to Point Reyes Peninsula (Bolinas, Ledum Swamp, Point Reyes). Very rarely the corollas may be pure white (Inverness Ridge). Var. *lanceolata* (Bart.) Fern., with small lilac or violet corollas, is much less common in Marin County, being restricted chiefly to wet meadows on Mount Tamalpais (Azalea Flat, Potrero Meadow, Willow Meadow). The first variety is endemic to the central California coast; the second, widely prevalent in North America, occurs even as far as eastern Asia and western Europe.

5. Lamium. HENBIT

1. **L. amplexicaule** L. This widespread garden weed is known from Marin County only from a Mill Valley record by Stacey. Native of Europe.

6. Stachys. HEDGE NETTLE

a. Corolla large, bright reddish-purple, the tube 1.5–2 cm. long .1. *S. Chamissonis*
a. Corolla whitish and more or less tinged or mottled with rose, the tube 1 cm.
    or less long ....b
b. Leaves cuneate at base, not rounded or cordate ....2. *S. ajugoides*
b. Leaves rounded or cordate at base c
c. Flower whorls not crowded above, all more or less distinct and evident; corolla
    rosy-tinged or mottled ....3. *S. rigida*
c. Flower whorls crowded above into a dense cylindric subcapitate spike, the
    lower whorls sometimes distinct; corolla whitish ....4. *S. pycnantha*

1. **S. Chamissonis** Benth. Rather common in coastal marshes and swales: Sausalito; Muir Beach; Stinson Beach; Point Reyes Peninsula (Inverness, Ledum Swamp, dunes near the radio station).
Overtopping low herbaceous and shrubby plants or growing rankly in willow and alder thickets, this is one of our most attractive flowering plants.
2. **S. ajugoides** Benth. Rather rare in clay soil of low valley lands: near San Rafael; Hicks Valley; laguna in Chileno Valley.
3. **S. rigida** Nutt. var. **quercetorum** (Hel.) Epl. Common and widespread in brushy or grassy places, generally on well-drained slopes in clayey, gravelly, or rocky soil: Sausalito, Angel Island, Tiburon, and Mount Tamalpais (Blithedale Canyon, Throckmorton Ridge, Cataract Gulch) to San Rafael Hills, Tomales, and Point Reyes Peninsula (Bolinas, Mud Lake, Inverness, Ledum Swamp).—*S. bullata* of Jepson's Manual.
4. **S. pycnantha** Benth. Restricted to springy areas on serpentine where it

may be locally abundant: Tiburon; Mount Tamalpais (Bootjack, Camp Handy); Carson country. These are the northernmost stations.

7. Salvia. SAGE

a. Plant annual ....1. *S. Columbariae*
a. Plant perennial, aerial stems arising from rootstocks ....2. *S. Verbenaca*

1. **S. Columbariae** Benth. CHIA. Occasional in gravelly soil in grassland or sunny open places in the chaparral: Sausalito Hills, acc. Sutliffe; Mount Tamalpais (Bootjack, acc. L. S. Rose); Bald Mountain west of San Anselmo; Carson country.
2. **S. Verbenaca** L. This Old World sage is listed by Stacey as established in Mill Valley.

8. Melissa

1. **M. officinalis** L. GARDEN BALM. Escaping from cultivation and occasionally adventive about habitations: Mill Valley; Ross; Lake Lagunitas road, acc. Sutliffe; San Rafael, acc. T. S. Brandegee; Ignacio. Introduced from Europe.

9. Lepechinia

1. **L. calycina** (Benth.) Epl. PITCHER SAGE. Occasional or sometimes locally common on brushy slopes, generally in the chaparral: Angel Island; Mount Tamalpais (Blithedale Canyon, West Point, Phoenix Lake); Fairfax Hills; Carson Ridge; San Geronimo Ridge.—*Sphacele calycina* Benth.
In chaparral fires the plants are killed but numerous seedlings replace the individuals destroyed and these flower and fruit within two years.

10. Satureja

1. **S. Douglasii** (Benth.) Briq. YERBA BUENA. Widespread in coastal hills and frequently common on slopes and flats more or less shaded by brush or trees: Rodeo Lagoon, Sausalito, and Mount Tamalpais (Corte Madera Ridge, Ocean View Trail, Rifle Camp) to Lagunitas Canyon, San Rafael Hills, and Point Reyes Peninsula (Bolinas, Mud Lake, Inverness).—*Micromeria Chamissonis* (Benth.) Greene.

11. Mentha. MINT

a. Inflorescence interrupted, the flower whorls distinct. Corollas hairy outside ..b
a. Inflorescence densely spicate above, more or less interrupted near the base ...c
b. Flower whorls borne in the axils of reduced or bractlike leaves; upper and
    lower calyx teeth dissimilar ....1. *M. Pulegium*
b. Flower whorls borne in the axils of well-developed leaves; upper and lower
    calyx teeth similar ....2. *M. arvensis*
c. Cauline leaves distinctly petiolate ....3. *M. citrata*
c. Cauline leaves sessile or nearly so ....d
d. Stems and leaves green and glabrous or nearly so; corollas glabrous outside
    ....4. *M. spicata*
d. Stems and leaves cinereous-pubescent; corollas hairy outside .5. *M. rotundifolia*

1. **M. Pulegium** L. PENNYROYAL. Common in low summer-dried fields and valley lands that have been marshy in the spring: Sausalito (acc. Stacey), Tamalpais Val-

ley, and Mount Tamalpais (Alpine Lake) to Bolinas, San Rafael, San Antonio
Creek, and Olema. Naturalized from the Old World.

2. **M. arvensis** L. MARSH MINT. Low coastal marshes on or adjacent to Point
Reyes Peninsula: Olema Marshes; McClure Beach road; Drakes Estero; near the
radio station. The Marin county plants are mostly referable to var. *typica* Stewart
f. *lanata* (Piper) Stewart.

3. **M. citrata** Ehrh. BERGAMOT MINT. Occasionally escaping from cultivation, as
at Muir Beach, Stinson Beach, and Mill Valley. Introduced from Europe where it
is generally regarded as a variant of *M. piperita* L. and both are treated as of
hybrid origin.

4. **M. spicata** L. SPEARMINT. Locally naturalized in low wet ground: Mill Valley
marshes; Bolinas; San Antonio Creek. Native of Europe.

5. **M. rotundifolia** (L.) Huds. This European mint is known in Marin County
only where it is locally naturalized in the Mill Valley marshes.

12. **Monardella.** WESTERN PENNYROYAL

a. Plants annual; leaf margins undulate ....3. *M. undulata*
a. Plants perennial; leaf margins plane or a little revolute, not crisped or undulate
   ....b
b. Upper leaf surface glabrous and shining, the lateral veins little or not at all
   impressed ....2. *M. neglecta*
b. Upper leaf surface puberulent or pilose, not shining, the lateral veins conspicuously impressed and extending to the leaf margin c
c. Stems and leaves puberulent or nearly glabrous, the hairs short and tending to
   be subappressed and retrorse ....1. *M. villosa*
c. Stems and leaves villous or subtomentose, the hairs elongate and slender
   ....1*a. M. villosa* var. *franciscana*

1. **M. villosa** Benth. Brushy canyonsides or open grassy or rocky slopes: Sausalito; Belvedere; Mount Tamalpais (Corte Madera Ridge, West Point, Steep
Ravine, Cataract Gulch); Fairfax; San Rafael Hills. On Tiburon is found a form
with glabrous stems and in Fern and Cascade canyons on Mount Tamalpais is a
form with very small leaves suggesting a hybrid between this species and *M.
neglecta*.

1*a*. **M. villosa** Benth. var. **franciscana** (Elmer) Jeps. Coastal slopes on the edge
of brush or in open rocky places: Rodeo Lagoon; Angel Island; Tiburon, the type
locality of *M. mollis* Heller; Muir Beach; Stinson Beach, acc. Stacey; Point Reyes
Peninsula (Inverness Ridge and near the lighthouse); Dillons Beach.

2. **M. neglecta** Greene. Not uncommon on open slopes of serpentine to which
it seems to be restricted: Tiburon; Mount Tamalpais (West Point, above Bootjack, Potrero Meadow, Berry Trail); Carson Ridge; San Geronimo Ridge—*M.
odoratissima* Benth. var. *neglecta* (Greene) Jeps.; *M. villosa* Benth. var. *neglecta*
(Greene) Jeps.

This neat pennyroyal, which appears entirely restricted to serpentine areas, is
only known from Marin County, the type locality, and from Sonoma County.

3. **M. undulata** Benth. Locally common on the sandy slopes and flats of the
Point Reyes dunes, the only occurrence known in Marin County.

SOLANACEAE. Nightshade Family

a. Corolla campanulate to broadly rotate; fruit a berry ....b
a. Corolla elongate, tubular or funnelform; fruit a capsule ....d
b. Corolla yellow with purplish center; calyx much enlarged in fruit ..3. *Physalis*
b. Corolla whitish to violet or purple; calyx not conspicuously enlarged in fruit c
c. Flowers in umbels; corolla rotate; stems erect or sometimes supported by other
    plants, not vinelike ....1. *Solanum*
c. Flowers solitary in leaf axils; corolla campanulate; stems trailing and vinelike
    ....2. *Salpichroa*
d. Stems and leaves glandular-pubescent; capsule not prickly ....4. *Nicotiana*
d. Stems and leaves not glandular; capsule prickly ....5. *Datura*

1. Solanum. Nightshade

a. Plants perennial, more or less shrubby; corollas 1–2 cm. broad ....b
a. Plants annual; corollas less than 1 cm. broad c
b. Stems and leaves spiny; leaves densely white-tomentose beneath
    ....1. *S. marginatum*
b. Stems and leaves not spiny; leaves pubescent but not white-tomentose
    ....2. *S. Xanti*
c. Stems and leaves glabrous or nearly so; fruit black ....3. *S. nigrum*
c. Stems and leaves densely glandular-pubescent; fruit green ....4. *S. sarachoides*

1. **S. marginatum** L. f. Occasional fugitive from cultivation, as in Sausalito (acc. Stacey) and near Point Bonita. A striking native of the Near East.

2. **S. Xanti** Gray var. **intermedium** Parish. Moist partly shaded woods or open brushy slopes in dry rocky soil: Mount Tamalpais (Blithedale Canyon, Steep Ravine, Potrero Meadow); Fairfax Hills; Chileno Valley; Inverness. This variety as it occurs in the hills from Marin and Napa counties northward was called *S. cupuliferum* Greene.

The violet saucershaped flowers are very attractive against the soft green foliage of the loosely branched plants and they are pleasantly fragrant. The closely related *S. umbelliferum* Esch. has been reported from Marin County but in that species the vesture is of branched nonglandular hairs and no plant corresponding to it has been seen.

3. **S. nigrum** L. BLACK NIGHTSHADE. Widespread and rather common in waste ground or in moist shaded places: Sausalito, Angel Island, Tiburon, and Mount Tamalpais (Blithedale Canyon, Phoenix Lake) to San Rafael Hills, Ignacio, and Point Reyes Peninsula (Mud Lake, Point Reyes).

4. **S. sarachoides** Sendt. Occasional weed of roadsides and waste ground: San Anselmo; San Rafael; Chileno Valley; near Inverness. According to several European botanists, *S. sarachoides* is the correct name of the American weed which has usually been called *S. villosum*. It is a native of South America.

2. Salpichroa

1. **S. rhomboidea** (Gill. & Hook.) Miers. LILY OF THE VALLEY VINE. This attractive plant can become a pernicious weed and as such it has been seen in Sausalito and Mill Valley. It is a native of South America.

3. Physalis

1. **P. ixocarpa** Brot. TOMATILLO. In the interior and southern parts of California, this weed is not uncommon, but from Marin County only one record has been seen, that by Mrs. Sutliffe near Camp Lilienthal above Fairfax. Introduced from Mexico.

4. Nicotiana. TOBACCO

a. Cauline leaves petiolate; limb of corolla mostly less than 1.5 cm. across
....1. *N. acuminata*
a. Cauline leaves mostly sessile; limb of corolla generally 2–4 cm. across
....2. *N. Bigelovii*

1. **N. acuminata** (Grah.) Hook. var. **multiflora** (Phil.) Reiche. This native of South America has been found on the gravelly bed of San Antonio Creek and at Point Reyes Station (acc. Stacey).—*N. attenuata* of Jepson's Manual.

2. **N. Bigelovii** (Torr.) Wats. The only Marin County record of this tobacco is one from San Rafael by Stacey.

5. Datura. THORNAPPLE

1. **D. Stramonium** L. JIMSON WEED. Occasional in disturbed or loose ground in clay or gravel: Fairfax Hills; Novato; San Antonio Creek; Olema Marshes; near Tomales. The form occurring in Marin County has lavender- or violet-tinged instead of white corollas and was originally named *D. Tatula* L.

SCROPHULARIACEAE. FIGWORT FAMILY

a. Stamens with anthers 5 ....1. *Verbascum*
a. Stamens with anthers 2 or 4 ....b
b. Corolla spurred or saccate at base c
b. Corolla not spurred or saccate at base ....d
c. Corolla spurred at base, the spur slender ....2. *Linaria*
c. Corolla saccate at base, the sac broad and blunt ....3. *Antirrhinum*
d. Corolla 2-lipped to nearly regular, the upper lip not hoodlike or beaked ....e
d. Corolla very irregular, the upper lip galeate, i.e., hoodlike or beaked ......o
e. Stamens with anthers 4 ....f
e. Stamens with anthers 2 l
f. Corolla nearly regular; flowers scapose ....9. *Limosella*
f. Corolla irregular, usually markedly so; flowers pedicellate ....g
g. Middle lobe of lower lip of corolla folded forming a keel ....4. *Collinsia*
g. Middle lobe of lower lip of corolla not folded or keel-like ....h
h. Flowers less than 5 mm. long ....5. *Tonella*
h. Flowers more than 5 mm. long i
i. Middle lobe of lower lip of corolla reflexed, the other four erect ....
....6. *Scrophularia*
i. Middle lobe of lower lip of corolla not reflexed ....j
j. Fifth stamen represented by a conspicuous filament ....7. *Penstemon*
j. Fifth stamen entirely lacking ....k
k. Calyx 5-parted, not tubular or prismatic; corolla tubular-campanulate, the lobes erect (i.e., extending straight out from the tube) ....14. *Digitalis*

k. Calyx tubular or prismatic; corolla not tubular-campanulate, the lobes not erect .... 8. *Mimulus*

l. Corolla and calyx 5-lobed or -parted; corolla tubular .... m

l. Corolla and calyx 4-lobed; corolla rotate or campanulate .... n

m. Flowers on filiform pedicels; corolla 2-lipped .... 10. *Lindernia*

m. Flowers on short stout pedicels; corolla nearly regular .... 11. *Gratiola*

n. Plants acaulescent or nearly so; corolla campanulate .... 12. *Synthyris*

n. Plants caulescent; corolla rotate .... 13. *Veronica*

o. Upper and lower lips of the corolla about equal or the lower lip more conspicuous than the upper .... p

o. Upper and lower lips of the corolla very unequal, the upper much larger than the lower .... r

p. Lips of corolla about equal; calyx 1-lobed, spathelike (or appearing 2-lobed when flower is subtended by a bract similar to calyx) ... 19. *Cordylanthus*

p. Lips of the corolla unequal; calyx tubular-campanulate .... q

q. Leaves alternate except near base of stem; anther cells unequal and separated .... 18. *Orthocarpus*

q. Leaves opposite; anther cells equal and together .... 15. *Bellardia*

r. Calyx lobes and floral bracts usually brightly colored; anther cells unequal and separated .... 17. *Castilleja*

r. Calyx lobes and floral bracts usually not brightly colored; anther cells equal and together .... 16. *Pedicularis*

1. Verbascum

1. **V. Blattaria** L. Moth Mullein. Occasional along roads and in weedy places about towns: Mill Valley, acc. Stacey; Alpine Lake; Woodacre; San Antonio Creek; Point Reyes Station, acc. Stacey. Native of Eurasia.

2. Linaria. Toad Flax

a. Corolla yellow with orange palate. Plants perennial, the erect leafy stems from rootstocks .... 1. *L. vulgaris*

a. Corolla lavender to deep violet .... b

b. Erect annual with sterile and fertile stems unlike; leaves entire, narrow; flowers racemose .... 2. *L. canadensis*

b. Trailing and climbing perennial (or annual) with roundish, palmately lobed leaves; flowers axillary .... 3. *L. Cymbalaria*

1. **L. vulgaris** Hill. Butter-and-Eggs. Rare in open ground on coastal hills: Point Reyes Station, acc. Eastwood; north of Dillons Beach. Introduced from Eurasia.

2. **L. canadensis** (L.) Dum.-Cours. var. **texana** (Scheele) Penn. Toad Flax. Occasional in gravelly soil of brushy or openly wooded slopes, becoming locally abundant following fires: Mount Tamalpais (Bolinas Ridge, *Eastwood,* Pipeline Trail); Carson country; San Rafael Hills; Shell Beach and near the lighthouse, Point Reyes Peninsula.

3. **L. Cymbalaria** (L.) Mill. Kenilworth Ivy. Escaping from cultivation and becoming naturalized on shady slopes in moist rocky ground: Sausalito; Tiburon; Belvedere. Introduced from Europe.

3. Antirrhinum. SNAPDRAGON

a. Plants supported by twining branchlets in the upper leaf axils and inflorescence
....1. *A. vexillo-calyculatum*
a. Plants supported by elongate twining pedicels, the branchlets not twining
....2. *A. Hookerianum*

1. **A. vexillo-calyculatum** Kell. Occasional in gravelly or rocky soil on open sunny slopes: Sausalito, acc. Stacey; Mount Tamalpais (Rock Spring and Eldridge Grade, acc. Sutliffe, Phoenix Lake); Fairfax Hills; Carson Ridge, *Sutliffe;* Forest Knolls, *Jussel;* San Rafael Hills.

The type locality of this species, which was described by Kellogg before Gray's *A. vagans,* is somewhere "near Point Reyes." Since the plant has never been definitely reported from Point Reyes Peninsula, it is likely that the type was found in the hill country between Mount Tamalpais and Tomales Bay. From that district also came the Bolander collection described as *A. vagans* var. *Bolanderi* Gray. *Antirrhinum vexillo-calyculatum* is apparently the first plant ever described from a Marin County collection.

2. **A. Hookerianum** Penn. Rather rare on warm brushy slopes in loose rocky soil, becoming locally abundant after fires: Sequoia Canyon (i.e. Muir Woods canyon), acc. Munz; Barths Retreat, Mount Tamalpais; Carson Ridge; San Rafael Hills; Inverness, *B. R. Jackson.—A. strictum* (H. & A.) Gray, not Sibth. & Smith.

4. Collinsia

a. Flowers solitary in the leaf axils, the lower pedicels as long as the calyx or longer; corolla mostly 1 cm. long or less ....1. *C. sparsiflora*
a. Flowers in whorls, the lower pedicels shorter than the calyx or almost lacking; corolla 1–2 cm. long ....b
b. Upper lip of corolla conspicuous and about as long as the lower lip; filaments of upper stamens with a linear appendage at the base
....2. *C. heterophylla*
b. Upper lip of corolla shorter than the lower lip, frequently inconspicuous; filaments of upper stamens with basal appendage rudimentary or none
....3. *C. corymbosa*

1. **C. sparsiflora** F. & M. var. **solitaria** (Kell.) Newsom. Widespread and sometimes locally abundant on open grassy slopes or on the edge of brush: Sausalito and Mount Tamalpais (Summit Meadow, Rock Spring) to San Rafael Hills, Big Rock Ridge, and Tomales. Sausalito Hills is the type locality of *C. divaricata* Kell., a synonym of *C. solitaria* Kell. which was described from the Oakland Hills.

Usually the flowers of var. *solitaria* are a pleasing violet or violet-purple, but at the west end of Mount Tamalpais the flowers are frequently an ivory- or creamy white.

2. **C. heterophylla** Grah. CHINESE HOUSES. Open, brushy or wooded slopes in partial shade: Tiburon; Phoenix Lake, Mount Tamalpais; Lagunitas Canyon; San Anselmo; San Rafael Hills; Point Reyes Station, acc. Stacey.—*C. bicolor* Benth.

3. **C. corymbosa** Herder. Coastal slopes, generally on sandy flats or dunes: Bolinas Bay, acc. Newsom; Point Reyes Peninsula, *Eastwood*.

5. Tonella

1. **T. tenella** (Benth.) Hel. Very rare in Marin County, known only from shaded, moss-covered rocks on the north side of Mount Tamalpais.

6. Scrophularia. FIGWORT

1. **S. californica** Cham. BEE-PLANT. Common and widespread in moist places in hills and valleys, especially abundant in brushy thickets of coastal slopes: Sausalito, Angel Island, Tiburon, and Mount Tamalpais (East Peak, Nora Trail, Cataract Gulch) to San Rafael, Ignacio, Tomales, and Point Reyes Peninsula (Bolinas, Inverness, road to Point Reyes).

The stems are usually not more than 5 or 6 feet tall but on Mount Tamalpais plants as much as 10 feet tall have been measured.

7. Penstemon

a. Stems woody forming low sprawling shrubs; corolla scarlet; anther lobes not
    saccate after dehiscence; sterile stamen hairy .... 1. *P. corymbosus*
a. Stems herbaceous above a woody caudex; corolla violet-blue; anther lobes
    saccate at base after dehiscence; sterile stamen glabrous ..2. *P. heterophyllus*

1. **P. corymbosus** Benth. Occasional on exposed rocks: Mount Tamalpais (Corte Madera Ridge, Bootjack, Mountain Theater, Northside Trail); San Anselmo Canyon; Bald Mountain west of San Anselmo.

2. **P. heterophyllus** Lindl. var. **Purdyi** (Keck) McMinn. Rare on high ridges in open shallow soil; near Potrero Meadow, Mount Tamalpais; Carson Ridge, acc. Sutliffe; San Rafael Hills.

8. Mimulus. MONKEYFLOWER

a. Shrubs with glutinous leaves .... 10. *M. aurantiacus*
a. Herbs with pubescent or glandular leaves .... b
b. Flowers red or scarlet c
b. Flowers yellow, the palate often dotted or spotted with reddish brown ....f
c. Perennials of wet places, more than 2.5 dm. tall; pedicels longer than the
    calyx .... 1. *M. cardinalis*
c. Annuals of dry open or brushy slopes, less than 2 dm. tall; pedicels shorter than
    the calyx .... d
d. Corolla about 1 cm. long, the lobes nearly equal; capsule beak strongly exserted
    from the fruiting calyx .... 7. *M. Rattanii*
d. Corolla mostly 2–5 cm. long, the lobes of the lower lip much shorter than those
    of the upper or obsolete; capsule entirely enclosed by the fruiting calyx ..e
e. Corolla less than 3 cm. long, the lower lip evident .... 8. *M. modestus*
e. Corolla more than 3 cm. long, the lower lip nearly obsolete ...9. *M. Douglasii*
f. Corolla not distinctly 2-lipped, the lobes nearly equal; flowers subtended by
    leaves; stems villous-tomentose .... 6. *M. moschatus*
f. Corolla distinctly 2-lipped; flowers subtended by reduced leaves or bracts;
    stems not villous-tomentose .... g

g. Stems 4-angled, sometimes winged. Plants annual ....h
g. Stems roundish. Bracts of the inflorescence usually not villous beneath or more
   or less villous in the perennials i
h. Bracts of the inflorescence conspicuously villous beneath; uppermost calyx lobe
   not much longer than the others; palate of corolla with many small reddish
   brown dots ....4. *M. arvensis*
h. Bracts of the inflorescence not villous beneath; uppermost calyx lobe much
   longer than the others; palate of corolla with one large reddish brown spot
   ....5. *M. nasutus*
i. Plants perennial, the stems slender or stout, generally more than 2 dm. tall
   ....2. *M. guttatus*
i. Plants annual, the stems slender, generally less than 2 dm. tall 3. *M. glareosus*

1. **M. cardinalis** Dougl. Occasional in wet ground along streams or about
springs: Sausalito, acc. Grant; Belvedere, acc. Stacey; Mount Tamalpais (Blithedale Canyon, Bootjack, Muir Woods, Alpine Dam); San Rafael Hills; between
Bolinas and Olema.

The plants of this handsome species are unusually robust as they grow in a
stream bed near Bootjack, where the stems growing up through the supporting
branches of western azalea attain a height of about 6 feet.

2. **M. guttatus** DC. Common and widespread in wet ground of coastal swales,
stream banks, and mountain meadows: Sausalito, Tiburon, and Mount Tamalpais (Rock Spring, Potrero Meadow, Camp Handy) to Dillons Beach and Point
Reyes Peninsula (Bolinas, Inverness, Ledum Swamp, McClure Beach).

The plant with very large flowers from springy places at lower elevations is
var. *grandis* Greene. In its lush way it is one of the most beautiful of our western
monkeyflowers. The smaller-flowered plant of upland meadows is quite different
and with its slender stems and its relatively few-flowered raceme it is reminiscent
of the montane *M. Tilingii* Regel.

The following three annuals, here thought distinct enough for specific recognition, are sometimes treated as varieties of *M. guttatus:*

3. **M. glareosus** Greene. Locally common on rocky or gravelly slopes where
water seeps in the spring: serpentine area above Bootjack and headwaters of
Cataract Creek, Mount Tamalpais; Carson Ridge.

4. **M. arvensis** Greene. Grassy or rocky slopes and flats where water seeps in the
spring: Tiburon; San Rafael; Tocaloma; Black Point; east of Aurora School.

5. **M. nasutus** Greene. Along ephemeral streamlets in the hills or on moist valley
lands: Phoenix Lake, Mount Tamalpais; Fairfax Hills; Tocaloma, acc. Grant;
Nicasio.

Usually the corolla is about 1.5 cm. long but occasionally it is much smaller,
not attaining 1 cm. in length. This small-flowered plant is var. *micranthus*
(Heller) Grant and has been seen from the San Rafael Hills and from Lagunitas
Canyon below Alpine Dam. The var. *eximius* (Greene) Grant with showy
corollas 2 cm. or more long has not been reported from Marin County, though
it may be expected since it has been found in the Coast Ranges just to the north.

6. **M. moschatus** Dougl. var. **sessilifolius** Gray. Moist slopes or flats, frequently
growing up among other herbaceous or shrubby plants: Blithedale and Cascade
canyons, Mount Tamalpais; Inverness, Point Reyes Peninsula.

In var. *sessilifolius* the leaves are nearly or quite sessile, but a second variety, var. *longiflorus* Gray, with leaves shortly petioled, has also been reported from Marin County: Inverness, acc. Grant; Point Reyes, acc. Stacey.

7. **M. Rattanii** Gray. Rare on open gravelly slopes or in brush on Mount Tamalpais: near the summit, acc. Grant; burned area south of Barths Retreat.

8. **M. modestus** Eastw. Gravelly soil on the edge of chaparral: Mount Tamalpais, *Eastwood;* burned area in the Carson country; Larkspur, acc. Stacey; Lagunitas, acc. Brandegee.

The type locality of this attractive little monkeyflower was on the old Bolinas Trail where it was discovered by Miss Eastwood.

9. **M. Douglasii** Gray. Occasional on open grassy or brushy slopes in gravelly soil: Tiburon; Mill Valley, acc. Stacey; Mount Tamalpais (Lone Tree, near Rock Spring, near West Point); on burn in the Carson country.

The earliest flowers of this plant have large showy corollas that are usually taller than the rest of the plant, and such queer chinless flowers as they are with the lower lip about obsolete! Later in the season the flowers no longer develop these showy corollas but are fertilized in bud and the abortive corolla remains closed, covering the top of the capsule like a little cap.

10. **M. aurantiacus** Curt. BUSH MONKEYFLOWER; DIPLACUS. Widespread on open or wooded canyonsides and flats, a common and characteristic constituent of brush and chaparral: Sausalito, Angel Island, Tiburon, and Mount Tamalpais (Blithedale Canyon, Cataract Gulch, Phoenix Lake) to San Rafael Hills, Black Point, Tomales, and Point Reyes Peninsula (Bolinas, Mud Lake, Inverness).—*Diplacus aurantiacus* (Curt.) Jeps.

When in full bloom, the bush monkeyflower is one of our most attractive shrubs. It is killed by fire but numerous seedlings come up the first winter following the conflagration, and, during the following summer, the vigorous young plants put on a spectacular flower show.

9. Limosella. MUDWORT

1. **L. aquatica** L. Rare in shallow water or on wet ground where water has stood: Lake Lagunits, Mount Tamalpais; Point Reyes Station, acc. Stacey; laguna in Chileno Valley.

10. Lindernia

1. **L. anagallidea** (Michx.) Penn. Rare along the strand of Lake Lagunitas, the only known station in Marin County.—*Ilysanthes dubia* of most California references, not *I. dubia* (L.) Barnh.

11. Gratiola

1. **G. ebracteata** Benth. Muddy flats or low fields where water has stood or on the drying strand of reservoirs: Mount Tamalpais (Hidden Lake, Lake Lagunitas, Portero Meadow); Ignacio; Hicks Valley; laguna in Chileno Valley; east of Aurora School.

12. Synthyris

1. **S. reniformis** (Dougl.) Benth var. **cordata** Gray. Deep shade of wooded canyonsides and flats: Cataract Gulch, Mount Tamalpais; Bolinas Ridge; Big Carson; Lagunitas.—*S. rotundifolia* of Jepson's Manual in part.

This little plant, which forms in Marin County such an attractive ground-cover under redwoods, is not known farther south.

13. Veronica. SPEEDWELL

a. Racemes terminating the stems, the uppermost leaves or leaflike bracts
   alternate ....b
a. Racemes axillary, the leaves all opposite ....e
b. Pedicels longer than the subtending bracts; corolla more than 5 mm. across
   ....1. *V. persica*
b. Pedicels shorter than the subtending bracts; corolla less than 5 mm. across ..c
c. Plants perennial, the stems creeping and rooting at the base; style elongate,
   exceeding the corolla ....2. *V. serpyllifolia*
c. Plants annual; style much shorter than the corolla ....d
d. Leaves elongate, oblongish to lanceolate, obscurely serrulate or entire; capsule
   shallowly emarginate ....3. *V. peregrina*
d. Leaves broad, ovate or roundish, conspicuously crenate; capsule deeply
   obcordate ....4. *V. arvensis*
e. Leaves broad, ovatish, shortly petiolate; capsule shallowly emarginate
   ....5. *V. americana*
e. Leaves elongate, oblongish to lanceolate, sessile; capsule deeply obcordate
   ....6. *V. scutellata*

1. **V. persica** Poir. Moist soil on open rocky slopes, Mill Valley. Native of Eurasia.—*V. Buxbaumii* Ten.

2. **V. serpyllifolia** L. In Marin County known only as a lawn weed in San Rafael. The species is native in many parts of California but it is undoubtedly adventive in its occurrence in Marin County.

3. **V. peregrina** L. var. **xalapensis** (H.B.K.) Penn. Low ground where water has seeped or stood: Lagunitas Meadows, Mount Tamalpais; Ignacio; east of Aurora School; laguna in Chileno Valley.

4. **V. arvensis** L. Weedy places in the hills and about habitations: Sausalito; Laurel Dell and Phoenix Lake, Mount Tamalpais; San Anselmo; San Rafael. Native of Europe.

5. **V. americana** Schwein. AMERICAN BROOKLIME. Rather common in wet ground near springs, streams, and marshes: Sausalito; Tiburon; Mount Tamalpais (Muir Woods, Cascade Canyon, Potrero Meadow); Bolinas and Ledum Swamp, Point Reyes Peninsula.

6. **V. scutellata** L. MARSH SPEEDWELL. Marshy ground near the coast: between Bolinas and Olema; Ledum Swamp, Point Reyes Peninsula. This attractive herb has not been reported south of the Golden Gate.

14. Digitalis. FOXGLOVE

1. **D. purpurea** L. Escaping from cultivation and becoming naturalized in moist shady places in coastal woods: Bolinas Ridge, acc. Stacey; Verde Canyon near Salmon Creek; Inverness. Native of Europe.

15. Bellardia

1. **B. Trixago** (L.) All. Locally common on open grassy hills: Tiburon; Boot-

jack and Dipsea Trail, Mount Tamalpais; Greenbrae; San Rafael Hills; Tocaloma.

This introduction from the Mediterranean region bears a superficial resemblance to owls clover and is one of the few plants naturalized in California
which assume the aspect of a native. Not only is it an attractive wildflower but
the glandular secretion of the leaves and inflorescence has the pleasing fragrance of
ripe apricots.

16. **Pedicularis.** LOUSEWORT

1. **P. densiflora** Benth. INDIAN WARRIOR. Moist slopes and flats in grassland,
brush, and woodland: Mount Tamalpais (Corte Madera Ridge, Throckmorton
Ridge, Lake Lagunitas); Fairfax Hills; Carson Ridge.

The Indian warrior with its colorful, full-flowered inflorescences is among the
most cherished of our spring flowers. Associated with blue hound's tongue or
ivory star zigadene, or dominating a meadowy flat by itself, this bright Californian presents pictures of choice beauty that are a joy to see and to remember.
Usually the corollas are crimson but on rare occasions they are white.

On the open serpentine barrens of the Carson country the plants are more
slender and less hairy and the corolla is rose instead of red. The differences
between it and the common widespread form are slight but the plant is distinctive
enough and undoubtedly represents a genetic variant adapted to the particular
habitat.

17. **Castilleja.** INDIAN PAINTBRUSH

a. Plants annual; bracts entire; lower lip of corolla bright red ....7. *C. stenantha*
a. Plants perennial; bracts generally lobed or parted; lower lip of corolla green
    or only the tips of the teethlike lobes yellowish or red ....b
b. Calyx more deeply cleft on the ventral (or outer) side than on the dorsal (or
    inner) side, the limb spathelike and standing erect at time of anthesis;
    corolla outwardly curved, the galea conspicuously exserted ....4. *C. affinis*
b. Calyx about equally cleft on the dorsal and ventral sides, or if somewhat
    unequal, the calyx limb not erect and spathelike; corolla nearly straight, the
    galea usually not conspicuous (except in *C. Douglasii*) c
c. Pubescence more or less viscidulous or capitate-glandular below the in-
    florescence ....1. *C. latifolia*
c. Pubescence not glandular below the inflorescence ....d
d. Stems and leaves densely felted with branched hairs ....6. *C. foliolosa*
d. Stems and leaves not felted-pubescent, the hairs simple ....e
e. Leaves velvety-pilose, the pubescence dense and cinereous; inflorescence 5–6
    cm. broad, rose-color, the bracts flabellate or broadly cuneate, 3–5-lobed
    ....5. *C. Leschkeana*
e. Leaves scabrous-hirsutulous, the pubescence sparse; inflorescence 5 cm. broad or
    less, generally red or yellow, the bracts narrowly cuneate or oblongish,
    entire or generally 3-lobed ....f
f. Inflorescence broader, 3–5 cm. broad, the flowers spreading, the galeas more or
    less exserted; bracts and calyx red ....2. *C. Douglasii*
f. Inflorescence strict, about 2 cm. broad, the flowers erect-ascending, the galeas
    not evident; bracts and calyx yellow, sometimes tinged with red
    ....3. *C. neglecta*

1. **C. latifolia** H. & A. Widespread in Marin County from coastal bluffs and mesas to wooded or brushy canyon slopes and ridges, exhibiting several distinctive forms. In exposed places on the immediate coast the typical form, with somewhat sprawling stems and more or less scattered oblong or oval leaves, is occasional at Stinson Beach (acc. Stacey), Dillons Beach, Point Reyes and McClure Beach, Point Reyes Peninsula. The form of this species that occurs at Point Reyes has been described as *C. inflata* by Pennell who regards it "a very localized and remarkable species." On wooded or brushy slopes somewhat back from the coast, the plants are bushy, more glandular, and densely leafy, var. *Wightii* (Elmer) Zeile. It is more common than the typical form and occurs from Rodeo Lagoon, Tennessee Cove, and Stinson Beach to Salmon Creek Canyon, Tomales, Inverness, and Point Reyes. In both these forms the inflorescence may be red or yellow, but red is more usual in the typical form and yellow in var. *Wightii*. Still farther inland, in moist shaded places on brushy or wooded hills, is a laxly branched, sparsely leafy form, quite unlike the maritime types but definitely related to them. This plant, var. *rubra* (Pennell) J. T. Howell, has been seen on Tiburon, Mount Tamalpais (Blithedale Canyon and Corte Madera Ridge), Fairfax Hills, and San Rafael Hills. It was first described as *C. Wightii* Elmer subsp. *rubra* Pennell from the Fairfax Hills. Very different, but also a part of this variable species, is a low densely leafy plant that has been found only on chaparral-covered slopes in the bishop pine woods on Inverness Ridge.

2. **C. Douglasii** Benth. Similar to *C. latifolia* var. *rubra* but with the leaves harshly pubescent and not viscidulous. It is rare in Marin County, being known only from Sausalito (acc. Stacey) and Angel Island. A related plant, which differs from the more typical form chiefly in its lower and more leafy habit, grows in exposed rocky places on Carson Ridge.—*C. parviflora* Bong. var. *Douglasii* (Benth.) Jeps.

3. **C. neglecta** Zeile. A remarkable endemic, known only from Tiburon Peninsula where it is restricted to open slopes on the serpentine.

4. **C. affinis** H. & A. Widespread on open, brushy, or wooded coastal hills from Sausalito, Tiburon, and Mount Tamalpais (Steep Ravine) to San Rafael Hills, Black Point, Tomales, and Point Reyes Peninsula (Bolinas, Inverness Ridge).

This is one of the most attractive and colorful plants in brushy places near the coast and it is well marked from the paintbrushes related to *C. latifolia* and *C. Douglasii* by the brilliant spathelike calyx. Although it is one of the most distinctive plants in this puzzling genus, there is some doubt as to its correct name: according to Pennell our plant should be called *C. franciscana* Pennell, but here the use of *C. affinis* is continued in its traditional sense.

5. **C. Leschkeana** J. T. Howell. Known only from swales behind the dunes on Point Reyes Peninsula, where its handsome inflorescence overtops the low dense tangle of herbaceous and woody plants. It is a member of a group of paintbrushes which extends southward along the coast from Alaska and which reaches its southernmost limit at this station.

6. **C. foliolosa** H. & A. Common on dry rocky slopes in open or brushy places: Tiburon; Mount Tamalpais (Blithedale Canyon, Throckmorton Ridge); Fairfax Hills; Carson Ridge; San Rafael Hills.

7. **C. stenantha** Gray. In wet soil along streamlets on the serpentine near

Bootjack, Mount Tamalpais. This is the only occurrence of this species known between the mountains of Lake and Napa counties to the north and the Santa Lucia Mountains in Monterey County to the south.

18. Orthocarpus

a. Anthers 2-celled; galea finely and densely hairy to the tip ....b
a. Anthers 1-celled; galea glabrous, at least near the tip, or if sparsely hairy at the tip, then the corolla less than 6 mm. long ....f
b. Spikes slender, usually less than 1.5 cm. wide, not very colorful; lower lip of corolla 2 mm. deep or less ....2. *O. attenuatus*
b. Spikes broader, usually more than 1.5 cm. wide, flowers or bracts colorful; lower lip of corolla more than 2 mm. deep c
c. Bracts green, not colored at the tips; lower lip of corolla about 6 mm. deep ....1. *O. lithospermoides*
c. Bracts colored at the tips with white, yellow, or red; lower lip of corolla about 3 mm. deep ....d
d. Galea hooked at the tip ....5. *O. purpurascens*
d. Galea straight at the tip ....e
e. Stems pilose to subglabrous, the hairs generally elongate and shining; leaves oblongish, not attenuate, generally not cinereous ..3. *O. castillejoides*
e. Stems and leaves mostly cinereous-pubescent, the hairs not translucent or shining; leaves linear-attenuate ....4. *O. densiflorus*
f. Corolla less than 6 mm. long, the galea sometimes pilose-hairy to the tip ....9. *O. pusillus*
f. Corolla 10 mm. long or more, the galea glabrous at least at the tip ....g
g. Stamens a little longer than the galea; corolla less than 1.5 cm. long ....8. *O. floribundus*
g. Stamens shorter than the galea; corolla 1.5–2.5 cm. long ....h
h. Stems glabrous; galea white, yellow, or pink ....6. *O. faucibarbatus*
h. Stems cinereous-pubescent; galea dark purple ....7. *O. erianthus*

1. **O. lithospermoides** Benth. CREAM SACS. In clay soil on open grassy slopes that are boggy in the spring: Tiburon; Carson country; Lucas Valley; road to Point Reyes.

2. **O. attenuatus** Gray. Rather widespread but not common on open grassy hills: Corte Madera, the type locality; Potrero Meadow and Corte Madera Ridge, Mount Tamalpais; Fairfax Hills; San Geronimo Ridge; San Rafael Hills; Big Rock Ridge; Black Point.

3. **O. castillejoides** Benth. Moist meadows in the hills and low ground along the upper reaches of the salt marshes, occasional: Rodeo Lagoon; Lagunitas Meadows, Mount Tamalpais; Greenbrae Marshes; San Rafael Hills; Hamilton Field; Drakes Estero and Point Reyes dunes, Point Reyes Peninsula.

This attractive species is variable in habit and foliage and is not always readily recognized. The stems may be erect or sprawling, the leaves broadly oblongish or narrow, and the spikes short and thick or elongate. A form with broad leaves and long inflorescences was described as *O. longispicatus* Elmer from Point Reyes Post Office on Point Reyes Peninsula.

4. **O. densiflorus** Benth. OWL'S CLOVER. Locally common on open grassy hills

and in valleys, frequently giving a rosy color to the landscape: Sausalito, Angel Island, Tiburon, and Mount Tamalpais (Corte Madera Ridge, Lagunitas Meadows) to the Carson country, San Rafael Hills, and Hamilton Field. The common name is given for the remarkable resemblance of the corolla to the image of an owl.

On open hills along the coast is a form with stouter spikes that are more colored with creamy white than rose. This is var. *noctuinus* (Eastw.) J. T. Howell which was originally described from Inverness and which has been found more recently above Drakes Estero on Point Reyes Peninsula, bluffs south of Tomales, *Leschke,* and on the rocks east of Dillons Beach.

5. **O. purpurascens** Benth. Rare in Marin County on coastal slopes and dunes: Stinson Beach, acc. Stacey; Point Reyes near the lighthouse. The plant from the latter station with its broader leaves and bracts is one of the showiest and most colorful in the wildflower gardens of that exposed headland and is referable to var. *latifolius* Wats.

6. **O. faucibarbatus** Gray. Wet grassy hills or swampy flats of valley lands or salt marsh borders: Tiburon; Corte Madera, the type locality; San Rafael; Olema; east of Aurora School.

This plant is often locally common and its clear attractive yellow may color extensive areas. Along roads near the coast it is particularly abundant where it forms narrow flowery strips on the graded borders for miles.

Generally the flowers are yellow but near the coast the flowers are white fading rose. This is var. *albidus* (Keck) J. T. Howell, and has been seen at Point Reyes Station, Tomales, Inverness, and near the lighthouse on Point Reyes. Near Tomales and Olema intermediates in color between the species and variety have been seen.

7. **O. erianthus** Benth. The typical form with flowers predominantly yellow is known in Marin County only from a flat field near the Laguna School in Chileno Valley.

On open rocky or sandy slopes near the ocean is the beautiful var. *roseus* Gray in which the fragrant flowers are at first white but fade rose. Memorable gardens of this plant have been seen on the rocks east of Dillons Beach, above Drakes Estero, and on Point Reyes near the lighthouse.

8. **O. floribundus** Benth. Locally common on open slopes and flats near the coast on Point Reyes Peninsula: Inverness; Drakes Estero; Point Reyes near the lighthouse. These are the northernmost stations for a rare plant known otherwise only from San Francisco and San Mateo counties.

9. **O. pusillus** Benth. Common and widespread on grassy slopes and flats: Sausalito, Angel Island, Tiburon, and Mount Tamalpais (Rock Spring, Phoenix Lake) to San Rafael Hills, Tomales, and Point Reyes Peninsula.

The lowly plants of this species are evident not from its tiny reddish-purple flowers but rather from the purplish-tinged herbage which contrasts with the green of other plants with which it grows. On Tiburon, however, some individuals of *O. pusillus* have been noted as green without any shading of purple. Far to the north in Washington, plants of *O. pusillus* with yellow-green herbage have been called var. *densiusculus* (Gdgr.) Keck.

Two instances of hybridization between *O. pusillus* and another species of *Orthocarpus* may be recorded: *O. pusillus* x *O. floribundus* has been reported by Keck from Point Reyes Peninsula near the radio station; *O. pusillus* x *O. erianthus* var. *roseus* is represented by a single plant from the rocks east of Dillons Beach.

19. Cordylanthus

a. Leaves linear, the uppermost tipped with a callous spot; flowers subtended by a linear leaflike bract and enclosed by 2 broader "calyx-leaves", the lower a floral bract, the upper the modified calyx; plants of dry hillsides 1. *C. pilosus*
a. Leaves elongate but lanceolate to oblong, not tipped with a callous spot; flowers without a subtending leaflike bract but enclosed by a floral bract and the modified calyx; plants of salt marshes ....b
b. Floral bracts shallowly 3-toothed at the summit; stamens 4 ....2. *C. maritimus*
b. Floral bracts shallowly pinnately lobed with 5–8 lobes; stamens 2 ..3. *C. mollis*

1. **C. pilosus** Gray. Occasional on dry grassy or brushy slopes: Phoenix Lake and Corte Madera Ridge, Mount Tamalpais; Fairfax; Greenbrae Hills; San Rafael, acc. Stacey; White Hill, *Eastwood*.

2. **C. maritimus** Nutt. *Salicornia* flats in salt marshes along the bay and ocean: Almonte and Greenbrae marshes; Stinson Beach; Inverness; Drakes Estero.

Both in this species and the following, the inflorescence frequently glistens with a coating of salt crystals.

3. **C. mollis** Gray. Rare, only known in salt marshes bordering the northern reaches of San Francisco Bay: San Rafael, acc. Ferris; Burdell Station; San Antonio Creek near San Antonio Station.

LENTIBULARIACEAE. BLADDERWORT FAMILY

1. Utricularia. BLADDERWORT

1. **U. vulgaris** L. Known in Marin County only from old records by K. and T. S. Brandegee who reported it from near Olema.

OROBANCHACEAE. BROOMRAPE FAMILY

a. Filaments densely hairy at the base; capsule usually opening in 4 parts
....1. *Boschniakia*
a. Filaments not hairy; capsule opening in 2 parts ....2. *Orobanche*

1. Boschniakia

a. Inflorescence 3 cm. in diameter or less; bracts of the inflorescence subacute; corolla 1.5 cm. long or less; filaments hairy only at the base; seeds less than 2 mm. long, the alveolae less than 0.5 mm. long ....1. *B. Hookeri*
a. Inflorescence more than 3 cm. in diameter; bracts of the inflorescence broadly rounded or nearly truncate; corolla 1.5 cm. long or more; filaments hairy both at the base and top; seeds generally 2 mm. long or more, the alveolae usually more than 0.5 mm. long ....2. *B. strobilacea*

1. **B. Hookeri** Walp. Rare on moist brushy slopes where it is parasitic on the roots of *Gaultheria:* Mill Valley, acc. Stacey; Pipeline Trail, *Papina*. The plant reported by R. G. Johnson from above Lily Lake is also probably this species.

2. **B. strobilacea** Gray. Occasional in chaparral on dry rocky slopes where it is

parasitic on the roots of *Arctostaphylos* and perhaps of *Arbutus:* Carson Ridge; possibly also the plant reported from the East Peak of Mount Tamalpais by Mary Courtright.—*B. tuberosa* (Hook.) Jeps. in large part.

The report that this species is parasitic on the roots of *Arbutus* needs verification.

2. Orobanche. BROOMRAPE

a. Flowers on pedicels shorter than the calyx, the pedicels with 2 bractlets just
  below the calyx ....b
a. Flowers on elongate, scapelike pedicels, the pedicels without bractlets c
b. Flowers 3–4 cm. long; inflorescence glandular-puberulent; calyx lobes linear-
  attenuate ....1. *O. Grayana*
b. Flowers less than 2 cm. long; inflorescence cinereous-puberulent, not glandular;
  calyx lobes oblong-lanceolate ....2. *O. bulbosa*
c. Pedicels 1–3, much longer than the stems ....3. *O. uniflora*
c. Pedicels usually more than 3, rarely only 2 or 3, about equaling the stem or
  shorter ....4. *O. fasciculata*

1. **O. Grayana** G. Beck. This broomrape, which was formerly called *O. comosa* Hook., is represented in Marin County by two varieties, both of which are rare. Var. *violacea* (Eastw.) Munz, in which the corollas are purplish and about 4 cm. long, is restricted to slopes near the ocean where it is parasitic on gumweed. In Marin County it has been found on Mount Vision, the type locality, and near Muir Beach. Var. *Jepsonii* Munz, in which the corollas are ivory-white or pinkish and about 3 cm. long, is found farther inland. It is known in Marin County only from Novato where it was discovered by Berta Kessel near California laurel and blue elderberry.

2. **O. bulbosa** G. Beck. Occasional on Mount Tamalpais, parasitic on the roots of shrubs in the chaparral, chiefly *Adenostoma fasciculatum:* Baltimore Ridge, *Sutliffe;* Throckmorton Ridge; Rock Spring Trail.—*O. tuberosa* (Gray) Heller, not Hook.

3. **O. uniflora** L. The inconspicuous plants of this broomrape may have been overlooked in Marin County but are apparently rare, being known only from three stations. In the Sausalito Hills and near Tomales the plants were parasitic on *Sedum spathulatum* and represent var. *Sedi* (Suksdorf) Achey. In the San Rafael Hills the plants were parasitic on a species of *Lithophragma* and are var. *minuta* (Suksdorf) G. Beck.

4. **O. fasciculata** Nutt. The typical variety with fewer flowers and purplish corollas has been collected on Tiburon Peninsula and on the Rock Spring Trail, Mount Tamalpais. Much more common on Mount Tamalpais is var. *franciscana* Achey in which the flowers are more numerous and the corolla yellowish: West Point; above Muir Woods; near Potrero Meadows. Mount Tamalpais is the type locality of the variety.

PLANTAGINACEAE. PLANTAIN FAMILY

1. Plantago. PLANTAIN

a. Corolla not opening, the lobes erect with the tips approximate, forming a
  conical cover over the capsule ....b

a. Corolla generally opening, the lobes ascending, spreading, or reflexed c

b. Plants perennial, robust; capsule 3-seeded ....4. *P. hirtella*

b. Plants annual, low and generally slender; capsule 2-seeded ....5. *P. firma*

c. Leaves divided into narrow spreading lobes; inflorescence nodding in bud

....2. *P. Coronopus*

c. Leaves entire or toothed, not divided; inflorescence not nodding ....d

d. Leaves fleshy-thickened; corolla tube hairy ....e

d. Leaves not fleshy-thickened; corolla tube not hairy ....f

e. Leaves erect, about equaling the erect scapes; seeds more than 2 mm. long

....3. *P. juncoides*

e. Leaves spreading, shorter than the spreading scapes; seeds 2 mm. long or less

....3a. *P. juncoides* var. *californica*

f. Leaves ovate; spike glabrous; seeds 6 or more ....1. *P. major*

f. Leaves elongate, linear to narrowly oblong; spike more or less hairy; seeds

2–7 ....g

g. Plants usually perennial; leaves lanceolate to oblong-lanceolate, with 3–5

prominent nerves. Stamens 4; seeds 2 ....6. *P. lanceolata*

g. Plants annual; leaves linear to linear-oblanceolate, without prominent nerves h

h. Corolla lobes about 2 mm. long; stamens 4; seeds 2, concave on the ventral face

....7. *P. erecta*

h. Corolla lobes less than 1 mm. long; stamens 2; seeds 4–7, irregularly pitted,

more or less convex on both faces ....8. *P. Bigelovii*

1. **P. major** L. Common Plantain. Occasional along roads and around towns in low places that are wet in the spring: Sausalito (acc. Lovegrove), Tiburon, and Mount Tamalpais (Blithedale Canyon, Alpine Lake) to San Rafael, Olema Marshes, and Point Reyes Peninsula (Bolinas, Inverness). Native of the Old World.

2. **P. Coronopus** L. Locally common in subsaline soil bordering the salt marshes near San Rafael. Native of the Old World.

3. **P. juncoides** Lamk. Occasional in salt marshes bordering the bay or ocean: Almonte; Escalle; San Antonio Creek; Stinson Beach; Inverness; Drakes Estero.

3a. **P. juncoides** Lamk. var. **californica** Fern. Rocky soil of exposed maritime slopes: Rodeo Lagoon; Point Reyes; McClure Beach.

Plantago juncoides and its distinctive variety have generally been treated collectively in California floras under the name *P. maritima.*

4. **P. hirtella** Kunth var. **Galeottiana** (Decne.) Pilger. Wet soil of springy slopes in the hills or marshy flats along the coast: Sausalito, Tiburon, and Mount Tamalpais (Rock Spring, Potrero Meadow) to China Camp, Dillons Beach, and Point Reyes Peninsula (Bolinas, Inverness, Ledum Swamp, McClure Beach).

Depending on the environment, the plants vary from small individuals only an inch or so tall to robust specimens over a foot tall.

5. **P. firma** Kunze. Open grassy slopes and meadows that are wet in the spring: Lone Tree and Lagunitas Meadows, Mount Tamalpais; Fish Grade; San Anselmo Canyon about Fairfax. This small weedy introduction from South America has been reported in California only from Sonoma and Marin counties.

6. **P. lanceolata** L. Ribwort. Common and widespread in hills and valleys and in waste ground along roads and in towns: Sausalito, Angel Island, Tiburon, and Mount Tamalpais (Bootjack, Rock Spring) to San Rafael, Tomales, and Point

Reyes Peninsula (Bolinas, Inverness, Point Reyes). Introduced from Europe.

7. **P. erecta** Morris. Common in loose sandy or gravelly soil from maritime bluffs and mesas to grassy hills and brushy or wooded canyonsides: Sausalito, Angel Island, Tiburon, and Mount Tamalpais (Lake Lagunitas) to San Rafael, Carson Ridge, Tomales, and Point Reyes Peninsula (Inverness Ridge, Point Reyes).

8. **P. Bigelovii** Gray. Locally common on moist sandy flats near the ocean: Rodeo Lagoon; Dillons Beach; near the radio station, Point Reyes Peninsula.

RUBIACEAE. MADDER FAMILY

a. Flowers in a small headlike cluster subtended by a deeply divided involucre; corolla funnelform, bright pink or lilac . . . . .1. *Sherardia*
a. Flowers in cymes or solitary, not subtended by an involucre; corolla rotate, white, buff, or purplish red . . . . .2. *Galium*

1. Sherardia

1. **S. arvensis** L. FIELD MADDER. Rather common on moist grassy slopes and flats in the spring, the bright pink flowers, although small, sometimes numerous enough to be attractive: Sausalito, Angel Island, Tiburon, and Mount Tamalpais (Rock Spring) to San Rafael and Point Reyes Peninsula (Mud Lake). Introduced from Europe.

2. Galium

a. Middle cauline leaves 5–8 in a whorl . . . . .b
a. Middle cauline leaves 4 in a whorl . . . . .g
b. Ovary and fruit glabrous c
b. Ovary and fruit hairy . . . . .d
c. Plants annual with capillary branchlets; fruit very small, less than 1 mm. long
   . . . . .2. *G. divaricatum*
c. Plants perennial, the branches less slender; fruit usually 1–2 mm. long
   . . . . .6. *G. tinctorium*
d. Fruit small, about 1 mm. long. Diffusely branched annual with slender branchlets . . . . .1. *G. parisiense*
d. Fruit larger, 1.5–5 mm. long . . . . .e
e. Plants perennial; leaves narrowly elliptic to obovate, 6 in a whorl
   . . . . .5. *G. triflorum*
e. Plants annual; leaves linear or linear-oblong to linear-oblanceolate, frequently 8 in a whorl . . . . .f
f. Fruit 4–5 mm. long . . . . .3. *G. Aparine*
f. Fruit 1.5–3 mm. long . . . . .4. *G. spurium*
g. Plants annual; fruits elongate, dry, hairy, the hairs uncinate and mostly tufted at the apex of the fruit . . . . .9. *G. murale*
g. Plants perennial; fruits round or broader than long, fleshy, glabrous or hairy, the hairs straight . . . . .h
h. Stems and leaves scabrous, the angles of the stems and the margins of the leaves bearing retrorse trichomes; fruit glabrous; plants clambering, the stems slender and woody below . . . . .7. *G. Nuttallii*
h. Stems and leaves pilose-hispidulous, the hairs straight and spreading; fruit generally hairy; plants mostly low and herbaceous . . . . . . .8. *G. californicum*

1. **G. parisiense** L. In Marin County known only from grassy slopes and flats in San Anselmo Canyon above Fairfax. Introduced from Europe.

2. **G. divaricatum** Lamk. Locally abundant in open grassland and along the edge of brush: Mount Tamalpais (Potrero Meadow, Rifle Camp, Lagunitas Meadows); Fairfax Hills; San Rafael Hills; Inverness Ridge. Native of Europe. —G. *parisiense* L. var. *anglicum* of Jepson's Manual.

3. **G. Aparine** L. Common and widespread on sunny or shaded slopes in grassland, brush, and woodland: Sausalito, Angel Island, Tiburon, and Mount Tamalpais (Blithedale Canyon, Rock Spring) to San Rafael, Tomales, and Point Reyes Peninsula (Mud Lake). Introduced from Europe.

4. **G. spurium** L. var. **Echinospermum** (Wallr.) Hay. Occasional in grassy. brushy, or wooded places: Lagunitas Meadows, Mount Tamalpais; Carson country; San Rafael Hills.

This seems scarcely more than a small-fruited variety of G. *Aparine* but the two are kept distinct by most botanists in Europe where the plants are native. *Galium Aparine* L. var. *Echinospermum* (Wallr.) Farwell, G. *Vaillantii* DC., and G. *Aparine* L. var. *Vaillantii* (DC.) Koch are other names that have applied to this plant.

5. **G. triflorum** Michx. Moist shady canyonsides and wooded flats: Sausalito; Phoenix Lake and Cataract Gulch, Mount Tamalpais; San Rafael Hills; Inverness and Mud Lake, Point Reyes Peninsula.

The foliage of this plant has the same pleasing fragrance as that of the waldmeister, *Asperula odorata* L., which is so esteemed for perfuming beverages in the Old World. The leaves of the *Galium* have proven a satisfactory substitute in America.

6. **G. tinctorium** L. Common and rather widespread in marshes and bogs: Mount Tamalpais (Potrero and Willow meadows, Lake Lagunitas, acc. Stacey) ; Chileno Valley; Point Reyes Peninsula (Bolinas, Inverness, Ledum Swamp).

This plant is part of a large and variable group that is sometimes treated collectively under the name G. *trifidum* L. Our Marin County plant is the particular form that has been described as G. *tinctorium* var. *diversifolium* Wight.

An anomalous plant with smooth fruits but with stems and leaves retrorsely hispidulous-scabrous has been found near Shell Beach on Point Reyes Peninsula. From the character of the plant, it would appear to have been derived from a cross between G. *tinctorium* and either G. *Aparine* or G. *spurium* var. *Echinospermum*.

7. **G. Nuttallii** Gray. Coastal mesas and wooded or brushy slopes of interior hills, widespread and common: Sausalito, Angel Island, Tiburon, and Mount Tamalpais (Blithedale Canyon, Throckmorton Ridge, Cataract Gulch) to San Rafael Hills, Tomales, and Point Reyes Peninsula (Mud Lake, Ledum Swamp).

8. **G. californicum** H. & A. Common in sandy, clayey, or gravelly soil of coastal bluffs, sheltered woodland, rocky ridges, and brushy hillsides: Sausalito, Tiburon, and Mount Tamalpais (Corte Madera Ridge, West Peak) to San Rafael Hills, Lagunitas, and Point Reyes Peninsula (Inverness Ridge, Ledum Swamp, Point Reyes).

9. **G. murale** (L.) All. Inconspicuous annual in open grassland: Tiburon; Laurel Dell, Mount Tamalpais. Introduced from Europe.

CAPRIFOLIACEAE. Honeysuckle Family

a. Leaves compound; corolla rotate or saucershaped .... 1. *Sambucus*
a. Leaves simple; corolla campanulate to funnelform .... b
   b. Corolla regular; berries white .... 2. *Symphoricarpos*
   b. Corolla irregular; berries red or black .... 3. *Lonicera*

1. Sambucus. Elderberry

a. Flowers in pyramidal or subglobose clusters; fruits red .... 1. *S. callicarpa*
a. Flowers in flat-topped clusters; fruits bluish purple, usually covered with a
   white bloom .... 2. *S. coerulea*

1. **S. callicarpa** Greene. Red Elderberry. Moist slopes and flats along the coast and in wooded canyons: Rodeo Lagoon; Sausalito; Steep Ravine, Mount Tamalpais; Mud Lake and Inverness, Point Reyes Peninsula.—*S. racemosa* L. var. *callicarpa* (Greene) Jeps.

2. **S. coerulea** Raf. Blue Elderberry. Widespread on open or wooded hills: Sausalito, Angel Island, Tiburon, and Mount Tamalpais (Cataract Gulch, Fish Grade) to San Rafael Hills, Salmon Creek, and Point Reyes Peninsula (Bolinas, Inverness).—*S. racemosa* Nutt.

2. Symphoricarpos. Snowberry

a. Stems erect, 1–2 m. tall; branchlets and upper surface of leaves glabrous or
   nearly so .... 1. *S. rivularis*
a. Stems low and trailing, mostly less than 3 dm. tall; branchlets and upper
   surface of leaves more or less hairy .... 2. *S. mollis*

1. **S. rivularis** Suksdorf. Common and widespread on moist partially shaded canyonsides and in woodland or open valley lands: Sausalito, Angel Island, Tiburon, and Mount Tamalpais (Blithedale and Cascade canyons, Cataract Gulch, Phoenix Lake) to San Rafael Hills, Tomales, and Point Reyes Peninsula (Mud Lake, Inverness).—*S. albus* of Jepson's Manual.

2. **S. mollis** Nutt. Rocky or gravelly slopes in brushy or wooded places: Tiburon; Mount Tamalpais (Throckmorton and Corte Madera ridges, Northside Trail); San Rafael Hills.

3. Lonicera. Honeysuckle

a. Flowers several in sessile whorls at the end of axillary or terminal branchlets;
   the pairs of leaves below the flowering branchlets connate and disklike.
   Stems twining and trailing .... 3. *L. hispidula*
a. Flowers 2 on short axillary peduncles; leaves not connate .... b
   b. Stems twining; bracts below the flowers small but green and leaflike
   .... 1. *L. japonica*
   b. Stems erect; bracts below the flowers yellowish or reddish-tinged, not leaflike
   .... 2. *L. Ledebourii*

1. **L. japonica** Thunb. The cherished garden honeysuckle, which has become such a rampant weed in the eastern United States, was seen as a rare escape

from cultivation in Marin County only at Belvedere and on Tiburon Peninsula. It is a native of eastern Asia.

2. **L. Ledebourii** Esch. TWINBERRY. Wet soil along streams and in coastal marshes: Rodeo Lagoon; Stinson Beach; Mill Valley, acc. Stacey; near Camp Taylor; Tomales; Inverness and road to Point Reyes, Point Reyes Peninsula.— *L. involucrata* Banks var. *Ledebourii* (Esch.) Jeps.

3. **L. hispidula** (Lindl.) T. & G. var. **vacillans** Gray. CALIFORNIA HONEYSUCKLE. Common and widespread in wooded canyons and on brushy slopes: Sausalito, Angel Island, Tiburon, and Mount Tamalpais (Blithedale Canyon, Throckmorton Ridge, Cataract Gulch, Phoenix Lake) to San Rafael Hills, Black Point, and Point Reyes Peninsula (Bolinas, Mud Lake, Inverness).—*L. hispidula* (Lindl.) T. & G. var. *californica* (Greene) Jeps.

The rosy flowers of our native honeysuckle lack the fragrance of the garden honeysuckle from Asia but they make up for their deficiency by producing bright red fruits that are one of the colorful attractions of our autumnal woodlands.

DIPSACACEAE. TEASEL FAMILY

a. Heads oblong, the flowers about equaling the conspicuous hooked, sharp-pointed bracts; lowest flowers in the head not very irregular ....1. *Dipsacus*
a. Heads broadly hemispheric, the flowers longer than the bracts; lowest or marginal flowers in the head very irregular, the outer lobes much larger than the inner and raylike ....2. *Scabiosa*

1. Dipsacus

1. **D. sativus** (L.) Honckeny. FULLERS' TEASEL. Common and widespread in low valley lands and on open slopes in deep soil: from Sausalito and Muir Beach to Chileno Valley and Point Reyes (acc. Stacey). Introduued from Europe.—*D. fullonum* of authors, not L.

2. Scabiosa

1. **S. atropurpurea** L. PINCUSHIONS. Occasionally escaping from cultivation and persisting for a longer or shorter time along roads or in waste places: Sausalito; Manzanita; Mill Valley; Bolinas. Native of the Mediterranean region.

VALERIANACEAE. VALERIAN FAMILY

a. Plants perennial; calyx segments plumose in fruit; corolla generally more than 1 cm. long ....1. *Centranthus*
a. Plants annual; calyx limb obsolete; corolla less than 1 cm. long ..2. *Plectritis*

1. Centranthus

1. **C. ruber** (L.) DC. JUPITER'S BEARD. Fugitive from cultivation and persisting in waste ground and along roads, especially on rocky slopes in shallow soil: Sausalito; Angel Island; Belvedere; Ross; San Anselmo; Fairfax; San Rafael; between Salmon Creek and Chileno Valley. Introduced from Europe.

The flowers occur commonly in three colors, a purplish red, a paler reddish rose, and a pure white, any one of which would have given Jupiter's beard a tone startling enough to be commemorated in fact or fiction, mythology or botany.

2. Plectritis

a. Fruits 3-sided, without wings or with wings much narrower than the nearly plane sides ....b
a. Fruits conspicuously winged, not obviously 3-sided, the back side more or less curving into the wings c
b. Fruits without wings ....1. *P. samolifolia*
b. Fruits with very narrow wings, the ventral face between the wings hirsutulous ....2. *P. aphanoptera*
c. Corolla 2–4 mm. long, the spur much shorter than the rest of the corolla ...d
c. Corolla 5–7 mm. long, the spur about equaling the length of the rest of the corolla ....g
d. Keel on back of fruits acute or rounded, not longitudinally grooved ....e
d. Keel on back of fruits shallowly grooved ....f
e. Keel on back of fruits acute ....1a. *P. samolifolia* var. *involuta*
e. Keel on back of fruits rounded ....3. *P. magna*
f. Corollas white or pale pink, the spur stout ....4. *P. macrocera*
f. Corollas deep pink or reddish, the spur slender ..5a. *P. californica* var. *rubens*
g. Wings on fruit inwardly curving, the fruit thus tending to be transversely thickish ....5. *P. californica*
g. Wings on fruit widely spreading, the fruit tending to be flattish ....
....6. *P. macroptera*

1. **P. samolifolia** (DC.) Hoeck. Rare on brushy or wooded slopes: near Point Reyes Station, *Eastwood;* Nicasio.

1a. **P. samolifolia** (DC.) Hoeck var. **involuta** (Suksdorf) Dyal. Wet grassy flat: Potrero Meadow, Mount Tamalpais.

2. **P. aphanoptera** (Gray) Suksdorf. On moist grassy slopes and along the edge of brush: Tiburon, *Berta Kessel;* Kentfield, *Eastwood;* Bald Mountain west of San Anselmo; head of Nicasio Creek.

3. **P. magna** (Greene) Suksdorf. Widespread and rather common on moist brushy or grassy slopes and flats: Sausalito, Tiburon, and Mount Tamalpais (Lagunitas Meadows) to San Rafael Hills, Dillons Beach, and Point Reyes Peninsula, *Eastwood.*

4. **P. macrocera** T. & G. Occasional in grassy places: Rock Spring, Mount Tamalpais; Bald Mountain; Big Rock Ridge.

5. **P. californica** (Suksdorf) Dyal. Grassy borders of brush and woods: Bald Mountain; San Rafael Hills; Big Rock Ridge.

This species with long-spurred corolla is one of several treated collectively by Jepson under the name of *P. ciliosa.* The present analysis of the Marin County entities of this complex genus follows the revision by Sarah Dyal Nielsen.

5a. **P. californica** (Suksdorf) Dyal var. **rubens** (Suksdorf) Dyal. In Marin County known only from gravelly slopes on Mount Tamalpais near Bootjack.

6. **P. macroptera** (Suksdorf) Rydb. Rare in Marin County, known only from Rock Spring, Mount Tamalpais, and from the Fairfax Hills.

CUCURBITACEAE. Gourd Family

1. Marah

a. Corolla of staminate flowers rotate; ovary round, not attenuate at apex into the beak; fruit densely spiny, seeds generally 4 ....1. *M. fabaceus*
a. Corolla of staminate flowers broadly campanulate; ovary attenuate at apex into the slender beak; fruit sparsely spiny or nearly smooth; seeds generally more than 4 ....2. *M. oregonus*

1. **M. fabaceus** (Naud.) Dunn. MANROOT. Widely distributed from coastal dunes and bluffs to interior valleys and hills: Sausalito, Angel Island, Tiburon, and Mount Tamalpais (Phoenix Lake) to San Rafael, Tomales, and Point Reyes Peninsula (Mud Lake, Inverness, McClure Beach, Point Reyes).—*Echinocystis fabacea* Naud.

2. **M. oregonus** (T. & G.) Howell. Not as common as the preceding but widespread, especially in coastal canyons and on moist wooded slopes: Sausalito, Angel Island, Tiburon, and Mount Tamalpais (Blithedale and Cascade canyons, Nora Trail, Phoenix Lake) to Tomales and Point Reyes Peninsula (Inverness Ridge).—*Echinocystis oregona* (T. & G.) Cogn.

CAMPANULACEAE. Bellflower Family

a. Flowers pedicellate, the pedicels sometimes short ....1. *Campanula*
a. Flowers sessile ....b
b. All flowers with corollas; calyx lobes linear to linear-oblong, in fruit about 1 cm. long or more ....2. *Githopsis*
b. Uppermost flowers with showy corollas, the lowest cleistogamous with inconspicuous corollas; calyx lobes narrowly deltoid to roundish-ovate, in fruit mostly 0.5 cm. long or less c
c. Calyx lobes linear-deltoid, entire; capsule narrowly oblong ......3. *Specularia*
c. Calyx lobes roundish-ovate, few-toothed; capsule subglobose, about as broad as long ....4. *Heterocodon*

1. Campanula

a. Plants annual; corolla about equaling the calyx lobes or shorter
....2. *C. angustiflora*
a. Plants perennial; corolla much longer than the calyx lobes ....b
b. Style shorter than the corolla; pedicels generally longer than the capsule
....1. *C. californica*
b. Style longer than the corolla; pedicels generally shorter than the capsule
....3. *C. prenanthoides*

1. **C. californica** (Kell.) Heller. Occasional in coastal swales on Point Reyes Peninsula. The delicate and attractive California bluebell has been reported only from Marin County to Mendocino County.—*C. linnaeifolia* Gray.

2. **C. angustiflora** Eastw. Widespread but usually not common in gravelly or rocky soil of high exposed ridges in the chaparral: Mount Tamalpais (Corte Madera Ridge, East and West peaks, Matt Davis and Berry trails, Barths Retreat); Carson country.—*C. exigua* of Jepson in part, not *C. exigua* Rattan.

The plants with stout stems, short internodes, and broad leaves are typical of the plants originally described from Mount Tamalpais by Miss Eastwood. Occasionally a slender, narrow-leaved form, var. *exilis* J. T. Howell, is found, as on Tiburon, Mount Tamalpais (Northside Trail), and Carson Ridge.

3. **C. prenanthoides** Durand. On Mount Tamalpais in open woods or on the edge of brush in deep soil: Pipeline Trail; Fern Canyon above Muir Woods; Matt Davis Trail; Northside Trail.

2. Githopsis

1. **G. specularioides** Nutt. Rather widespread but not common in gravelly or clayey soil of open woodland and brushy slopes: Mount Tamalpais (Laurel Dell Trail, acc. Sutliffe, Rifle Camp); Fairfax Hills; San Rafael Hills.

3. Specularia

1. **S. biflora** (R. & P.) F. & M. Rare on brushy ridges along the coast and in the interior: San Anselmo Canyon above Fairfax; Carson Ridge; Tomales. In these Marin County plants, all the flowers were cleistogamous and none produced the attrative open bellflower-like corollas that are generally found in the later flowers of this species.

4. Heterocodon

1. **H. rariflorum** Nutt. Moist grassy meadows and open streambanks: Mount Tamalpais (Rock Spring, Potrero Meadow, Rocky Ridge Fire Trail); Lagunitas Canyon; Dillons Beach. This is a lovely delicate bellflower generally occurring in the most pleasant places.

LOBELIACEAE. Lobelia Family

1. Downingia

1. **D. concolor** Greene. In Marin County known only from low fields and pastures east of Aurora School that are inundated during the rainy season.

COMPOSITAE. Sunflower Family

Key to the Tribes

a. Corollas all ligulate, 5-toothed at the end; plants usually with milky juice
....Tribe 11. *Cichorieae*
a. Corollas all tubular and regular, or the inner flowers (disk flowers) tubular and the outer flowers (ray flowers) ligulate and 3-toothed at the end; plants without milky juice. (In *Lessingia* and *Centaurea,* the heads are discoid but the outermost flowers are sometimes palmately enlarged.) ....b
b. Receptacle naked, without chaffy bracts or abundant long hairs c
b. Receptacle bearing chaffy bracts or abundant long hairs ....k
c. Bracts of the involucre well imbricated in several series ....d
c. Bracts of the involucre in 1 or 2 series (disregarding much-reduced bracts at the base of the involucre in some species) ....g
d. Bracts of the involucre without a scarious margin ....e
d. Bracts of the involucre with a conspicuous dry or scarious margin ....f
e. Leaves alternate or basal, not fleshy-thickened ....Tribe 2. *Astereae*
e. Leaves opposite, fleshy-thickened ....*Jaumea,* p. 279
f. Pappus of capillary bristles ....Tribe 3. *Inuleae*

f. Pappus none or rarely present as a minute border or crown
....Tribe 8. *Anthemideae*

g. Pappus none, or if present, paleaceous or coarsely bristly ....h

g. Pappus present, of capillary bristles i

h. Outer achenes not much enlarged; leaves not broadly triangular
....Tribe 7. *Helenieae*

h. Outer fertile achenes much longer than the involucre, the inner sterile and
   abortive; leaves mostly basal, broadly triangular, undulate-dentate, green
   above and white below ....*Adenocaulon,* p. 274

i. Style branches clavate; shrubby plant with opposite, 3-nerved, deltoid-ovate
   leaves; heads discoid; corollas white ....Tribe 1. *Eupatorieae*

i. Style branches not clavate; herbs or suffrutescent plants with leaves generally
   alternate (opposite in *Arnica*) and with disk flowers generally yellow (whitish
   in *Petasites*) ....j

j. Bracts of the involucre equal and in a single series (except for very short outer
   ones) ....Tribe 9. *Senecioneae*

j. Bracts of the involucre a little unequal, in about 2 series ....*Erigeron,* p. 270

k. Receptacle with abundant long hairs; thistles and thistlelike plants, the
   leaves frequently spiny ....Tribe 10. *Cynareae*

k. Receptacle with bracts, generally not hairy; leaves not spiny l

l. Anthers distinct or scarcely united; heads unisexual (or in *Iva,* with nodding
   heads solitary in the axils of the leaves, the outer flowers pistillate and the
   inner perfect); fruit commonly a bur ....Tribe 4. *Ambrosieae*

l. Anthers united; heads not unisexual; fruit not a bur ....m

m. Bracts of the involucre scarious or coriaceous, at least on the margin ....n

m. Bracts of the involucre green, not scarious or coriaceous ....o

n. Heads without rays ....Tribe 3. *Inuleae*

n. Heads with white rays ....Tribe 8. *Anthemideae*

o. Bracts of the involucre not embracing or enfolding the ray achenes
....Tribe 5. *Heliantheae*

o. Bracts of the involucre embracing or enfolding the ray achenes
....Tribe 6. *Madieae*

Tribe 1. **Eupatorieae.** Eupatory Tribe

Represented by only one genus ....1. *Eupatorium*

Tribe 2. **Astereae.** Aster Tribe

a. Heads radiate, the ligules sometimes small but always evident ....b

a. Heads discoid or the outermost flowers with very tiny inconspicuous or
   obsolete ligules. (In *Lessingia* the heads are discoid but the marginal flowers
   are frequently enlarged and palmately spreading.) l

b. Rays yellow c

b. Rays white, pink, or blue, never yellow ....h

c. Heads medium-sized, 1 cm. broad or more, generally rather few ....d

c. Heads small, less than 1 cm. broad, generally numerous ....f

d. Pappus of few readily deciduous awns or slender scales in one series
....2. *Grindelia*

d. Pappus of numerous persistent bristles in 2 series, the outer short and inspicuous ....e

e. Annual herb, usually 1 m. tall or more; leaves serrate .... 4. *Heterotheca*

e. Perennial herb, 3 dm. tall or less; leaves entire .... 5. *Chrysopsis*

f. Pappus of rather few short scales .... 3. *Gutierrezia*

f. Pappus of numerous bristles ....g

g. Stems herbaceous, low or tall; leaves more than 1 cm. long, not fascicled, broad or linear .... 6. *Solidago*

g. Stems low and woody; leaves linear, resinous-dotted, fascicled, less than 1 cm. long .... 7. *Aplopappus*

h. Leaves basal, the heads solitary on slender scapes; pappus none .... 9. *Bellis*

h. Leaves chiefly cauline; pappus bristles few or many i

i. Plants annual with solitary heads on slender peduncles; pappus bristles 3–5 .... 8. *Pentachaeta*

i. Plants annual or perennial, if annual the heads small and clustered on the branchlets; pappus bristles numerous ....j

j. Plants grayish green with a close tomentum; ray flowers sterile .... 11. *Corethrogyne*

j. Plants glabrous, pilose-hairy, or glandular, not tomentose; ray flowers fertile .k

k. Involucral bracts mostly oblong or wider, more or less foliaceous in texture, at least at the tip; style branches sharply acute .... 12. *Aster*

k. Involucral bracts linear or narrowly oblong, generally not foliaceous; style branches obtuse .... 13. *Erigeron*

l. Plants dioecious; corollas whitish. Perennial herbs and shrubs with glandular-punctate leaves .... 14. *Baccharis*

l. Plants monoecious, the flowers perfect or unisexual; corollas yellow, lavender, or purplish ....m

m. Plants annual ....n

m. Plants perennial. Corollas yellow; pappus bristles numerous ....p

n. Pappus bristles 3 .... 8. *Pentachaeta*

n. Pappus bristles numerous ....o

o. Corollas lavender or purplish, in the outermost flowers more or less enlarged .... 10. *Lessingia*

o. Corollas yellowish, in the outermost flowers reduced to a filiform tube and minute ligule .... 13. *Erigeron*

p. Leaves oblong to ovate; upper stems and leaves capitate-glandular 5. *Chrysopsis*

p. Leaves linear or oblong-linear, glabrous, glandular-punctate, or villous, not capitate-glandular ....q

q. Tall shrub; heads very numerous .... 7. *Aplopappus*

q. Low herbs; heads rather few .... 13. *Erigeron*

TRIBE 3. Inuleae. CUDWEED TRIBE

a. Pistillate (outer) flowers closely subtended or enclosed by concave or saclike bracts of the receptacle. Involucral bracts few ....b

a. Pistillate (outer) flowers not subtended or enclosed by bracts of the receptacle f

b. Heads a little longer than wide, cylindric to conical or broadly ovate; outer pistillate and inner perfect flowers separated by a circle of 5 persistent bracts c

b. Heads globose or a little broader than long; inner and outer flowers not separated by bracts, or if present, then the bracts very small and deciduous d

c. Heads conical or broadly ovate, the bracts of the receptacle subtending or enclosing the pistillate flowers, the inner bracts stellate-spreading in age from the periphery of the mushroom-shaped receptacle ....15. *Filago*

c. Heads cylindric, the bracts of the receptacle only subtending the pistillate flowers, the inner bracts spreading at the top of the slender, upwardly prolonged receptacle ....16. *Evax*

d. Leaves opposite; upper dorsal part of the bracts of the receptacle becoming gibbous and hoodlike in fruit. Receptacle broadly turbinate 19. *Psilocarphus*

d. Leaves alternate; bracts of the receptacle becoming gibbous dorsally ....e

e. Receptacle flat or depressed, more or less parted into short raylike projections each bearing a pistillate flower and bract; bracts of the receptacle radially arranged in one plane, becoming bony, hyaline only at the beaklike top of the aperture ....17. *Micropus*

e. Receptacle columnar with an erect elongate axis; bracts of the receptacle spirally arranged and imbricate, herbaceous, the aperture bordered by a winglike hyaline membrane ....18. *Stylocline*

f. Involucral bracts usually 4 or 5, united at the base, much shorter than the enlarged fertile achenes; perennial herb with broad, triangular or 3-lobed, bicolored leaf blades ....22. *Adenocaulon*

f. Involucral bracts few to numerous, distinct, longer than the small fertile achenes; annual or perennial plants with linear or oblong entire leaves ....g

g. Plants monoecious, the heads with both pistillate (outer) and perfect (inner) flowers; involucral bracts usually shining and hyaline, not milky-white ....20. *Gnaphalium*

g. Plants dioecious or nearly so, the heads of pistillate (fertile) or perfect (sterile) flowers mostly occurring on different plants; involucral bracts milky- or chalky-white ....21. *Anaphalis*

TRIBE 4. **Ambrosieae.** RAGWEED TRIBE

a. Heads containing both pistillate (marginal) and perfect (central) flowers, the pistillate flowers with corollas; achene not enclosed in the fruiting involucre; leaves entire ....23. *Iva*

a. Heads either staminate and many-flowered or pistillate and 1- to several-flowered, corolla of pistillate flowers generally lacking; achenes enclosed in the fruiting involucre; leaves serrate, lobed, or bipinnatifid ....b

b. Fruiting involucre not spiny; perennial herb with erect stems from elongate rootstocks ...../..24. *Ambrosia*

b. Fruiting involucre spiny; annual or perennial herbs without rootstocks c

c. Plants annual; stems erect; spines hooked ....25. *Xanthium*

c. Plants perennial; stems spreading or prostrate; spines straight ...26. *Franseria*

TRIBE 5. **Heliantheae.** SUNFLOWER TRIBE

a. Heads solitary at the ends of leafy stems or peduncles ....b

a. Heads loosely clustered at the ends of stems or branches c

b. Leaves alternate on the lower part of the stem; achenes thick and angled, not flattened ....28. *Wyethia*

b. Leaves opposite on the lower part of the stem; achenes strongly flattened
....29. *Helianthella*

c. Leaves scabrous-pubescent, alternate, more or less petioled .....27. *Helianthus*

c. Leaves smooth and glabrous, opposite, sessile ....30. *Bidens*

TRIBE 6. Madieae. TARWEED TRIBE

a. Involucral bracts only half enclosing the achenes of the ray flowers ....b

a. Involucral bracts completely enclosing the achenes of the ray flowers ......d

b. Uppermost leaves and involucral bracts terminated by a broad sessile gland;
disk flowers subtended by conspicuous gland-bearing bracts on the receptacle;
heads aggregated into compact subglobose clusters at the ends of branches
....31. *Holocarpha*

b. Uppermost leaves and involucral bracts not terminated by a broad sessile
gland; receptacle with or without bracts; heads scattered or aggregated c

c. Heads broadly campanulate or hemispheric; rays 5 or more, yellow or white;
capitate glands not tackshaped ....32. *Hemizonia*

c. Heads cylindric or narrowly campanulate; rays few, usually 1–3; white or pink
tinged; capitate glands of uppermost leaves and involucral bracts tackshaped ....33. *Calycadenia*

d. Achenes of ray flowers laterally compressed or turgid, not compressed dorso-
ventrally; involcral bracts very glandular ....34. *Madia*

d. Achenes of ray flowers dorso-ventrally compressed; involucral bracts bearing
scattered glands or the glands inconspicuous or lacking ....e

e. Pappus of disk achenes very conspicuous in two series of broad, shining paleae,
the inner series becoming longer in fruit than the disk corollas; ray achenes
conspicuously ribbed ....38. *Achyrachaena*

e. Pappus of disk achenes none or if present of slender or paleaceous bristles
usually shorter than the disk corollas; ray achenes not conspicuously
ribbed ....f

f. Heads large, 2–4 cm. across, or if only about 1 cm. across, then the disk
achenes with numerous plumose paleaceous pappus bristles ......35. *Layia*

f. Heads small or inconspicuous, less than 2 cm. across; disk pappus usually
lacking or sometimes present as slender deciduous bristles ....g

g. Plants annual; heads sessile, subtended by reduced leaves simulating an outer
series of involucral bracts; rays inconspicuous, pale yellow ..36. *Lagophylla*

g. Plants perennial; heads on slender peduncles, not subtended by leaves; rays
conspicuous, white, veined and tinged with reddish purple ....37. *Holozonia*

TRIBE 7. Helenieae. SNEEZEWEED TRIBE

a. Leaves opposite ....b

a. Leaves alternate, at least above the base ....d

b. Bracts of the involucre imbricate in several series; perennials with fleshy
thickened leaves ....39. *Jaumea*

b. Bracts of the involucre in one series; annuals (or one species perennial) with
usually thin leaves c

c. Bracts of the involucre distinct ....40. *Baeria*

c. Bracts of the involucre united into a cup ....41. *Lasthenia*

d. Leaves entire or denticulate .... e

d. Leaves lobed and divided .... g

e. Leaves glandular-dotted; disk becoming subglobose .... 45. *Helenium*

e. Leaves not glandular-dotted; disk flat or nearly so .... f

f. Bracts of the involucre united into a cup; rays showy; leaves broadly lanceolate
.... 42. *Monolopia*

f. Bracts of the involucre distinct; rays inconspicuous; leaves linear
.... 44. *Rigiopappus*

g. Plants perennial, woody at the base or above; leaves more or less tomentose;
disk achenes fertile .... 43. *Eriophyllum*

g. Plants annual; leaves not tomentose; disk achenes sterile ..46. *Blennosperma*

TRIBE 8. **Anthemideae.** MAYWEED TRIBE

a. Heads with rays .... b

a. Heads without rays .... d

b. Heads less than 1.5 cm. across, closely clustered in a flat-topped corymb
.... 48. *Achillea*

b. Heads 1.5 cm. across or more, loosely clustered or solitary on slender
peduncles c

c. Receptacle conical, the upper part bearing bracts .... 47. *Anthemis*

c. Receptacle flat or hemispheric, without bracts .... 50. *Chrysanthemum*

d. Heads solitary and sessile along the stems or in the forks of the branches;
achenes tipped by the persistent spine-like style .... 52. *Soliva*

d. Heads with stalks or clustered in spikes or panicles; achenes without spine-
like style .... e

e. Receptacle conical .... 49. *Matricaria*

e. Receptacle flat or convex .... f

f. Marginal pistillate flowers pedicellate, without a corolla .... 51. *Cotula*

f. Marginal flowers pistillate or perfect, sessile, with a coralla .... g

g. Heads more than 1 cm. across, few, the inflorescence not elongate
.... 53. *Tanacetum*

g. Heads less than 1 cm. across, many, the inflorescence elongate ....54. *Artemisia*

TRIBE 9. **Senecioneae.** GROUNDSEL TRIBE

a. Leaves opposite, at least below the inflorescence; pappus bristles subplumose .... 56. *Arnica*

a. Leaves alternate; pappus bristles slender, not plumose .... b

b. Foliage leaves basal, roundish in outline, palmately cleft; flowers white or
purplish .... 55. *Petasites*

b. Foliage leaves cauline or basal and cauline, elongate (or roundish in *Senecio
mikanioides*); flowers yellow c

c. Heads discoid, the outer flowers pistillate, the inner perfect ..57. *Erechtites*

c. Heads discoid or radiate, if discoid, all of the flowers perfect ......58. *Senecio*

TRIBE 10. **Cynareae.** THISTLE TRIBE

a. Leaves not spiny .... 63. *Centaurea*

a. Leaves spiny (only slightly spiny in *Cynara*) .... b

b. Pappus of short scales in several series; outer bracts of the involucre leaflike ....64. *Carthamus*

b. Pappus of slender elongate bristles, the bristles sometimes paleaceous and
united at the base; outer bracts of the involucre not leaflike (the heads
subtended by leaves in some species of *Cirsium*) c

c. Involucral bracts numerous, broad, imbricate, with or without a terminal
spine, the margins not spiny; heads large; flowers violet; receptacle fleshy;
leaves very large ....61. *Cynara*

c. Involucral bracts usually narrower; heads small or large; flowers white, pink,
or purplish; receptacle not fleshy; leaves large or small ....d

d. Leaves blotched with white along the veins ....62. *Silybum*

d. Leaves not blotched with white ....e

e. Pappus bristles plumose ....59. *Cirsium*

e. Pappus bristles setose but not plumose ....60. *Carduus*

TRIBE 11. **Cichorieae.** CHICORY TRIBE

a. Pappus none ....65. *Lapsana*

a. Pappus present (sometimes short and inconspicuous or sometimes deciduous) ....b

b. Pappus paleaceous, the paleae frequently bearing a conspicuous slender
bristle c

b. Pappus of slender bristles, not paleaceous (or slightly thickened at the base
in *Stephanomeria*) ....g

c. Heads sessile or nearly so; corollas bright blue (rarely white); pappus entirely
paleaceous ....66. *Cichorium*

c. Heads on slender scapes or peduncles; corollas yellow or buff, sometimes tinged
with rose; pappus paleae bearing bristles at least on the innermost
achenes ....d

d. Inner bracts of the involucre becoming corky-thickened in fruit and enclosing
the outermost achenes. Pappus of outer achenes entirely paleaceous, some
of the paleae of the inner achenes bearing bristles ....70. *Hedypnois*

d. Inner bracts of the involucre not becoming corky in fruit or enclosing the
outermost achenes ....e

e. Pappus of outer achenes entirely paleaceous, the paleae of the inner achenes
bearing conspicuously plumose bristles ....69. *Leontodon*

e. Pappus paleae of all the achenes bearing bristles, the bristles scabrous, setose,
or barely subplumose ....f

f. Pappus paleae gradually narrowed into the elongate bristle; heads nodding in
bud ....67. *Microseris*

f. Pappus paleae sharply notched at the apex, the bristle arising from the notch;
heads erect in bud ....68. *Uropappus*

g. Pappus bristles conspicuously plumose ....h

g. Pappus bristles scabrous or smooth, not plumose l

h. Achenes not beaked; heads terminating leafy-bracteate branches or subsessile
along them; corollas rose ....73. *Stephanomeria*

h. Achenes beaked, or the outermost achenes rarely beakless; heads on short or
long peduncles, the peduncles naked or bracteate; corollas yellow, white,
or lavender-purple i

i. Corollas yellow ....j
i. Corollas white or lavender-purple ....k
j. Involucral bracts not corky-thickened, the outer shorter but not very different from the inner; herbage glabrous or pubescent, the hairs not barbed ....71. *Hypochoeris*
j. Involucral bracts becoming corky-thickened in fruit, the inner and outer bracts very unlike, the outer broadly ovate and nearly as long as the inner lanceolate ones; herbage rough with barbed hairs ....72. *Picris*
k. Plants annual; heads about 2 cm. long; corollas white, sometimes purplish-tinged ....74. *Rafinesquia*
k. Plants perennial; heads about 5 cm. long; corolla lavender-purple ....75. *Tragopogon*
l. Achenes strongly flattened ....m
l. Achenes terete ....n
m. Involucres campanulate in flower; achenes truncate, not beaked ..78. *Sonchus*
m. Involucres cylindric in flower; achenes beaked, or if only contracted at the summit, then the pappus brownish ....79. *Lactuca*
n. Achenes beaked ....o
n. Achenes not beaked ....q
o. Leaves basal and cauline; heads many in a loose corymbose cluster ..81. *Crepis*
o. Leaves basal or nearly so; heads solitary at the end of naked scapes ....p
p. Leaf lobes prominently retrorse; involucral bracts in two distinct series, the outer very short; achenes 4- or 5- ribbed ....77. *Taraxacum*
p. Leaf lobes not prominently retrorse; involucral bracts not in two distinct series; achenes 10-ribbed ....80. *Agoseris*
q. Pappus bristles stiffish, sordid or brownish; plants perennial with leaves chiefly basal. Corollas whitish or tinged with buff ....82. *Hieracium*
q. Pappus bristles fragile, white; plants annual with leaves basal or cauline ....r
r. Pappus bristles more or less united at the base and falling from the achene; leaves chiefly basal; heads nodding in bud ....76. *Malacothrix*
r. Pappus bristles distinct and persistent; leaves chiefly cauline; heads erect in bud ....81. *Crepis*

Tribe 1. Eupatorieae. EUPATORY TRIBE

1. Eupatorium

1. **E. adenophorum** Spreng. Naturalized on moist slopes or in waste ground near habitations: hills overlooking the Golden Gate; Sausalito. Introduced as a garden plant from Mexico.

Tribe 2. Astereae. ASTER TRIBE

2. Grindelia. GUMWEED

a. Stems shrubby or suffrutescent above the base; plants of salt or brackish marshes ....b
a. Stems herbaceous or if woody above the base, the stems prostrate or nearly so; plants not of salt or brackish marshes c
b. Leaves not resinous; heads clustered; rays 1–1.5 cm. long ....1. *G. humilis*

b. Leaves a little resinous; heads on more elongate slender branchlets; rays
    1.5–2 cm. long ....2. *G. stricta*

c. Leaves very densely resinous-punctate, glabrous, the margin of the uppermost
    leaves not scabrous-ciliate ....3. *G. camporum*

c. Leaves moderately or not at all resinous-punctate, the blades tomentulose,
    hirsutulous-hairy, or glabrous, the margin more or less ciliate or scabrous-
    ciliate ....d

d. Bracts of the involucre erect and flattened, acute but usually not narrowed
    into a spreading or recurved tip. Stems, leaves, and involucral bracts more
    or less tomentulose or hirsutulous ....4. *G. hirsutula*

d. Bracts of the involucre narrowed into a slender or filiform tip that is strongly
    recurved or revolute ....e

e. Stems erect, herbaceous ....5. *G. rubricaulis*

e. Stems prostrate, frequently woody above the base ....6. *G. arenicola*

1. **G. humilis** H. & A. Common and widespread in salt marshes bordering San Francisco and San Pablo bays: Sausalito and Tiburon to San Rafael, McNears Landing, and Black Point. According to Steyermark, this robust species also ranges to Bolinas and Tomales bays along the ocean but he cites no collections from there. An occasional form of *G. humilis* in which the involucral bracts are more prominently reflexed or even revolute is f. *reflexa* Steyerm.

2. **G. stricta** DC. Salt marshes near the head of Tomales Bay and perhaps also along Bolinas Bay. This species is most common on the coast of Oregon, Washington, and British Columbia where it is represented by many forms and varieties, but it is rather rare in California and is not known south of Marin County. Steyermark cites an Elmer collection from Point Reyes as f. *venulosa* (Jeps.) Steyerm. (*G. venulosa* Jeps.), a plant originally described from Humboldt County.

3. **G. camporum** Greene. Near San Rafael and along the railroad in Sausalito where it has perhaps been introduced. On Angel Island is var. *Davyi* (Jeps.) Steyerm., a variant in which slender elongate branches are terminated by few or solitary heads and bear narrower thinner leaves.

In both typical *G. camporum* and var. *Davyi,* the heads are relatively large, 2.5 cm. or more in diameter. More common than either of these forms in Marin County is var. *parviflora* Steyerm. in which the heads are about 1.5 cm. across. This variety has been found near San Rafael, Hamilton Field, Ignacio, and Black Point.

4. **G. hirsutula** H. & A. In typical form and several variations, this is the most common and widely distributed species of *Grindelia* in Marin County, being characteristic of open or brushy hills and valleys both along the coast and in the interior. The typical form in which the involucral bracts have looser spreading tips is restricted in Marin County to coastal hills from Sausalito to Tennessee Cove. The extreme form, f. *patens* (Greene) Steyerm., with the outermost involucral bracts foliaceous, has been collected at Olema.

The widespread form of *G. hirsutula* in Marin County is var. *brevisquama* Steyerm. and it is known from Angel Island, Tiburon, and Mount Tamalpais (Lagunitas Meadows) to San Rafael, Hamilton Field, Black Point, and Chileno Valley. The most conspicuously hairy individuals (as from Fairfax Hills and Olema) are f. *tomentulosa* Steyerm.

5. **G. rubricaulis** DC. Open hills near the coast and rocky or grassy slopes on

Mount Tamalpais: Muir Beach; Stinson Beach; Bolinas; Point Reyes Station, *Leschke;* Inverness Ridge; McClure Beach road. The plants from Bootjack on Mount Tamalpais have more slender stems and Steyermark cites a broad-leaved plant from the West Point Road as var. *platyphylla* (Greene) Steyerm.

In *G. rubricaulis* the stems and leaves are usually glabrous but occasionally they are tomentulose, as at Stinson Beach, on Inverness Ridge, and near Olema. These plants may be hybrids between *G. rubricaulis* and *G. hirsutula.*

6. **G. arenicola** Steyerm. Exposed coastal slopes, mesas, and dunes on Point Reyes Peninsula. The form from McClure Beach with larger, thinner leaves is typical of the species but the usual form near the radio station and at Point Reyes is the small-, thick-leaved var. *pachyphylla* Steyerm. Although *G. arenicola* presents a very distinctive appearance, it seems to be only an extreme maritime form of variable *G. rubricaulis.*

3. Gutierrezia

1. **G. californica** (DC.) T. & G. Rare and local on rocky hills: Sausalito Hills; Angel Island. These are the only stations known north of the Golden Gate.

4. Heterotheca

1. **H. grandiflora** Nutt. Sporadic along roads and undoubtedly introduced from central and southern California: Sausalito; northern part of Tiburon Peninsula; Ignacio.

5. Chrysopsis

a. Heads radiate; leaves villous, scarcely glandular . . . . .1. *C. Bolanderi*
a. Heads discoid; leaves conspicuously glandular . . . . .2. *C. oregona*

1. **C. Bolanderi** Gray. Shallow soil of open grassy or brushy hills: Sausalito, Angel Island, Tiburon, and Mount Tamalpais (Lagunitas Meadows) to San Rafael Hills, Tomales, and Point Reyes Peninsula, the type locality of *C. arenaria* Elmer (Mount Vision, McClure Beach road).—*C. villosa* Nutt. var. *Bolanderi* (Gray) Gray.

2. **C. oregona** (Nutt.) Gray. Occasional on gravelly fills along the railroad, probably introduced from the more northern Coast Ranges where it grows on the floodbeds of streams: south of Mill Valley; Corte Madera; north of San Rafael.

6. Solidago. GOLDENROD

a. Leaves linear to linear-lanceolate, entire; inflorescence loosely branched
. . . . .4. *S. occidentalis*
a. Leaves elliptic-lanceolate to oblanceolate and obovate, entire or usually few-
toothed or serrate; inflorescence compactly branched . . . . .b
b. Upper and lower leaf surfaces densely grayish-puberulent. Involucre 4–5 mm.
long . . . . .1. *S. californica*
b. Upper and lower leaf surfaces glabrous or sparsely hairy c
c. Leaves acute, herbaceous, not glutinous, scabrous-ciliate; involucre 3–4 mm.
long . . . . .2. *S. lepida*
c. Leaves obtuse to acute, coriaceous, glutinous, sparsely, if at all, ciliate; involucre
6–7 mm. long . . . . .3. *S. spathulata*

1. **S. californica** Nutt. Occasional on grassy or brushy slopes: Rodeo Lagoon; Angel Island; Tiburon; Azalea Flat, Mount Tamalpais; San Rafael Hills; Chileno Valley; Point Reyes Peninsula (Bolinas, McClure Beach road).

2. **S. lepida** DC. var. **elongata** (Nutt.) Fern. Rather rare on moist brushy slopes near the coast: Rodeo Lagoon; near Inverness and Ledum Swamp, Point Reyes Peninsula.—*S. elongata* Nutt.

3. **S. spathulata** DC. Shallow soil of rocky slopes or sandy mesas near the coast, rare: Tomales; Point Reyes Peninsula (Inverness and Point Reyes, acc. Stacey, Drakes Estero).

4. **S. occidentalis** (Nutt.) T. & G. Low wet places along roads or the edge of marshes: Rodeo Lagoon; Tamalpais Valley; Olema Marshes; Tomales Bay near Inverness.

7. Aplopappus

a. Heads radiate; plants of coastal dunes, less than 1 m. tall ....1. *A. ericoides*
a. Heads discoid; plants of the chaparral, generally more than 1 m. tall
....2. *A. arborescens*

1. **A. ericoides** (Less.) H. & A. Occasional on coastal slopes and dunes: Angel Island; Stinson Beach; Dillons Beach; dunes on Point Reyes Peninsula.—*Ericameria ericoides* (Less.) Jeps.

2. **A. arborescens** (Gray) Hall. Rather common in the chaparral on Mount Tamalpais and the ridges to the north: Mount Tamalpais (Corte Madera Ridge, East Peak, Throckmorton Ridge, Berry Trail); San Geronimo Ridge.—*Ericameria arborescens* (Gray) Greene.

8. Pentachaeta

a. Heads on slender peduncles exceeding the leaves, radiate, the rays white and conspicuous ....1. *P. bellidiflora*
a. Heads on peduncles scarcely exceeding the leaves, apparently discoid, the outer pistillate flowers without a ligule ....2. *P. alsinoides*

1. **P. bellidiflora** Greene. Grassy slopes on the edge of woods or brush, rare but locally common: Corte Madera, the type locality; Larkspur, *Eastwood;* Greenbrae Hills; Marin City, *Peter Raven.*—*P. exilis* Gray var. *Grayi* Jeps.

This is a rare but attractive little daisy occurring locally from Marin County south to the Santa Lucia Mountains.

2. **P. alsinoides** Greene. Occasional in grassy or gravelly places, inconspicuous and perhaps overlooked: Sausalito; Angel Island; Tiburon; Laurel Dell, Mount Tamalpais; Bald Mountain west of San Anselmo; Carson Ridge; San Rafael Hills.

9. Bellis

1. **B. perennis** L. ENGLISH DAISY. Naturalized in moist grassy places and in lawns: Sausalito, acc. Stacey; Mill Valley; ridge southeast of Muir Woods; San Rafael. Introduced from Europe.

10. Lessingia

a. Heads narrowly turbinate; outer corollas in the head scarcely or not at all enlarged; pappus bristles more or less united and tending to form 5 pale-aceous awns ....1. *L. ramulosa*

a. Heads broadly turbinate; outer corollas in the head enlarged and simulating rays; pappus bristles mostly distinct and not paleaceous ......2. *L. hololeuca*

1. **L. ramulosa** Gray var. **micradenia** (Greene) J. T. Howell. Shallow gravelly soil, occasional on shale or sandstone, common on serpentine: Mount Tamalpais (Phoenix Lake, Camp Handy, Rocky Ridge Fire Trail); Liberty Spring; Carson Ridge; Big Rock Ridge.

In typical form this variety occurs only in Marin County, although in the Russian River district, a plant is found that is somewhat intermediate between the present small-headed variety and the larger-headed plants characteristic of the species found near Santa Rosa. The type collection of *L. micradenia* Greene was made on the north side of Mount Tamalpais.

2. **L. hololeuca** Greene. Low fields and grassy hillsides, becoming locally abundant and attractive: Tamalpais Valley; Corte Madera; between San Anselmo and Fairfax; San Rafael Hills; west of Santa Venetia; Ignacio.—*L. leptoclada* Gray var. *hololeuca* (Greene) Jeps.

This is an attractive species with its ample panicle of lavender flower-heads (which rarely may be pure white). In grass fires, the plants may be burned to the ground, but from the root-crown that survives, a full-flowered rosette is produced that sets fruit before the winter rains.

11. Corethrogyne

1. **C. californica** DC. Shallow rocky soil or open coastal hills and mesas: Forest Knolls, *Gordon True;* Lagunitas, acc. Stacey; Point Reyes Peninsula (Mud Lake, Inverness, Ledum Swamp).

12. Aster

a. Plants annual; heads small, less than 1 cm. wide, the rays less than 5 mm. long ....3. *A. exilis*
a. Plants perennial; heads showy, more than 1 cm. wide, the rays more than 5 mm. long ....b
b. Leaves elongate, linear-oblanceolate to oblong-oblanceolate; backs of the involucral bracts glabrous ....1. *A. chilensis*
b. Leaves broadly oblanceolate-obovate to obovate; backs of the involucral bracts hairy ....2. *A. radulinus*

1. **A. chilensis** Nees. Common and widespread from salt marshes and coastal swales to low valleys, brushy canyonsides, and wooded slopes: Sausalito, Angel Island, Tiburon, and Ross to San Rafael, Chileno Valley, Olema Marshes, and Point Reyes Peninsula (Inverness Ridge, swale near radio station).

The plants of this species are almost as diverse in appearance as the places where they grow. When well developed and floriferous, the plants resemble the garden asters that are cultivated under the name of Michaelmas daisies.

2. **A. radulinus** Gray. Widespread but not very common on brushy or wooded slopes in shallow soil: Sausalito; Tiburon; Mount Tamalpais (Corte Madera Ridge, Potrero Meadow, Phoenix Lake, acc. Sutliffe); Fairfax, acc. Stacey; Tomales.

Near Potrero Meadow on Mount Tamalpais a peculiar form without rays and with unusually copious pappus was found growing with the normal plant. The discoid plant did not set good fruit but the adjacent radiate one did.

3. **A. exilis** Ell. In Marin County known only along the strand of Lake Lagunitas. The species is to be expected in low ground in the northern and eastern part of the county but no specimens have been seen from there.

13. Erigeron. FLEABANE

a. Heads with conspicuous rays ....b
a. Heads with inconspicuous rays or discoid ....d
b. Leaves less than 5 mm. wide, elongate, linear-oblong or linear-oblanceolate, entire ....3. *E. foliosus*
b. Leaves (at least the basal and middle cauline) more than 5 mm. wide, broadly oblanceolate to obovate, entire, crenulate-dentate, or saliently few-toothed c
c. Lower leaves acute; head about 1.5 cm. broad ....1. *E. Karvinskianus*
c. Lower leaves obtuse; head 2 cm. broad or more ....2. *E. glaucus*
d. Plants annual; marginal flowers in the head pistillate, the ligules shorter than the corolla tube and either very inconspicuous or apparent ....e
d. Plants perennial; heads entirely discoid, the flowers perfect with 5-toothed corollas ....f
e. Corolla of pistillate flowers with a small but apparent ligule that is longer than the style branches; stem and leaves thinly subappressed-hirsutulous or nearly glabrous ....4. *E. canadensis*
e. Corolla of pistillate flowers filiform, with a minute ligule that is shorter than the style branches; stem and leaves appressed-hirsutulous, grayish ....5. *E. crispus*
f. Stems and leaves villous ....6. *E. petrophilus*
f. Stems (above the base) and leaves glabrous ....7. *E. inornatus*

1. **E. Karvinskianus** DC. Escaping from cultivation and persisting along streets and in waste ground: Belvedere; San Rafael; Inverness. Native of Mexico.

2. **E. glaucus** Ker. SEASIDE DAISY. Coastal bluffs and sandy flats: Sausalito, acc. Lovegrove; Angel Island; Stinson Beach, acc. Stacey; Point Reyes Peinsula (Bolinas, Shell Beach, McClure Beach, Point Reyes).

3. **E. foliosus** Nutt. var. **Hartwegii** (Greene) Jeps. Grassy places on brushy or wooded slopes, not common: Sausalito; Tiburon; Mount Tamalpais, acc. Cronquist; San Rafael Hills; near Chinese Camp.

4. **E. canadensis** L. HORSEWEED. Common and widespread weed of roadsides and fields: Sausalito, Angel Island, and Mount Tamalpais (Cascade Canyon) to Ignacio, Tomales, and Point Reyes Peninsula (Bolinas, Mud Lake, Ledum Swamp).

5. **E. crispus** Pourret. Occasional weed about habitations and along roads: Sausalito; Angel Island; Mill Valley, acc. Stacey; San Anselmo; San Rafael. Naturalized from the Mediterranean region.—*E. linifolius* Willd.

6. **E. petrophilus** Greene. Of local occurrence on rocky bluffs on Mount Tamalpais: Corte Madera Ridge; East Peak; Northside Trail.

7. **E. inornatus** (Gray) Gray var. **angustatus** Gray. Rocky soil on Mount Tamalpais, frequently in areas of serpentine: West Point; Bootjack; Barths Retreat; Northside Trail.

In var. *angustatus* the leaves are less than 2 mm. wide and are almost needle-like in appearance. On Point Reyes Peninsula near Point Reyes Post Office, Elmer collected var. *Biolettii* (Greene) Jeps. in which the leaves are more than 2 mm. wide.

14. Baccharis

a. Plants herbaceous, the substrictly erect stems 1–2 m. tall; leaves lanceolate to broadly ovate-lanceolate, the margin entire or serrulate ....1. *B. Douglasii*
a. Plants shrubby, the stems prostrate to erect, branching widely; leaves obovate, coarsely few-toothed or entire ....b
b. Plants matlike, the stems prostrate or nearly so ....2. *B. pilularis*
b. Plants bushy, the stems erect ....2a. *B. pilularis* var. *consanguinea*

1. **B. Douglasii** DC. Low wet valley lands and about springs in the hills: Tiburon; Mill Valley marshes, acc. Sutliffe; Mount Tamalpais (Blythedale and Cascade canyons, Willow Meadow, acc. Leschke); Stinson Beach; west of Fairfax; Novato Creek; Olema Marshes; McClure Beach road, Point Reyes Peninsula.
2. **B. pilularis** DC. Open hills, mesas, and dunes near the coast: Sausalito; Angel Island; Tennessee Cove; Tomales; Point Reyes and dunes, Point Reyes Peninsula.
2a. **B. pilularis** DC. var. **consanguinea** (DC.) Ktze. COYOTE BRUSH. Common and widespread on coastal and interior hills and on open flats and valley lands: Sausalito, Angel Island, Tiburon, and Mount Tamalpais (Blithedale and Cascade canyons, Azalea Flat, Phoenix Lake) to San Rafael, Black Point, Tomales, and Point Reyes Peninsula.

Hybrid intermediates are frequent where typical *B. pilularis* and var. *consanguinea* grow near together but usually the two forms are very distinct. The common name, fuzzy-wuzzy, is sometimes given to the plants because of the abundant pappus produced by fruiting pistillate plants.

TRIBE 3. Inuleae. CUDWEED TRIBE

15. Filago

a. Uppermost leaves oblong, about equaling the heads; outer bracts of the receptacle only partly enclosing the achenes ....1. *F. californica*
a. Uppermost leaves linear-lanceolate, conspicuously longer than the heads; outer bracts of the receptacle entirely enclosing the achenes ....2. *F. gallica*

1. **F. californica** Nutt. Occasional in loose clayey or gravelly soil of brushy ridges and canyonsides: Mount Tamalpais (Corte Madera Ridge, West Point, Bootjack); Carson country; Fairfax and San Rafael hills; Tomales.
2. **F. gallica** (L.) L. Common and widespread in grassland, brush, or open woodland: Sausalito, Angel Island, Tiburon, and Mount Tamalpais (Corte Madera Ridge, Bootjack, Potrero Meadow) to San Rafael Hills and Point Reyes Peninsula (Bolinas, Mud Lake). Native of the Old World.

16. Evax

1. **E. sparsiflora** (Gray) Jeps. Clay soil of open or brushy hills: Tiburon; Bootjack, Mount Tamalpais; Carson Ridge; Dillons Beach. On Mount Vision, Point

Reyes Peninsula, the plants have shorter stems and smaller leaves, the form described as var. *brevifolia* (Gray) Jeps.

17. Micropus

1. **M. californicus** F. & M. Common on open or brushy hills: Angel Island; Tiburon; Bootjack and Lagunitas Meadows, Mount Tamalpais; Carson Ridge; San Rafael Hills; Hamilton Field.

18. Stylocline

a. Hyaline membrane on the ventral face of the fruiting bract narrower than the herbaceous part enclosing the flower; receptacle about 0.5 mm. long
....1. *S. amphibola*
a. Hyaline membrane on the ventral face of the fruiting bract wider than the herbaceous part enclosing the flower; receptacle about 1 mm. long
....2. *S. gnaphalioides*

1. **S. amphibola** (Gray) J. T. Howell. Occasional on open grassy slopes or in the chaparral: Angel Island; Tiburon; Mount Tamalpais, acc. Jepson; Fairfax, Greenbrae, and San Rafael hills; Big Rock Ridge.—*Micropus amphibolus* Gray.
2. **S. gnaphalioides** Nutt. In Marin County known only from gravelly soil on a steep canyonside in the San Rafael Hills. The only other station for this species that has been reported in the North Coast Ranges is in Lake County.

19. Psilocarphus

1. **P. tenellus** Nutt. Occasional in clay soil on flats or slopes where water stands or seeps in the spring: Sausalito, Angel Island, and Mount Tamalpais (Berry Trail) to San Rafael Hills and Tomales.

20. Gnaphalium. CUDWEED

a. Heads small, the involucre about 3–3.5 mm. long; plants annual ....b
a. Heads larger, the involucre mostly 4–7 mm. long; plants annual or perennial c
b. Capitate clusters of heads leafy-bracteate; involucral bracts tomentose at the base, hyaline only at the tip; stems usually low ....1. *G. palustre*
b. Capitate clusters of heads not leafy-bracteate; involucral bracts hyaline to the base, scarcely tomentose; stems usually 2–3 dm. tall, slender
....2. *G. luteo-album*
c. Involucre turbinate, cuneate at the base, the bracts bright pink
....3. *G. ramosissimum*
c. Involucre hemispheric or campanulate, not cuneate at the base, the bracts usually white, straw-color, or brownish ....d
d. Leaves linear or linear-oblong; heads in small clusters in an openly branched panicle. Involucral bracts white ....4. *G. microcephalum*
d. Leaves oblong or oblanceolate; heads in compact clusters, the clusters spicate, capitate, or at the end of branchlets ....e
e. Involucral bracts pearly, white or nearly so; upper stems and leaves more or less glandular, the tomentum tending to be deciduous ...5. *G. californicum*
e. Involucral bracts straw color, purplish, or brownish; stems and leaves not glandular, the tomentum mostly persistent ....f

f. Heads in compact capitate clusters, the clusters solitary or on short branches; involucral bracts straw color or tawny .... 6. *G. chilense*

f. Heads in leafy-bracteate clusters, the clusters arranged in an elongate, ovate to oblong spike, the spike compact or interrupted; involucral bracts purplish or brownish .... 7. *G. purpureum*

1. **G. palustre** Nutt. Occasional in low ground where water stands in the spring: Potrero and Lagunitas meadows, Mount Tamalpais; Carson Ridge; laguna in Chileno Valley; east of Aurora School; Dillons Beach; Point Reyes Peninsula (Bolinas, Inverness, Ledum Swamp).

2. **G. luteo-album** L. Widespread along roads and trails and in grassland on the edge of brush or woodland: Sausalito, Angel Island, Tiburon, and Mount Tamalpais (Corte Madera Ridge, Bootjack, Lake Lagunitas) to San Rafael, Ignacio, and Point Reyes Peninsula (Inverness, Ledum Swamp). Introduced from the Old World.

3. **G. ramosissimum** Nutt. Brushy slopes generally near the coast: Rodeo Lagoon; Stinson Beach; Throckmorton Ridge, Mount Tamalpais; Bolinas and Inverness, Point Reyes Peninsula. This attractive pink cudweed has not been reported north of Marin County.

4. **G. microcephalum** Nutt. Rocky slopes and ridges in brush or chaparral: Sausalito, Tiburon, and Mount Tamalpais (Blithedale Canyon, Throckmorton Ridge) to San Rafael Hills, Black Point, and Point Reyes Peninsula (Inverness Ridge).

5. **G. californicum** DC. Common and widespread on moist brushy slopes and maritime headlands: Sausalito, Angel Island, Tiburon, and Mount Tamalpais (Corte Madera Ridge, Bootjack) to San Rafael Hills and Point Reyes Peninsula (Mud Lake, Inverness).—*G. decurrens* Ives var. *californicum* (DC.) Gray.

Usually the upper stems and leaves lack tomentum and are more or less glandular but near the coast a rather common form is persistently and uniformly tomentose up to the shining involucres. In the character of the vestiture, this form resembles *G. chilense* but it can be readily distinguished by the color and texture of the involucre. It has been noted as follows: Tiburon; Elk Valley; Tennessee Cove; Bolinas and Point Reyes, Point Reyes Peninsula.

6. **G. chilense** Spreng. Widely distributed on coastal slopes, brushy and wooded hills, and open valleys: Sausalito, Angel Island, Tiburon, and Mount Tamalpais (Corte Madera Ridge, Azalea Flat, Phoenix Lake) to Novato, Dillons Beach, and Point Reyes Peninsula (Bolinas, Inverness, Point Reyes, McClure Beach).

7. **G. purpureum** L. Common in grassy places on the hills and in the valleys: Sausalito, Angel Island, Tiburon, and Mount Tamalpais (Corte Madera Ridge, Rock Spring) to San Rafael Hills, Dillons Beach, and Point Reyes Peninsula (Bolinas, Inverness, Ledum Swamp, Point Reyes).

21. Anaphalis

1. **A. margaritacea** (L.) Benth. & Hook. f. PEARLY EVERLASTING. Occasional on moist brushy slopes: Sausalito; Tiburon; Muir Beach; Mount Tamalpais, *Eastwood;* Bolinas and Inverness, Point Reyes Peninsula.

22. Adenocaulon

1. **A. bicolor** Hook. Moist flats in shaded canyons: Muir Woods; Blithedale Canyon and Cataract Gulch, Mount Tamalpais; Ross; Lagunitas, acc. Stacey.

TRIBE 4. Ambrosieae. RAGWEED TRIBE

23. Iva

1. **I. axillaris** Pursh. POVERTY WEED. Gravelly banks along the railroad at Ignacio where it may have been introduced. Sterile plants seen on Angel Island were perhaps this species which is widespread through the lowlands of California.

24. Ambrosia. RAGWEED

1. **A. psilostachya** DC. var. **californica** (Rydb.) Blake. Occasional in low fields and along roads: north of Stinson Beach; Greenbrae; San Rafael; Black Point; Point Reyes, acc. Stacey.

25. Xanthium

a. Stems spiny; leaves ovate-lanceolate ....1. *X. spinosum*
a. Stems not spiny; leaves broadly ovate ....b
b. Fruits subglabrous or glandular-puberulent; spines on the fruits mostly sparse
   ....2. *X. calvum*
b. Fruits and the bases of the spines bearing coarse trichomes as well as a finer
   puberulence; spines on the fruits numerous and dense ......3. *X. italicum*

1. **X. spinosum** L. SPINY CLOTBUR. Occasional but widespread in waste ground about farms and along roads: Sausalito, Tiburon, and Mill Valley (acc. Sutliffe) to Burdell, Chileno Valley, and Point Reyes Peninsula (Bolinas, Inverness). Naturalized from Europe.

2. **X. calvum** Millsp. & Sherff. Autumnal plants on the dried bottom of Alpine Lake near the east end.

3. **X. italicum** Mor. ITALIAN COCKLEBUR. Occasional in low ground where water has stood during the rainy season: Mill Valley, acc. Stacey; Kentfield; Phoenix Lake; San Rafael; Ignacio; Black Point.—*X. canadense* of Jepson's Manual in part.

The common American cocklebur, *X. pennsylvanicum* Wallr., is to be expected but no specimen has been seen from Marin County. It has the densely spiny fruits of *X. italicum* but they are without the coarse trichomes that distinguish the latter species.

26. Franseria

1. **F. Chamissonis** Less. Occasional in deep sand of ocean beaches and dunes: Rodeo Lagoon; Stinson Beach; Dillons Beach; Inverness and McClure Beach, Point Reyes Peninsula.

Much more common than *F. Chamissonis* is var. *bipinnatisecta* Less. (*F. bipinnatifida* Nutt.) which is found on Angel Island and near Almonte as well as on most of the beaches named above. In the species the leaves are prominently serrate or shallowly lobed, while in the variety the leaves are twice or thrice pinnatifid. Typical plants of the two are very distinct and would deserve specific recognition if they did not completely intergrade in foliage characters wherever they occur together. In such places, the number of plants exhibiting various intermediate leaf forms is usually greater than the number of either of the extremes.

TRIBE 5. **Heliantheae.** SUNFLOWER TRIBE

27. Helianthus

a. Plants annual; bracts of the involucre broadly ovate, abruptly narrowed above
....1. *H. annuus*
a. Plants perennial; bracts of the involucre lanceolate, gradually narrowed above
....2. *H. californicus*

1. **H. annuus** L. COMMON SUNFLOWER. Rare as a roadside weed: Novato; San Antonio Creek. Native of the central United States.

2. **H. californicus** DC. Rocky banks and flood beds of streams, reported from San Rafael by Stacey.

28. Wyethia

a. Leaves lanceolate to narrowly ovate, acute; outer bracts of the involucre not broad and foliaceous ....1. *W. angustifolia*
a. Leaves broadly oblong or ovate, obtuse; outer bracts of the involucre broad and foliaceous ....2. *W. glabra*

1. **W. angustifolia** Nutt. Common and widespread on open hills in brush and grassland: Sausalito, Angel Island, Tiburon, and Mount Tamalpais (Potrero Meadow, Phoenix Lake) to San Rafael Hills, Nicasio, and Point Reyes Peninsula (Bolinas, Point Reyes).

2. **W. glabra** Gray. MULE-EARS. Occasional but rather widely distributed on wooded or brushy slopes: Mount Tamalpais (Bootjack, Phoenix Lake) and San Rafael Hills to Novato, Salmon Creek, and Tomales.

29. Helianthella

1. **H. californica** Gray. Wooded or brushy slopes of high rocky ridges, reported from Marin County without definite stations by both Greene and Jepson.

30. Bidens

1. **B. laevis** (L.) B.S.P. BUR MARIGOLD. In shallow water along the banks of streams and margins of ponds, in Marin County known only from the Olema Marshes.

TRIBE 6. **Madieae.** TARWEED TRIBE

31. Holocarpha

1. **H. macradenia** (DC.) Greene. Known in Marin County only from a collection made in Ross Valley in 1883 by Mary Katharine Curran. This is a very rare plant ranging from Monterey County north to Marin County and recorded from only a few localities.

32. Hemizonia

a. Basal and lower cauline leaves pinnately divided, the divisions linear-oblong; disk flowers not subtended by bracts ....3. *H. corymbosa*
a. Basal and lower cauline leaves entire or toothed, not lobed or divided; disk flowers subtended by bracts ....b

b. Flat free tip of the involucral bracts usually shorter than the body of the
bract; rays yellow (varying to cream or white in hybrids)
....1. *H. luzulaefolia*
b. Flat free tip of the involucral bracts equaling or longer than the body of the
bract; rays white, veined or tinged with pink ....2. *H. congesta*

1. **H. luzulaefolia** DC. var. **lutescens** Greene. HAYFIELD TARWEED. Common and
widespread in valleys and on open hills: Rodeo Lagoon, Tiburon, and Mount
Tamalpais (Phoenix and Alpine lakes) to Hamilton Field, Nicasio, Dillons Beach,
and Point Reyes Peninsula (near Mud Lake).—*H. congesta* DC. var. *lutescens*
(Greene) Jeps.

Although this tarweed with its bright yellow flowers is usually easy enough to
place, it is variable in many characters and some individuals may be puzzling if
they are derived from crosses with the following species. There are two forms
recognized depending on the time of flowering, a spring-blooming form and a
summer- and fall-blooming form. Scarcely have the coastal hills begun to lose
their vernal greenness when the spring form will recover the hills with green and
gold. On maritime bluffs and slopes at Dillons Beach this spring bloomer forms
low widely branching plants with unusually broad leaves, a type that was described from the "northern part of Marin County" by Greene as *H. citrina*. In
summer and autumn, after the hills have become dried and tawny, the late
blooming form transforms broad areas into tarweed gardens and there is the floral
effect of a second spring that lasts until the rains begin in October or November.
This abundant and often-maligned plant is the golden link that binds with
flowers and verdure the end and beginning of successive rainy seasons.

2. **H. congesta** DC. Occasional in low fields or flats in clay soil where water
stands or seeps in the spring: Bolinas; San Rafael; Hamilton Field; east of Aurora
School; Tomales, acc. Hall.

When *H. congesta* and *H. luzulaefolia* var. *lutescens* grow together, hybridization generally takes place on a limited scale. Frequently the hybrids can be detected by intermediate color shades in the rays but at other times plants with
white rays attest mixed parentage by their habit, size, vestiture, and shape of the
involucral bracts. A series of hybrid intermediates has been found near Hamilton
Field, and white-rayed plants from San Rafael and Stinson Beach may be hybrids.

3. **H. corymbosa** (DC.) T. & G. Occasional in valleys and on open hillsides:
Stinson Beach; Mount Tamalpais, acc. Stacey; San Rafael Hills; Black Point; Bolinas, Point Reyes Peninsula.

When fresh, as well as after drying, the herbage of this species has a fragrance
that is sweet and pleasant.

33. Calycadenia

1. **C. multiglandulosa** DC. var. **cephalotes** (DC.) Jeps. Locally common on open
gravelly or rocky slopes in clay soil: Tiburon; Mount Tamalpais (Bootjack, Laurel
Dell, Rocky Ridge Fire Trail, Alpine Lake); San Geronimo Ridge; south of
Olema. This variety, according to D. D. Keck, is not known south of Marin
County.

34. Madia. TARWEED

a. Rays showy and much longer than disk, generally more than 5 mm. long ..b

a. Rays inconspicuous and little longer than the disk, generally less than 5 mm. long c
b. Plants perennial; lower leaves serrate, the serrations few but sometimes salient; rays about 0.5 cm. long ....1. *M. madioides*
b. Plants annual; lower leaves entire or inconspicuously toothed; rays usually 1 cm. long or more ....2. *M. elegans*
c. Involucre less than 5 mm. long ....7. *M. exigua*
c. Involucre 5 mm. long or more ....d
d. Achenes of ray flowers turgid, not compressed or laterally flattened ....6. *M. anomala*
d. Achenes of ray flowers strongly compressed ....e
e. Stems slender, more or less openly branched; herbage not very glandular, the secretion with an agreeable smell; plants of wooded or brushy hills, blooming in spring or early summer ....5. *M. gracilis*
e. Stems stout, the flowering branches usually compactly branched; herbage heavily glandular; weedy plants of valleys and roadsides ....f
f. Leaves subtending the heads oblong-lanceolate; heads generally 0.5 cm. long; glandular secretion dark brown and strong smelling; summer and fall flowering plants ....3. *M. sativa*
f. Leaves subténding the heads ovate-lanceolate; heads generally 1–1.3 cm. long; glandular secretion light brown, not strong smelling; spring flowering plants ....4. *M. capitata*

1. **M. madioides** (Nutt.) Greene. Moist wooded hills and canyons: Mount Tamalpais (Blithedale and Cascade canyons, Cataract Gulch, Rifle Camp); Ross; Fairfax Hills; San Rafael Hills.

2. **M. elegans** Don. Rocky slopes on Mount Tamalpais near Laurel Dell and Barths Retreat; a slender, few-flowered plant blooming in the spring.

Most of the plants in Marin County belong to var. *densifolia* (Greene) Jeps. and are much more robust and leafy and are summer- and fall-blooming plants: Sausalito; Phoenix Lake, Mount Tamalpais; Lagunitas; Ignacio; Chileno Valley; Point Reyes Station. With its well-branched inflorescence bearing many yellow daisies, this is one of the most attractive of the native *Compositae*. The flowers are open in the morning but close by midday.

3. **M. sativa** Mol. Common and widespread in fields and along roads in valleys and hills: Sausalito, Angel Island, Tiburon, and Mill Valley to San Rafael and Point Reyes Peninsula (Bolinas, Inverness, Point Reyes). In low rich ground near the head of Bolinas Lagoon, plants measured 8 feet in height.

4. **M. capitata** Nutt. Occasional in low fields or on slopes wet in the spring: Tiburon; Mill Valley and Mount Tamalpais, acc. Stacey; Fairfax; Greenbrae Hills. Very unlike the preceding with which it has been confused.—*M. sativa* Mol. var. *congesta* T. & G.

5. **M. gracilis** (Smith) Keck. Common and widespread on wooded or brushy hills: Sausalito, Angel Island, Tiburon, and Mount Tamalpais (Cascade Canyon, West Point, Bootjack) to San Rafael Hills, Tomales, and Point Reyes Peninsula (Inverness Ridge, Shell Beach).—*M. dissitiflora* (Nutt.) T. & G.

6. **M. anomala** Greene. Occasional in grassy opens on wooded or brushy hills: Mill Valley, *Campbell;* Mount Tamalpais, acc. Stacey; Fairfax Hills; San Anselmo

Canyon, *Sutliffe;* Bald Mountain west of San Anselmo; Greenbrae Hills; Black Point.—*M. dissitiflora* (Nutt.) T. & G. var. *anomala* (Greene) Jeps.

This distinctive California tarweed is not known from any localities south of Marin County.

7. **M. exigua** (Smith) Greene. Shallow gravelly soil of rocky slopes and ridges: Tiburon; Mount Tamalpais (Corte Madera Ridge, West Point, Berry Trail); Carson country; San Rafael Hills.

The plants of this species are usually low and less than 6 inches tall, but after the 1945 fire on Carson Ridge, individuals were quite robust and as much as 15 inches tall.

35. Layia

a. Rays white, inconspicuous, little exceeding the disk ....4. *L. carnosa*
a. Rays yellow or yellow tipped with creamy white, showy, much longer than the
   disk ....b
b. Back of the involucral bracts bearing coarse trichomes, these especially numerous near the fold of the bract and looking like teeth, capitate glands lacking ....3. *L. chrysanthemoides*
b. Back of the involucral bracts villous to hirsutulous, the top of the peduncle
   and the involucral bracts bearing scattered capitate glands c
c. Stems hirsutulous- or tomentulose-pubescent, the hairs not arising from dark
   spots; pappus bristles white, setulose, not plumose ....1. *L. platyglossa*
c. Stems somewhat hirsute, the hairs arising from dark spots; pappus bristles
   sordid, plumose-hairy at the base ....2. *L. gaillardioides*

1. **L. platyglossa** (F. & M.) Gray. Widespread and sometimes locally abundant on sandy coastal mesas, gravelly ridges, or grassy slopes: Rodeo Lagoon, acc. Sutliffe; Tiburon; Laurel Dell, Mount Tamalpais; Point Reyes and Inverness Ridge, Point Reyes Peninsula.

The beauty of this daisy is generally enhanced by the creamy-white tips of the yellow rays, hence the common name, tidy-tips. Not infrequently in Marin County, however, the tips are not tidy but are yellow throughout.

2. **L. gaillardioides** H. & A. Occasional in grassland of wooded or brushy hills: Mount Tamalpais (above Mill Valley, acc. Sutliffe, south of Phoenix Lake); Stinson Beach, acc. Stacey; Fairfax Hills; Carson Ridge; San Rafael Hills; Tomales; Mud Lake, Point Reyes Peninsula.

Mount Tamalpais is the type locality of *Blepharipappus nemorosus* Greene which does not differ enough from *L. gaillardioides* to warrant even the varietal status accorded it by Jepson.

3. **L. chrysanthemoides** (DC.) Gray. Clay soil of low valleys and bounding hills, locally common on flats bordering the salt marshes: Ignacio; Novato; Lucas Valley; Chileno Valley.

Depending on the presence or absence of pappus on the disk achenes, two species have usually been recognized for this plant which is here regarded as a single species. *Layia Calliglossa* Gray and its var. *oligochaeta* Gray are the forms in which pappus is present.

4. **L. carnosa** (Nutt.) T. & G. Occasional on fixed dunes on Point Reyes Peninsula.

36. Lagophylla

a. Stems openly branched above; involucre about 0.5 cm. long ..1. *L. ramosissima*
a. Stems narrowly and compactly branched above; involucre about 1 cm. long
....2. *L. congesta*

1. **L. ramosissima** Nutt. Occasional on dry grassy hills and flats: Greenbrae Hills; San Rafael Hills; Santa Venetia.
2. **L. congesta** Greene. In Marin County this is apparently known only from the type collection which was made on Mount Tamalpais by Mary Katharine Curran in 1883.—*L. ramosissima* Nutt. var. *congesta* (Greene) Jeps.

37. Holozonia

1. **H. filipes** (H. & A.) Greene. Rocky, gravelly bed of summer-dried water courses on Mount Tamalpais: Cataract Creek near Rock Spring; Lagunitas Creek below Lake Lagunitas.

38. Achyrachaena

1. **A. mollis** Schauer. Open grassland of hills and valleys, widespread but not very common in Marin County: Angel Island, Tiburon, and Mount Tamalpais (Rock Spring, Lagunitas Meadows) to Fairfax Hills, Hamilton Field, and Black Point.

TRIBE 7. **Helenieae.** SNEEZEWEED TRIBE

39. Jaumea

1. **J. carnosa** (Less.) Gray. Common with *Salicornia* in salt marshes along the bay and ocean: Almonte and Tiburon to McNears and Black Point; Rodeo Lagoon and Stinson Beach to Inverness and Drakes Estero.

40. Baeria

a. Leaves (at least the lower) with prominent lobes or divisions; pappus, if present, of both slender awns and shorter broad scales ......4. *B. uliginosa*
a. Leaves entire; pappus, if present, of slender awns or bristles ....b
b. Plants perennial, the roots thickened and arising from a short rhizome or narrow caudex; heads mostly more than 2.5 cm. across .....3. *B. macrantha*
b. Plants annual; heads mostly less than 2.5 cm. across c
c. Leaves linear; bracts of the involucre ovate-lanceolate or elliptic-ovate
....1. *B. chrysostoma*
c. Leaves oblong or linear-oblong; bracts of the involucre broadly ovate or obovate
....2. *B. hirsutula*

1. **B. chrysostoma** F. & M. GOLD FIELDS; SUNSHINE. Common and widespread on open grassy or rocky slopes and valleys: Sausalito, Angel Island, Tiburon, and Mount Tamalpais (Rock Spring, Lagunitas Meadows) to Carson Ridge, Dillons Beach, and Point Reyes Peninsula (Drakes Estero).

How appropriate are the common names for this slender yellow-flowered annual which clothes with golden brightness so many hills and valleys in California! The individual heads may be small but through a common purpose in legions of plants, color and beauty are imparted to many a vernal landscape. In Marin County this color tumbles in avalanches over grassy coastal slopes from

Sausalito to Tomales, and on the Tiburon hills the golden yellow contrasts with the blue of *Lupinus nanus* in unforgettable wildflower pictures.

Depending on whether or not the achenes bear pappus, various names have been applied to this widespread plant. Here all are combined under the name *B. chrysostoma*, the name originally applied to epappose plants collected by the Russians. *Baeria chrysostoma* var. *gracilis* (DC.) Hall is one of the names given to plants with pappus.

2. **B. hirsutula** (Greene) Greene. Sandy soil of open coastal slopes and mesas: Dillons Beach; near Point Reyes Lighthouse. Marin County is the type locality of this maritime species.

3. **B. macrantha** (Gray) Gray. Coastal slopes and mesas: near Point Reyes, the type locality; hills north of Dillons Beach.—*B. macrantha* (Gray) Gray var. *littoralis* Jeps.

The typical form has usually erect stems and narrowly linear leaves. On sandy bluffs overlooking the ocean is a thick-stemmed sprawling plant with broad some-what fleshy leaves. This is var. *thalassophila* J. T. Howell (type locality, Dillons Beach). South of Marin County, *B. macrantha* apparently occurs only on the San Luis Obispo County coast where it has been found by R. F. Hoover.

4. **B. uliginosa** (Nutt.) Gray. Occasional on sandy slopes and flats near the ocean: Dillons Beach; Point Reyes Peninsula (McClure Beach, dunes near the radio station, Point Reyes).

41. Lasthenia

a. Heads conspicuously radiate, the rays 5–10 mm. long . . . . .1. *L. glabrata*
a. Heads apparently discoid, the rays not exceeding the involucre 2. *L. glaberrima*

1. **L. glabrata** Lindl. Low wet clay flat at Burdell, the only station known in Marin County.

2. **L. glaberrima** DC. Occasional in wet clay soil of low fields, ponds, or ditches: Lagunitas Meadows, Mount Tamalpais; Ignacio; Hicks Valley; Burdell; east of Aurora School.

42. Monolopia

1. **M. major** DC. Open grassy hills in clay soil: Sausalito, acc. Stacey; San Rafael, acc. Crum.

43. Eriophyllum

a. Heads solitary at the ends of slender leafless stems, more than 1.5 cm. across
. . . . .1. *E. lanatum*
a. Heads in small clusters at the ends of leafy stems, less than 1.5 cm. across . .b
b. Bracts of the involucre overlapping, mostly less than 5 mm. long
. . . . .2. *E. confertiflorum*
b. Bracts of the involucre not overlapping, about 5–6 mm. long
. . . . .3. *E. staechadifolium*

1. **E. lanatum** (Pursh) Forbes var. **arachnoideum** (F. & L.) Jeps. Widespread and common on open slopes along the ocean or on brushy or wooded canyonsides in the interior, frequently abundant and very attractive: Sausalito, Muir Beach, and Mount Tamalpais (Cascade Canyon, Cataract Gulch, Phoenix Lake) to San Rafael Hills, Tomales, and Point Reyes Peninsula (Mud Lake, Point Reyes).

2. **E. confertiflorum** (DC.) Gray. Common on moist brushy hillsides and on dry slopes in the chaparral: Sausalito, Angel Island, Tiburon, and Mount Tamalpais (Throckmorton Ridge, Phoenix Lake) to San Rafael Hills and Black Point.

3. **E. staechadifolium** Lag. var. **artemisiaefolium** (Less.) Macbr. Common on rocky maritime slopes or sandy flats behind the dunes: Sausalito, Angel Island, Stinson Beach to Dillons Beach and Point Reyes Peninsula (Bolinas, Point Reyes).

The low leafy densely branched plants when in flower are frequently most showy, seemingly encrusted with purest gold—low golden mats or mounds set off by blue of sky and ocean.

44. Rigiopappus

1. **R. leptocladus** Gray. Occasional in gravelly soil on high rocky ridges: Mount Tamalpais (near West Point, Potrero Meadow); Carson Ridge.

45. Helenium

a. Leaves linear, 1–2 mm. wide, not decurrent ....1. *H. tenuifolium*
a. Leaves oblanceolate or narrowly lanceolate, more than 2 mm. wide, decurrent b
 b. Rays short, 4–6 mm. long ....2. *H. puberulum*
 b. Rays longer, mostly 10–15 mm. long ....3. *H. Bigelovii*

1. **H. tenuifolium** Nutt. Low field along Redwood Highway north of Hamilton Field. Perhaps introduced by airplane from the eastern or southern United States where it is native.

2. **H. puberulum** DC. Marshy ground or moist beds of streams, occasional: Sausalito; Stinson Beach; Mount Tamalpais (Cascade Canyon, Alpine Dam, Phoenix Lake); Ross; Bolinas and Drakes Estero, Point Reyes Peninsula.

3. **H. Bigelovii** Gray. Rare in wet ground in serpentine areas: Tiburon; west of Little Carson Falls.

46. Blennosperma

a. Stems slender; involucres about 5 mm. long, the bracts ovate; achenes less than
   3 mm. long ....1. *B. nanum*
a. Stems somewhat thickened and fistulous; involucres about 7 mm. long, the
   bracts ovate-lanceolate; achenes 3–4.5 mm. long 1a. *B. nanum* var. *robustum*

1. **B. nanum** (Hook.) Blake. Low fields and roadside hollows where water collects in the spring, in Marin County known only from Lucas Valley where it was found by Leschke.—*B. californicum* (DC.) T. & G.

1a. **B. nanum** (Hook.) Blake var. **robustum** J. T. Howell. A rare plant known only on open coastal hills in sandy soil near McClure Beach on Point Reyes Peninsula, the type locality. In some plants the achenes are papillate as they are in the typical form, but in other plants the achenes are entirely glabrous.

Tribe 8. Anthemideae. Mayweed Tribe

47. Anthemis

1. **A. Cotula** L. Mayweed. Common and widespread in hills and valleys, along roads and near habitations: Sausalito, Tiburon, and Mount Tamalpais (acc. Stacey) to San Rafael Hills and Point Reyes Peninsula (Bolinas, Mud Lake, Inverness). Naturalized from the Old World.

48. Achillea. YARROW

1. **A. borealis** Bong. Two varieties of this northern species are found in Marin County, var. *arenicola* (Hel.) J. T. Howell, a heavy-leaved tomentose plant of maritime dunes, and var. *californica* (Poll.) J. T. Howell, a thinner-leaved arachnoid or glabrate plant of hills and valleys. The former has been found only on Point Reyes Peninsula: Point Reyes, dunes near the radio station, McClure Beach. The latter is widely distributed from coastal bluffs and salt marsh borders to wooded or brushy canyons and slopes: Sausalito, Tiburon, Angel Island, and Mount Tamalpais (Bootjack, Lagunitas Meadows) to San Rafael Hills, Burdell marshes, Tomales, and Point Reyes Peninsula (Bolinas, Mud Lake, Inverness).— *A. Millefolium* and varieties *maritima* Jeps. and *californica* (Poll.) Jeps., at least in part.

49. Matricaria

1. **M. matricarioides** (Less.) Porter. PINEAPPLE WEED. Occasional in waste ground or along roads: Sausalito, Angel Island, Tiburon, and Mill Valley to San Rafael, Bolinas, and Tomales.—*M. suaveolens* of authors, not L. Probably not native.

50. Chrysanthemum

a. Heads 1–1.5 cm. across; rays white ....3. *C. Parthenium*
a. Heads 2.5–4 cm. across; rays yellow ....b
b. Leaves serrate or pinnately lobed ....1. *C. segetum*
b. Leaves bipinnatifid ....2. *C. coronarium*

1. **C. segetum** L. CORN CHRYSANTHEMUM. Occasional but becoming locally common on coastal hills: near Stinson Beach; Point Reyes Station; near Marshalls; north of Tomales; Chileno Valley; Inverness and road to Point Reyes, Point Reyes Peninsula. Native of Eurasia.

2. **C. coronarium** L. Fugitive from cultivation on grassy slopes near homes: Sausalito; Bolinas. Introduced from the Mediterranean region.

3. **C. Parthenium** (L.) Bernh. FEVERFEW. Escaping from cultivation and becoming established in towns: Sausalito; Tamalpais Valley; Mill Valley; Ross; Point Reyes, acce. Stacey. Native of Eurasia.

51. Cotula

a. Leaves glabrous, entire or pinnately lobed, the rachis broad; heads bright
   yellow ....1. *C. coronopifolia*
a. Leaves thinly villous-pubescent, once or twice pinnatifid, the rachis narrow;
   heads pale dull yellow ....2. *C. australis*

1. **C. coronopifolia** L. Wet ground of springy slopes, low fields, and salt marsh borders: Sausalito, Angel Island, Tiburon, and Mount Tamalpais (acc. Stacey) to San Rafael, Dillons Beach, and Point Reyes Peninsula (Bolinas, Mud Lake, Ledum Swamp, McClure Beach). Naturalized from South Africa.

2. **C. australis** (Less.) Hook. f. Occasional weed of gardens and waste places: Sausalito, Angel Island, Belvedere, and Mill Valley to Stinson Beach (*L. S. Rose*), San Rafael, and Inverness. Introduced from Australia.

52. Soliva

1. **S. daucifolia** Nutt. Common and widespread in grassy places: Sausalito, Angel Island, Tiburon, and Mount Tamalpais (Bootjack, Potrero Meadow, Lagunitas Meadows) to San Rafael Hills, Ignacio, Dillons Beach, and Point Reyes Peninsula (Mud Lake, Ledum Swamp, Point Reyes).—*S. sessilis* of California references.

53. Tanacetum

1. **T. camphoratum** Less. In Marin County known only from sandy maritime slopes near Rodeo Lagoon.

54. Artemisia

a. Plants shrubby ....1. *A. californica*
a. Plants herbaceous, sometimes woody at the base ....b
b. Leaves glabrous or nearly so; plants annual or biennial ....2. *A. biennis*
b. Leaves villous or tomentose, at least on the lower side; plants perennial c
c. Leaves divided into linear segments, silky-villous ....3. *A. pycnocephala*
c. Leaves entire or pinnately lobed, tomentose ....4. *A. Douglasiana*

1. **A. californica** Less. CALIFORNIA SAGEBRUSH. Widely distributed on coastal bluffs, steep canyonsides, and rocky ridges, generally on dry, windy, or sunny slopes: Sausalito, Angel Island, Tiburon, and Mount Tamalpais (Cascade Canyon, Bootjack, Phoenix Lake) to San Rafael, Big Rock Ridge, Black Point, Tomales, and Point Reyes Peninsula (Bolinas, road to Point Reyes).

2. **A. biennis** Willd. In places where water has stood, rare in Marin County: Lake Lagunitas; Point Reyes, acc. Stacey. Native from the Rocky Mountains north to British Columbia.

3. **A. pycnocephala** (Less.) DC. Maritime bluffs and coastal dunes: Sausalito Hills; Angel Island; Bolinas sand spit; Dillons Beach; Point Reyes and McClure Beach, Point Reyes Peninsula.

4. **A. Douglasiana** Bess. CALIFORNIA MUGWORT. Common and widespread throughout the hills and valleys along watercourses and in places wet in the spring: Sausalito, Angel Island, Tiburon, and Mount Tamalpais (Phoenix Lake) to San Rafael, Big Rock Ridge, Tomales, and Point Reyes Peninsula (Inverness, road to Point Reyes). On moist flats at the north end of Bolinas Lagoon the plants form herbaceous thickets with stems 8 feet tall.

TRIBE 9. Senecioneae. GROUNDSEL TRIBE

55. Petasites

1. **P. palmatus** (Ait.) Gray. Occasional in low wet places in shaded canyons or woods: Mount Tamalpais, *Brandegee;* Lagunitas Canyon; near Camp Taylor. Our plant is closely related to the boreal *P. frigidus* (L.) Fries and should perhaps be referred to that species as var. *palmatus* (Ait.) Cronquist.

56. Arnica

1. **A. discoidea** Benth. Rocky or gravelly slopes under brush or chaparral, not common: Blithedale Canyon, Mount Tamalpais; Carson Ridge.

Arnica parviflora Gray subsp. *alata* (Rydb.) Maguire has been reported from Marin County by Maguire but all specimens that have been examined seem to be *A. discoidea.*

57. Erechtites

a. Leaves dentate; pappus longer than the corollas ....1. *E. prenanthoides*
a. Leaves pinnately divided; pappus about equaling the corollas ....2. *E. arguta*

1. **E. prenanthoides** DC. Common along roads, in waste ground, and in woods or brush: Sausalito, Almonte marshes, and Mount Tamalpais (Steep Ravine, Alpine Dam, Fern Canyon) to Chinese Camp, Chileno Valley, Tomales, and Point Reyes Peninsula (Bolinas, Mud Lake, Ledum Swamp). Introduced from Australia.

2. **E. arguta** DC. Common around towns and on wooded or brushy slopes in the hills: Sausalito, Angel Island, Tiburon, and Mount Tamalpais (Blithedale Canyon, Throckmorton Ridge) to Chinese Camp, Chileno Valley, Tomales, and Point Reyes Peninsula (Bolinas, Inverness). Native of Australia and New Zealand.

58. Senecio. GROUNDSEL

a. Stems twining and climbing; leaves roundish, palmately veined and lobed
   ....5. *S. mikanioides*
a. Stems erect, not twining; leaves elongate, not palmate ....b
b. Plants perennial; leaves toothed or entire c
b. Plants annual; leaves once or twice pinnatifid ....d
c. Herbage glabrous, glaucous, and somewhat succulent; leaves entire or nearly
   so ....1. *S. hydrophilus*
c. Herbage thinly tomentulose or glabrate, not glaucous or succulent; leaves
   mostly dentate or crenate ....2. *S. aronicoides*
d. Involucre bearing small short bracts at the base; rays none ......3. *S. vulgaris*
d. Involucre almost without short bracts at the base; rays present but small and
   inconspicuous ....4. *S. sylvaticus*

1. **S. hydrophilus** Nutt. var. **pacificus** Greene. In Marin County known only from the marshy borders of the laguna in Chileno Valley.

2. **S. aronicoides** DC. Occasional in moist shaded woods or on brushy slopes or flats: Sausalito, Angel Island, Tiburon, and Mount Tamalpais (Corte Madera Ridge, Berry Trail) to San Rafael Hills, Big Rock Ridge, Tomales, and Point Reyes Peninsula (near the radio station).

3. **S. vulgaris** L. COMMON GROUNDSEL. Common weed of gardens and waste ground, occasional in woods or brush: Sausalito, Angel Island, and Mill Valley to San Rafael, Tomales, and Inverness. Native of the Old World.

4. **S. sylvaticus** L. Widespread along streets and roads and on rocky slopes in woods or brush, becoming very common after fires: Sausalito, Angel Island, Tiburon, and Mount Tamalpais (Cascade Canyon, West Peak, Berry Trail) to San Rafael Hills and Point Reyes Peninsula (Inverness Ridge). Introduced from Europe.

5. **S. mikanioides** Otto. GERMAN IVY. Forming dense tangles in shaded canyons or on moist open slopes: Belvedere; Mill Valley; Stinson Beach; Bolinas; Inverness. Introduced from South Africa.

Tribe 10. Cynareae. Thistle Tribe

59. Cirsium

a. Plants dioecious, perennial, the stems from rootstocks; heads 2–2.5 cm. high
....10. *C. arvense*

a. Plants monoecious, the flowers perfect; plants annual, biennial, or perennial;
   heads usually more than 2.5 cm. high ....b

b. Leaves decurrent as spiny wings on the stem ....1. *C. vulgare*

b. Leaves not decurrent or very shortly so c

c. Involucral bracts without a glutinous ridge ....d

c. Involucral bracts with an elongate glutinous ridge along the midrib i

d. Involucral bracts nearly glabrous, usually not conspicuously archnoid or
   tomentose; plants perennial, the stems from rootstocks or crowns; corollas
   sordid white or buff ....e

d. Involucral bracts tomentose or arachnoid; plants annual or biennial; corollas
   reddish or purplish, rarely white ....f

e. Plants usually about 1 m. tall, the stems slender; involucres generally 2–2.5
   cm. long, the bracts closely or loosely imbricate, the margins of the bracts
   lacerate or fimbriate and sometimes expanded at the top 2. *C. remotifolium*

e. Plants low, the stout stems mostly less than 3 dm. tall; involucres generally
   3–4 cm. long, the bracts closely and smoothly imbricate, the margins of the
   bracts coriaceous but not lacerate or fimbriate ....3. *C. quercetorum*

f. Heads subtended by more or less reduced leaves ....g

f. Heads not subtended by leaves ....h

g. Flowers about equaling the involucre; corolla lobes very narrowly linear;
   anther column shorter than the corolla, the appendages not conspicuous
   ....4. *C. edule*

g. Flowers exceeding the involucre; corolla lobes oblong-linear; anther column
   longer than the corolla, the appendages conspicuous ....5. *C. Andrewsii*

h. Outer bracts of the involucre much shorter than the inner, the base appressed
   and the tips spreading or reflexed, arachnoid-tomentose and more or less
   glabrate; flowers conspicuously exceeding the involucre ......6. *C. Coulteri*

h. Outer bracts of the involucre somewhat shorter than the inner, the whole
   straight, ascending or erect, and not much appressed, spreading-arachnoid;
   flowers scarcely exceeding the involucre ....7. *C. occidentale*

i. Stems and leaves densely white-tomentose; corollas dark reddish-purple
   ....8. *C. Breweri*

i. Stems and upper leaves glabrous (lower side of basal leaves tomentose); corollas
   light purplish-rose ....9. *C. Vaseyi*

1. **C. vulgare** (Savi) Airy-Shaw. BULL THISTLE. Common weed in hills and
valleys, around habitations and along roads or trails: Sausalito, Angel Island,
Tiburon, and Mount Tamalpais (Cascade Canyon, Laurel Dell Trail) to San
Rafael Hills, Tomales, and Point Reyes Peninsula (Bolinas, Mud Lake, Inverness
Ridge). Introduced from the Old World.—*C. lanceolatum* (L.) Scop., not Hill.

2. **C. remotifolium** (Hook.) DC. Widespread through the hills on brushy slopes,
in open woods, or on the edge of grassland: Angel Island, Tiburon, and Mount
Tamalpais (Lake Lagunitas) to San Rafael Hills, Lucas Valley, and Salmon Creek
summit.

This species is quite variable as to the form and texture of the involucral bracts and three names have been given to the plants on Mount Tamalpais alone by Petrak. Most of the Marin County specimens are referable to var. *odontolepis* in which the margin of the bracts is conspicuously fimbriate-toothed. In this variety the bracts are glabrous but in subsp. *pseudocarlinoides* the bracts are arachnoid-tomentose and more elongate-oblong. The third variant was described as a species, *C. amblylepis*, but it cannot be distinguished from those forms in which the bracts are more nearly entire. It may have originated as a hybrid between *C. remotifolium* and *C. quercetorum.* At the type locality of of *C. amblylepis* on Summit Avenue above Mill Valley no two plants were found to be the same. Plants which also appear to be hybrids of these two species have been found on Angel Island and near Lone Tree on Mount Tamalpais but in both of those places the plants were different again and none was like *C. amblylepis.*

3. **C. quercetorum** (Gray) Jeps. Widespread on open grassy or brushy slopes near the coast: Sausalito, Angel Island, and San Geronimo Ridge to Dillons Beach and Point Reyes Peninsula (Bolinas, Inverness Ridge, Ledum Swamp, Point Reyes). The plants are usually quite low and almost stemless but occasional forms may be a foot or two tall. Suspected hybrids between *C. quercetorum* and *C. Andrewsii, C. Coulteri,* and *C. remotifolium* are described under those species.

4. **C. edule** Nutt. Common on moist brushy or wooded slopes and flats: Sausalito, Angel Island, Tiburon, and Mount Tamalpais (Blithedale Canyon, Rock Spring) to Lagunitas Canyon, Tomales, and Point Reyes Peninsula (Bolinas and Mud Lake). On Inverness Ridge white-flowered plants were common with the usual purplish-rose form.

5. **C. Andrewsii** (Gray) Jeps. Locally common at a few places on maritime bluffs and in moist coastal canyons: Rodeo Lagoon; Tennessee Cove; Dillons Beach; Ledum Swamp and Point Reyes, Point Reyes Peninsula. This is a rare thistle found only from San Mateo County north to Sonoma County. The type locality is probably in Marin County.

Near the lighthouse on Point Reyes, hybrids between *C. Andrewsii* and *C. quercetorum* were found where the two species grew together. Plants otherwise like *C. Andrewsii* had the involucre of *C. quercetorum,* and another plant was entirely like *C. Andrewsii* except that the flowers were sordid white as in *C. quercetorum.* A hybrid between these two species was described by Petrak from Tennessee Cove.

6. **C. Coulteri** Harv. & Gray. Dry sunny slopes in loose or gravelly soil, in brush or open woods, rather common: Angel Island, Tiburon, and Mount Tamalpais (Corte Madera Ridge, Summit Avenue, below Rock Spring, Northside Trail) to Carson country, Fairfax Hills, and Black Point.—*C. occidentale* (Nutt.) Jeps. var. *Coulteri* (Harv. & Gray) Jeps.

This is a beautiful thistle, usually with dark crimson flowers, though occasionally white-flowered forms occur, either with the others or in pure colonies. On two occasions hybrids with other species have been noted: in the Sausalito Hills, with *C. quercetorum;* and on Summit Avenue above Mill Valley, with *C. remotifolium* (or the form of that species that grows there). Petrak described a hybrid between *C. occidentale* and *C. quercetorum* from the ridge north of Tennessee Cove and his plant may be the same as the one from the Sausalito Hills.

7. **C. occidentale** (Nutt.) Jeps. In Marin County known in typical form only from the dunes and beaches on Point Reyes Peninsula. A form in which the involucral bracts are a little more graduated and a little more spreading grows at the west end of Mount Tamalpais (Bootjack and above Summit Meadow). These plants are much nearer *C. occidentale* than *C. Coulteri* but they may represent a hybrid strain between those usually very distinct species.

8. **C. Breweri** (Gray) Jeps. var. **Wrangelii** Petrak. A beautiful thistle known in Marin County only from a wet willow-choked gully on the ocean bluffs north of Dillons Beach.

9. **C. Vaseyi** (Gray) Jeps. Around springs, along brooks, or in marshy ground in serpentine areas: Mount Tamalpais, the type locality (Azalea Flat, Bootjack, headwaters of Cataract Creek, Potrero Meadow, Camp Handy); Liberty Spring southwest of Carson Ridge.

This thistle is not known beyond Marin County. It is *C. Vaseyi* var. *hydrophilum* of Jepson's Manual in part but not *Carduus hydrophilus* Greene, a plant of the Suisun Marshes. Petrak described the plant from Mount Tamalpais (west of West Point) as *C. montigenum*.

10. **C. arvense** (L.) Scop. CANADA THISTLE. Along the Petaluma road east of Tomales, the only occurrence known in Marin County. Native of the Old World.

60. Carduus

a. Heads usually few (1–5) at the ends of the branches; bracts of the involucre with small rough trichomes on the margin and back ....1. *C. pycnocephalus*
a. Heads usually numerous (5–20) at the ends of the branches; bracts of the involucre glabrous or subciliate on the margin ....2. *C. tenuiflorus*

1. **C. pycnocephalus** L. Occasional on grassy flats and slopes and on the edge of woods or brush: coast road between Muir and Stinson beaches; Corte Madera; Phoenix Lake, Mount Tamalpais; near Fairfax; Tocaloma, *Leschke;* Burdell; Tomales; Inverness, Point Reyes Peninsula. Native of the Mediterranean region.

2. **C. tenuiflorus** Curt. Rare on wooded or brushy slopes: Chileno Valley; near Mud Lake, Point Reyes Peninsula. Introduced from Europe.

61. Cynara

1. **C. Scolymus** L. ARTICHOKE. Rare fugitive from cultivation: Sausalito; Mill Valley. Native of Europe.

62. Silybum

1. **S. Marianum** (L.) Gaertn. MILK THISTLE. Common and widespread weed in fields and pastures, along roads and about habitations: Sausalito, Angel Island, Tiburon, and Mount Tamalpais (Cascade Canyon, near Rock Spring, Phoenix Lake) to San Rafael, Dillons Beach, and Point Reyes Peninsula (Bolinas, Mud Lake, Inverness). Native of the Old World.

Although this is one of the weeds to be termed "bad," it has a certain beauty which, though scarcely recommending it for cultivation in the herbaceous border, does perhaps redeem it from the obnoxious. The large basal leaves are attractive in form and marbling and the fiercely spiny heads would be ornamental enough if it were not for the reddish-purple color of the flowers. Its common occurrence coupled with a vulgar prejudice against thistles is not likely to encourage genetic studies for the improvement of its color as an ornamental.

63. Centaurea

a. Middle involucral bracts toothed at the apex and along the sides; outer flowers with enlarged raylike corollas ....4. *C. Cyanus*
a. Middle involucral bracts tipped by a stiff spreading spine; outer flowers in the head not much enlarged ....b
b. Flowers purple; leaves not decurrent ....3. *C. calcitrapa*
b. Flowers yellow; leaves prominently decurrent c
c. Spines slender, about 1 cm. long; flowers exceeding the involucre by 3–5 mm.
   ....1. *C. melitensis*
c. Spines stouter, 1–2 cm. long; flowers exceeding the involucre by 7–10 mm.
   ....2. *C. solstitialis*

1. **C. melitensis** L. NAPA THISTLE; TOCALOTE. Common in grassy and brushy places in hills and valleys: Sausalito, Tiburon, Angel Island, and Mount Tamalpais (Blithedale Canyon, West Point, Rock Spring, Phoenix Lake) to San Rafael and Bolinas. Naturalized from the Mediterranean region.

2. **C. solstitialis** L. BARNABYS-THISTLE. Widespread along roads, in fields, and in waste places: Sausalito, Tiburon, and Mill Valley to Santa Venetia, Ignacio, and Lagunitas. Naturalized from the Mediterranean region.

3. **C. calcitrapa** L. PURPLE STARTHISTLE. Occasional in low fields and along roads: Sausalito, Tiburon, and Mill Valley to San Rafael and Novato. Native of Eurasia.

4. **C. Cyanus** L. CORNFLOWER. Occasionally escaping from cultivation and becoming locally established: Mill Valley; San Anselmo; Lagunitas. Native of Eurasia.

64. Carthamus

1. **C. lanatus** L. Open hills and along roads in Chileno Valley. Native of the Old World.

TRIBE 11. Cichorieae. CHICORY TRIBE

65. Lapsana

1. **L. communis** L. NIPPLEWORT. Occasional along roads in shady places: Mill Valley; Ross; Inverness. Introduced from the Old World.

66. Cichorium

1. **C. Intybus** L. CHICORY. Along roads and in waste places about habitations and in the hills: Sausalito; Mill Valley and San Geronimo, acc. Stacey; Ross; Fairfax; Santa Venetia; near Mud Lake, Point Reyes Peninsula. Native of the Old World.

The root of chicory furnishes a coffee-like beverage esteemed by many people; the leaf, in certain strains, is esculent and is a most palatable substitute for lettuce; the flower is tinctured with one of the most beautiful blues among plants. But also, Nature has decreed that its floral beauty be brief: the lovely caerulean disks that open to greet the dawn are all too quickly closed by the garish light of day, and only ugly twiggy shoots are left to mark the spot. Surely this is the Cinderella among our weeds!

67. Microseris

a. Plants perennial; involucre about 15 mm. long ....1. *M. paludosa*
a. Plants annual; involucre 12 mm. long or less ....b
b. Pappus paleae about 1 mm. long ....2. *M. aphantocarpha*
b. Pappus paleae 1.5–4 mm. long c
c. Achene a little constricted just below the top ....3. *M. Douglasii*
c. Achene not constricted below the top ....4. *M. Bigelovii*

1. **M. paludosa** (Greene) J. T. Howell. Open grassy slopes or on the edge of brush, not common: Blithedale Canyon, Mount Tamalpais; Corte Madera, the type locality; hills south of Salmon Creek; Dillons Beach; Drakes Estero and dunes, Point Reyes Peninsula.—*Scorzonella paludosa* Greene; *S. sylvatica* Benth. var. *Stillmanii* (Gray) Jeps.
2. **M. aphantocarpha** (Gray) Gray. In Marin County known only on Tiburon Peninsula. Common in the Inner Coast Ranges to the north and south.
3. **M. Douglasii** (DC.) Gray. Common on open grassy hills in clayey or gravelly soil: Tiburon and Mount Tamalpais (Rock Spring, Potrero and Lagunitas meadows) to the Carson country, San Rafael Hills, Hamilton Field, and Lucas Valley.
4. **M. Bigelovii** (Gray) Gray. Coastal hills and mesas, especially in sandy soil: Sausalito; Angel Island; Dillons Beach; Point Reyes Peninsula (Mount Vision, Drakes Estero, Point Reyes Lighthouse). The type came from Corte Madera.

68. Uropappus

a. Pappus buff, the bristles finely hairy; achenes only slightly narrowed upward ....1. *U. Lindleyi*
a. Pappus white or whitish, the bristles smooth or a little scabrous; achenes narrowed into a stout beak ....2. *U. linearifolius*

1. **U. Lindleyi** (DC.) Nutt. Occasional in loose clayey soil of open hills and canyonsides: Tiburon; between Muir and Stinson beaches; Mount Tamalpais (Blithedale Canyon, Phoenix Lake); Fairfax Hills; San Rafael Hills.
2. **U. linearifolius** (DC.) Nutt. Gravelly or clayey soil of open or brushy slopes: Sausalito; Angel Island; Tiburon; Phoenix Lake, Mount Tamalpais; Carson country; San Rafael Hills; Inverness Ridge.

69. Leontodon

1. **L. nudicaulis** (L.) Banks. Occasional in the hills and in open fields: Stinson Beach; Carson country; San Rafael Hills; near Santa Venetia; Bolinas. Native in Europe and North Africa.

The Marin County plants of this variable European species belong to subsp. *taraxacoides* (Vill.) Schinz & Thell. In some plants the involucres are glabrous, in others hirsutulous. These and several other forms have been given names in Europe.

70. Hedypnois

1. **H. cretica** (L.) Willd. Rather rare in clayey or sandy soil: Cascade Canyon, Mount Tamalpais; Ross Valley; San Rafael Hills; Dillons Beach. Naturalized from the Mediterranean region.—*Rhagadiolus Hedypnois* of California references.

71. Hypochoeris

a. Leaves glabrous; flower heads about 1 cm. across or less; outer achenes beakless, inner beaked ....1. *H. glabra*
a. Leaves hairy; flowering heads about 2 cm. across or more; all achenes beaked ....2. *H. radicata*

1. **H. glabra** L. SMOOTH CATS-EAR. Widespread and common through the hills and valleys in woods, brush, or grassland: Sausalito, Angel Island, Tiburon, and Mount Tamalpais (Corte Madera Ridge, Blithedale Canyon) to San Rafael Hills, Big Rock Ridge, and Point Reyes Peninsula (Inverness, Mud Lake). Naturalized from the Old World.

2. **H. radicata** L. HAIRY CATS-EAR. Usually occurring in grassland but also in brush and woods, common: Sausalito, Angel Island, Tiburon, and Mount Tamalpais (Cascade Canyon) to San Rafael Hills, Tomales, and Point Reyes Peninsula (Bolinas, Mud Lake, Point Reyes). Native in the Old World.

This is an attractive weed with its large bright yellow dandelion-like flowers, especially where it occurs with *Rumex Acetosella* on coastal slopes and mesas above the blue ocean.

72. Picris

1. **P. echioides** L. BRISTLY OX-TONGUE. Common weed of waste ground about towns, also along roads and in pastures: Sausalito, Angel Island, Tiburon, and Mount Tamalpais (Blithedale and Cascade canyons) to San Rafael, Bolinas, and Point Reyes (acc. Stacey). Introduced from the Mediterranean region.

73. Stephanomeria

a. Achenes buff, the ribs prominent; pappus completely deciduous ..1. *S. virgata*
a. Achenes grayish brown, angled but the ribs not prominent; pappus bristles breaking above the very short paleaceous base which remains attached to the achene ....2. *S. coronaria*

1. **S. virgata** Benth. Open slopes in loose clayey or gravelly soil: Angel Island, Tiburon, and Mount Tamalpais (Alpine Lake) to San Rafael Hills and Point Reyes Peninsula (Mud Lake, road to Point Reyes).

2. **S. coronaria** Greene. Occasional on open gravelly slopes: Matt Davis Trail and Phoenix Lake, Mount Tamalpais; Fairfax; San Rafael Hills; Bolinas Lagoon.

At the west end of Mount Tamalpais near Bootjack, plants are found in which the flowers are larger and the pappus more persistent. These plants seem more closely related to *S. coronaria* than to *S. virgata* but they may represent an undescribed entity.

74. Rafinesquia

1. **R. californica** Nutt. Occasional on open or brushy canyonsides, generally in

loose soil: Angel Island; Lagunitas Canyon, Mount Tamalpais; San Rafael Hills; Carson country; Shell Beach, Point Reyes Peninsula.

75. Tragopogon. SALSIFY

1. **T. porrifolius** L. OYSTER-ROOT. Open grassy places around towns and along roads: Sausalito; Tiburon; Mill Valley; San Rafael. Naturalized from the Old World where the root is widely used for food.

76. Malacothrix

1. **M. floccifera** (DC.) Blake. Gravelly rock slides in canyons or rocky slopes of hills: Marin County, acc. Behr. Seedlings noted on Corte Madera Ridge, Mount Tamalpais, may have belonged to this widely distributed Coast Range species.— *M. obtusa* Benth.

77. Taraxacum. DANDELION

a. Achenes pale grayish brown ....1. *T. officinale*
a. Achenes rose-tinged ....2. *T. laevigatum*

1. **T. officinale** Weber. Lawns and weedy places in the hills and about habitations: Sausalito; Angel Island; Mill Valley; Ross; Bolinas Lagoon; Olema Marshes. Naturalized from the Old World.—*T. vulgare* Schrank.

2. **T. laevigatum** (Willd.) DC. Open grassy flats near habitations in San Rafael. Introduced from Europe.

78. Sonchus. SOW-THISTLE

a. Lobes at base of leaf pointed; achenes finely transversely wrinkled or
    roughened, not thin-margined ....1. *S. oleraceus*
a. Lobes at base of leaf rounded; achenes thin-margined, not transversely wrin-
    kled or roughened ....2. *S. asper*

1. **S. oleraceus** L. Widespread and common weed about towns and in the hills and valleys: Sausalito, Angel Island, Tiburon, and Mount Tamalpais (Blithedale Canyon) to San Rafael, Burdell Marshes, and Point Reyes Peninsula (Bolinas, Inverness, McClure Beach). Naturalized from the Old World.

2. **S. asper** (L.) Hill. Occasional and ruderal about towns and along roads and trails: Sausalito (acc. Lovegrove), Angel Island, Tiburon, and Mount Tamalpais (Berry Trail) to San Rafael, Carson Ridge, and Inverness. Introduced from the Old World.

79. Lactuca. LETTUCE

a. Achenes narrowed above but scarcely beaked; pappus brownish ..1. *L. spicata*
a. Achenes with a slender beak; pappus white ....b
b. Leaves linear-lanceolate, or if divided, the divisions linear, margin of blade
    and divisions entire and smooth; panicle narrow and spikelike ..2. *L. saligna*
b. Leaves oblanceolate or oblong to elliptic, the margin dentate, spinulose-den-
    tate, or spinulose-ciliate; panicle openly and rather widely branching c
c. Achenes blackish or purple-black, minutely transversely roughened
    ....3. *L. virosa*
c. Achenes gray or brown, not transversely roughened ....4. *L. Serriola*

1. **L. spicata** (Lamk.) Hitchc. Moist coastal woods: near Olema, where the plant was collected a half century ago by T. S. Brandegee; Bear Valley, according to Miss Eastwood in 1898. This is the southernmost station known in the Coast Ranges.

2. **L. saligna** L. Common in waste ground around towns, along roads, and in fields: Sausalito, Mill Valley, and Mount Tamalpais (Bootjack, Laurel Dell, Lagunitas Meadows) to Fairfax, San Rafael, and Bolinas Lagoon. Naturalized from the Old World.

3. **L. virosa** L. Moist, shaded, brushy or wooded slopes, Sausalito and Mill Valley, acc. Stacey. This Old World lettuce is a conspicuous plant in the Oakland and Berkeley hills across San Francisco Bay.

4. **L. Serriola** L. Common about towns and along roads in weedy places: Sausalito, Tiburon, and Mill Valley to San Rafael and Point Reyes Station. Naturalized from the Old World.—*L. Scariola* L.

In the species the leaves are broadly pinnately lobed but frequently growing with it and differing from it only in leaf shape is f. *integrifolia* Bogenhard. This form has been noted in Sausalito, Tiburon, Mill Valley, and Point Reyes Station. —*L. Scariola* L. var. *integrata* Gren. & Godr.

80. Agoseris

a. Plants annual . . . . .4. *A. heterophylla*
a. Plants perennial . . . . .b
b. Ligules about equaling the involucral bracts; involucral bracts mostly in two series, the outer short and ovate-lanceolate, the inner equaling the pappus of the mature achenes and linear-lanceolate; beak of the achene 2–3 times longer than the achene . . . . .3. *A. plebeia*
b. Ligules much longer than the involucral bracts; involucral bracts rather evenly graduated, pilose and a little viscidulous; beak of the achene shorter to about twice longer than the achene c
c. Beak of the achene generally longer than the achene; plants of open grassy coastal hills . . . . .1. *A. hirsuta*
c. Beak of the achene shorter than the achene or about equaling it; plants of maritime bluffs and dunes . . . . .2. *A. apargioides*

1. **A. hirsuta** (Hook.) Greene. Widely distributed and rather common on open grassy slopes, especially towards the coast: Sausalito, Angel Island, Tiburon, and Mount Tamalpais (Dipsea Trail, Rock Spring, Potrero and Lagunitas meadows) to Carson country, San Rafael Hills, and Point Reyes Peninsula (Inverness Park).

2. **A. apargioides** (Less.) Greene. Maritime bluffs, mesas, and dunes, occasional: Dillons Beach; Point Reyes Peninsula (Point Reyes, dunes near the radio station, McClure Beach).

3. **A. plebeia** (Greene) Greene. Widespread and rather common on grassy, brushy, or wooded hills: Sausalito, Angel Island, Tiburon, and Mount Tamalpais (Blithedale Canyon, Bootjack, Rock Spring, Kent Trail) to San Rafael Hills, Black Point, Tomales, and Point Reyes Peninsula (Shell Beach).

Certain plants with more slender, fewer-flowered heads have been referred to *A. gracilens* (Gray) Ktze., but that is typically a more northern species and the Marin plants seem to be but a slender form of *A. plebeia*.

4. **A. heterophylla** (Nutt.) Greene. This is a variable species in the markings on the outer achenes. The typical form of the species, in which the ribs on the achenes are straight or nearly so, is rather rare in Marin County, having been seen only from the San Rafael Hills and Chileno Valley. In this form the outer achenes are also hirsutulous. The rarest form in Marin County is var. *cryptopleura* Greene in which the outer achenes are nearly smooth and appear as if inflated. It has been found in Marin County only in Potrero Meadow on Mount Tamalpais, apparently the southernmost station in California. Var. *kymapleura* Greene, in which the ribs of the achenes are well developed and strongly undulate, is the commonest form in brush and grassland: Mount Tamalpais (between Barths Retreat and Potrero Meadow, Lagunitas Meadows, Fish Grade); Fairfax Hills; Carson country; San Rafael Hills.

81. Crepis

a. Involucre 5–7 mm. long; achenes not beaked ....1. *C. capillaris*
a. Involucre about 1 cm. long; achenes shortly beaked ....2. *C. vesicaria*

1. **C. capillaris** (L.) Wallr. Rare in moist brushy places: Lagunitas, *Eastwood;* Bolinas; south end of Tomales Bay. Native of Europe.
2. **C. vesicaria** L. This European introduction, which is more common farther north along the Pacific coast, has been listed from Point Reyes by Stacey.

82. Hieracium

1. **H. albiflorum** Hook. Partially shaded places in woods or on the edge of brush: Sausalito, Tiburon, and Mount Tamalpais (Cascade Canyon, Bootjack, Phoenix Lake) to San Rafael Hills and Point Reyes Peninsula (Mud Lake, Inverness).

Additions and Corrections

Page 54. *Pilularia americana* A. Br. was found on the strand of Lake Lagunitas in July, 1949, the first known Marin County record in over fifty years.

Page 134. *Silene multinervia* Wats. has been collected in Sonoma County by M. S. Baker.

Page 194. According to Pierre Dansereau, the naturalized variant of *Cistus villosus* L. in Marin County is var. *tauricus* (Dun.) Grosser, the typical variety in this complex species.

Page 196. *Dirca occidentalis* Gray has been found near Bodega, Sonoma County, by M. S. Baker. The subtraction of this species and *Silene multinervia* from the group of plants reaching a northern limit in Marin County will change slightly the figures given in that section of the introduction where the relationship of the Marin County flora to the flora of the Coast Ranges is discussed (p. 25). The changes in the figures, however, do not alter the conclusions reached.

Page 216. *Pityopus californicus* (Eastw.) Copeland f. has been collected in 1949 near Mendocino, Mendocino County, by Jean Boyd and Marie Kelly.

Page 237. According to Stebbins and Paddock, *Solanum nigrum* L. probably does not occur in California. The plant so called in Marin County is *S. nodiflorum* Jacq. This species, which may grow as either an annual or perennial, is native in tropical America. A related South American species with large flowers (over 1 cm. across) is *S. furcatum* Dunal; it is to be watched for on slopes along the coast.

Principal References

ABRAMS, LEROY.

1923–1944. Illustrated Flora of the Pacific States. Vol. 1, xii + 558 pages; vol. 2, viii + 636 pages. Stanford University Press, Stanford University, California.

BOWERMAN, MARY L.

1944. The Flowering Plants and Ferns of Mount Diablo, California. xi + 290 pages. The Gillick Press, Berkeley.

BRANDEGEE, KATHARINE.

1892. Catalogue of the Flowering Plants and Ferns Growing Spontaneously in the City of San Francisco. Zoe 2: 334–386.

HITCHCOCK, A. S.

1935. Manual of the Grasses of the United States. 1040 pages. United States Government Printing Office, Washington, D.C.

JEPSON, WILLIS LINN.

1909–1943. A Flora of California. Vol. 1, 32–578 pages; vol. 2, 1–684 pages; vol. 3, 17–464 pages. Associated Students Store, University of California, Berkeley.

1923–1925. A Manual of the Flowering Plants of California. 1238 pages. Associated Students Store, University of California, Berkeley.

KEARNEY, THOMAS H., ROBERT H. PEEBLES, and collaborators.

1942. Flowering Plants and Ferns of Arizona. 1069 pages. United States Government Printing Office, Washington, D.C.

LAWSON, ANDREW C.

1914. Geologic Atlas of the United States, No. 193: San Francisco Folio. 24 pages. United States Geological Survey, Washington, D.C.

MARTIN, R. J., editor.

———. Climatic Summary of the United States. Section 15—Northwestern California. 24 pages.

MASON, H. L.

1934. Pleistocene Flora of the Tomales Formation. Carnegie Institution of Washington Publication No. 415, pages 81–179.

MUNZ, PHILIP A.

1935. A Manual of Southern California Botany. xxxix + 642 pages. Claremont Colleges, Claremont, California.

PECK, MORTON EATON.

1941. A Manual of the Higher Plants of Oregon. 866 pages. Binfords & Mort, Portland, Oregon.

RUSSELL, R. J.

1926. Climates of California. Univ. Calif. Publ. Geog. 2: 73–84.

SHARSMITH, HELEN K.

1945. Flora of the Mount Hamilton Range of California. Amer. Midl. Nat. 34: 289–367.

Glossary

Accessory, something added.

Achene, a dry, 1-celled, 1-seeded, indehiscent fruit.

Acuminate, gradually narrowing to the top.

Acute, sharp-pointed but not drawn out.

Adnate, grown to.

Adventive, applied to recently introduced plants.

Aerial, living above the ground or water.

Alternate, having one leaf or branch at a given level on the stem; having one part of a flower between, or alternating with, other parts.

Alveola, a pit or cavity.

Amphibious, growing both on dry land and in water.

Ampliate, enlarged.

Androgynous, having staminate and pistillate flowers in the same clusters; in *Carex* applied to inflorescences in which the staminate flowers are above the pistillate.

Angiosperm, a plant in which the seeds are borne in a closed case or ovary.

Anther, the part of a stamen containing the pollen, usually 2-celled and borne on a stalk or filament.

Anthesis, the time of flowering when fertilization takes place.

Apetalous, without petals.

Apiculate, having a small, sharp point.

Arachnoid, resembling a cobweb.

Arborescent, becoming tree-like.

Assurgent, ascending, rising upward.

Attenuate, gradually narrowed.

Auricle, a small earlike lobe; adj., auriculate.

Awn, a bristle-like appendage.

Axil, the upper angle formed by the stem and leaf; adj., axillary.

Banner, the uppermost petal in a pea flower.

Barb, a hooked hair.

Beak, a pointed projection.

Berry, a fleshy, many-seeded, indehiscent fruit.

Bi-, prefix meaning twice or two times.

Bifid, divided halfway into two.

Blade, the expanded part of a leaf or petal; the stem structure in *Lemnaceae.*

Bloom, a white waxy covering.

Bract, a reduced leaf, particularly the kind occurring in the inflorescence; adj., bracteate.

Bractlet, a small bract; a secondary bract; small appendages attached to the calyx and alternating with the lobes or sepals.

Bulb, a leaf bud with fleshy scales or leaf bases, frequently underground.

Bulbous, having bulbs or the structure of bulbs.

Bulblet, a small bulb.

Bur, a fruit bearing prickles.

Callous, hardened and thickened.

Callus, an abnormally thickened part; the base of the florets in grasses.

Calyx (pl. calyces), the outer series of floral leaves or envelopes, usually green.

Campanulate, bell-shaped.

Canescent, becoming gray.

Capillary, slender like a hair.

Capitate, with a head like that of a pin; growing in a globular structure or head.

Capsule, a dry dehiscent fruit of a compound pistil.

Carinate, having a keel.

Carpel, a simple pistil or element of a compound pistil.

Catkin, a deciduous spike.

Caudex, a short trunk, particularly the woody base of a perennial herb.

Cauline, belonging to, or arising from, the stem.

Cell, the unit of plant and animal tissues; the cavity of an anther or ovary.

Cellular, consisting of cells.

Cespitose, growing in tufts.

Chaparral, a plant formation consisting of hardwood shrubs that are frequently rigid and spiny.

Chartaceous, papery.

Chlorophyll, the green coloring matter of plants.

Ciliate, fringed with hairs.

Ciliolate, finely ciliate.

Circumscissile, dehiscing by a circular line around the fruit.

Clavate, club-shaped, thickened towards the apex.

Claw, the narrowed base of some petals, as in *Brassica.*

Cleft, cut about to the middle.

Cleistogamous, with fertilization taking place in an unopened flower.

Collar, in grass leaves, the junction of blade and sheath.

Compound leaf, one divided into separate blades or leaflets.

Connate, united or grown together from the beginning.

Cordate, heart-shaped, the notch at the broad lower end.

Coriaceous, leathery.

Corm, a fleshy bulb-like underground stem, a "solid bulb."

Corolla, the inner floral leaves or envelopes, usually of some color besides green.

Corymb, a flat-topped flower cluster in which the outermost flowers bloom first; adj., corymbose.

Cotyledon, the seed leaves or first leaves of the embryo, one in the monocotyledons, two in the dicotyledons.

Crenate, edged with rounded teeth or scallops.

Crest, an elevation or ridge on the top of a structure.

Crisped, curled or twisted.

Cucullus, a hood; adj., cucullate.

Culm, a straw; the stem of grasses, sedges, and rushes.

Cuneate, wedge-shaped or triangular widening upward; cuneiform.

Cuspidate, tipped with a sharp stiff point.

Cyme, a flower cluster, generally flattened, in which the central flower blooms first.

Deciduous, falling in season after a function is performed.

Decumbent, reclining, but with the tip ascending.

Decurrent (leaves), prolonged on the stem as wings below the point of insertion.

Deflexed, bent abruptly downwards.

Dehiscence, the opening of a mature anther cell or fruit in a definite way or along a definite line; adj., dehiscent.

Deltoid, triangular with the sides nearly equal.

Dentate, toothed with the teeth extending straight out.

Diadelphous, with two groups of stamens; in the Pea Family, with one stamen free and the rest united.

Digitate, arranged like the fingers of a hand.

Dimorphic, having 2 unlike forms.

Dioecious, unisexual with the male and female elements in different individuals.

Disarticulate, to separate or break at a joint.

Discoid, having a head or capitate inflorescence of flowers with only tubular corollas.

Discrete, separate, distinct.

Dissected, deeply cut into many segments.

Divided, cleft to the base or midrib.

Dorsal, pertaining to or attached to the back.

Drupe, a stone fruit, like a plum, in which the outer part is fleshy and the inner part is hard.

Drupelet, a small drupe, as an element in a blackberry.

Edaphic, pertaining to the soil and its influence on plant growth.

Elliptic, oblong with flowing outline and similar ends.

Elongate, drawn out in length.

Emarginate, with a notch at the apex.

Entire, with an even margin.

Epidermis, the cellular skin or covering of a plant.

Erose, eroded, as if gnawed.

Evanescent, soon disappearing.

Exfoliate, to fall away in flakes or scales.

Fascicle, a close cluster; adj., fasciculate.

Filament, the stalk of an anther; a threadlike structure.

Filamentose, consisting of filaments or resembling a filament.

Filiform, threadlike.

Fimbriate, fringed.

Fistulous, hollow.

Flaccid, limp, flabby.

Flexuous, bending in a zigzag way.

Floret, a small flower; the flower, lemma, and palea in the Grass Family.

Foliaceous, resembling a leaf in texture or shape.

Foliolate, having leaflets.

Follicle, a simple 1-celled pod opening by the inner (ventral) suture.

Foveola, a small pit or cavity.

Frond, the leaf of ferns.

Galea, a structure shaped like a helmet, frequently drawn out like a beak.

Gamopetalous, with united petals.

Glabrescent, becoming glabrous.

Glabrous, devoid of pubescence.

Gland, a small secreting structure; a small protuberance of similar form which does not secrete.

Glandular, furnished with glands.

Glaucous, covered with a white waxy powder or bloom.

Globose, spherical.

Glumes, the two lowest empty bracts in the spikelet of the grass inflorescence.

Glutinous, with a sticky cover.

Gymnosperm, a naked seed; a plant in which the seeds are borne on a scale and not in a closed case or ovary.

Gynaecandrous, with pistillate flowers above the staminate flowers in the inflorescence of *Carex*.

Habitat, the kind of locality in which a plant grows.

Hastate, with a spreading lobe on each side at the base.

Head, a round or roundish flower cluster in which the flowers are sessile or subsessile on a short or flattened axis or receptacle.

Hirsute, with moderately stiff hairs of medium length.

Hispid, with stiff harsh hairs of medium length.

Hood, the appendage borne on the anthers in the Milkweed Family.

Hyaline, translucent.

Hybrid, a plant derived from cross-breeding two related species.

Hypanthium, an enlargement, and particularly a cuplike development, of the receptacle below the calyx.

Hypogynous, inserted on a low- or high-conical receptacle beneath the pistil.

Imbricate, overlapping as shingles on a roof.

Imperfect, lacking a part usually present; an imperfect flower, lacking either stamens or pistil.

Impressed, pressed into, marked with slight depressions.

Indehiscent, not opening by valves or along a definite line.

Indigenous, native to the region, not introduced.

Indurate, hardened.

Indusium (pl. indusia), an epidermal outgrowth covering the sorus of ferns.

Inferior, below some other organ; inferior ovary appears to be below the calyx.

Inflorescence, the arrangement of flowers on the stem.

Internode, the part of the stem between two nodes or joints.

Involucral, pertaining to an involucre.

Involucrate, furnished with an involucre.

Involucre, a ring or imbricate series of bracts around a flower or flower cluster.

Involute, having the edges rolled inwards.

Irregular, lacking regularity in form or size of similar parts.

Keel, a ridge like the keel of a boat; in a pea flower the two lowest petals that are united along their lower margin.

Lacerate, with the margin appearing as if torn.

Laciniate, slashed, cut into narrow lobes.

Lanceolate, several times longer than wide with the widest part below the middle.

Leaf, the lateral organ borne on the stem at nodes.

Leaflet, the division or part of a compound leaf.

Legume, the fruit in the Pea Family which is a simple pod opening in 2 pieces.

Lemma, the lower of the 2 bracts enclosing the flower in the Grass Family.

Ligule, the appendage between the sheath and blade in a grass leaf; the strap-shaped corolla or ray in the Sunflower Family; adj., ligulate.

Limb, the expanded part of a gamopetalous corolla.

Linear, several times longer than wide with the same width throughout.

Lip, the largest lobe in the perianth of an orchid; one of the two principal parts of an irregular corolla.

Lobe, a division of an organ, especially one that is rounded and usually not deep.

Loculicidal, splitting down the middle of the back of the cell of a capsule.

Maquis, a xerophilous shrub formation of the Mediterranean region in which the plants are mostly rigid and small-leaved. It is generally believed that the formation has developed in areas that were once wooded.

Membranaceous, thin and soft in texture and more or less translucent.

Mephitic, having an offensive smell.

Midrib, the middle or main vein of a leaf.

Monadelphous, having the stamens united one to another.

Monocotyledon, a plant with 1 seed leaf or cotyledon.

Monoecious, with the male and female elements in the same individual; monoecious flowering plants may have the flowers perfect or unisexual.

Mucro, an abrupt sharp terminal point; adj., mucronate.

Nerve, a simple unbranched vein.

Node, the joint of a stem, that part of the stem which bears a leaf or leaves.

Nut, a hard, 1-seeded, indehiscent fruit.

Nutlet, a small nut.

Ob-, as a prefix, means inversely or oppositely.

Oblong, oval, but much lengthened, with the sides parallel.

Obsolete, rudimentary or wanting.

Obtuse, blunt or rounded at the end.

Opposite, having 2 leaves or branches at a node; having one part placed before another in a flower.

Ochroleucous, yellowish white.

Odd-pinnate, with a terminal unpaired leaflet.

Oval, broadly elliptic.

Ovary, the part of the pistil containing the ovules or immature seeds.

Ovate, shaped like the longitudinal section of an egg with the broader end downwards.

Ovoid, an egg-shaped solid.

Ovule, the immature seed in the ovary.

Palate, the projection of the lower lip into the throat of a 2-lipped corolla.

Palea, the upper of the 2 bracts enclosing the flower in the Grass Family.

Paleaceous, chaffy or chafflike in texture.

Palmate, having the divisions or lobes spreading like the outspread fingers of the hand.

Panicle, a loose, branched flower cluster; adj., paniculate.

Papilionaceous, having an irregular butterfly-shaped corolla which consists of a banner petal, 2 wing petals, and 2 keel petals united along the lower side.

Papilla (plural, papillae), a small, nippleshaped protuberance; adj., papillate.

Parted, separated into parts almost to the base.

Pectinate, pinnately divided into narrow and close segments, like the teeth of a comb.

Pedicel, the stalk of a flower in a flower cluster; adj., pedicellate.

Peduncle, the stalk of a single flower or flower cluster; adj., pedunculate.

Peltate, shield-shaped, attached to a stalk by the lower surface and not by the margin.

Penicillate, tipped with a tuft of fine hairs, like a painter's brush.

Perfect flower, one having both stamens and pistils.

Perianth, the floral envelopes, the calyx or corolla, or both.

Perigynium, the sac-like or bottle-like bract that encloses the pistillate flower and fruit in *Carex.*

Perigynous, with organs inserted on the perianth or on the hypanthium.

Petal, a division of the corolla.

Petiole, the stalk of a leaf; adj., petiolate.

Petiolule, the stalk of a leaflet.

Phyllode, a petiole taking on the form and function of a leaf blade.

Pilose, hairy, the hairs usually soft.

Pinna (pl. pinnae), the primary division of a pinnate leaf.

Pinnate, with the leaflets arranged on each side of a common axis or rachis.

Pinnatifid, deeply cut in a pinnate manner.

Pinnatisect, divided to the rachis in a pinnate manner.

Pinnule, a segment of a pinnately compound pinna.

Pistil, the female or seed-bearing structure of a flower, when complete consisting of ovary, style, and stigma.

Pistillate, furnished with a pistil or pistils.

Placenta, that part of the ovary which bears the ovules.

Plano-convex, 2-sided and flat on one side, outwardly curved on the other.

Plica, a plait or fold.

Plumose, having fine hairs on each side, as in a feather.

Polygamous, with perfect and unisexual flowers on the same individual.

Polypetalous, having distinct petals.

Pome, apple or other similar fleshy fruit developed from an inferior ovary with several cells.

Prismatic, with flat faces separated by angles.

Process, a projecting appendage.

Procumbent, lying along the ground.

Prostrate, lying flat on the ground.

Puberulent, minutely pubescent.

Pubescence, the hairiness or vestiture of plants.

Pubescent, clothed with hair, the hair usually fine and soft.

Punctate, marked with dots, depressions, or translucent glands.

Pungent, ending in a sharp rigid point; acrid to taste or smell.

Pustulate, with a blisterlike swelling.

Quadrate, nearly square.

Raceme, a simple flower cluster in which pedicellate flowers are borne on a common and usually elongate axis.

Racemose, in racemes or resembling a raceme.

Rachilla, the axis that bears the florets in the spikelet of grasses.

Rachis, the axis of an inflorescence or a compound leaf.

Radiate, spreading from a common center; bearing ray flowers.

Ray, the primary branch of an umbel; one of the marginal flowers in the head of the Sunflower Family when these are distinct from the disk.

Receptacle, the part of the flower that bears the sepals, petals, stamens, and pistils; the expanded summit of an axis or stalk which bears the collected flowers of a head.

Recurved, curved backward or downward.

Reflexed, bent abruptly backward or downward.

Regular, having the similar parts alike in size and shape.

Relict, a species belonging to a plant association or flora formerly more widespread and abundant.

Reniform, kidney-shaped.

Reticulate, netted; with a network.

Retrorse, pointing backward or downward.

Retuse, with a shallow notch at the apex.

Revolute, rolled inward or backward from the margin or apex.

Rhizome, a rootstock, a prostrate stem on or under the ground.

Rhomboidal, obliquely 4-sided.

Rib, a primary or prominent vein.

Rootstock, a rhizome.

Rosette, a circular cluster of leaves.

Rosulate, forming a rosette.

Rotate, wheel-shaped, flat and circular in outline.

Ruderal, growing in waste places or among rubbish.

Rudiment, an imperfectly or partially developed organ, a vestige; adj. rudimentary.

Rugose, wrinkled.

Saccate, bag-shaped.

Sagittate, shaped like an arrowhead.

Salient, projecting prominently.

Salverform, having a slender tube abruptly expanded into a flat limb.

Samara, an indehiscent winged fruit, like that of the maple or ash.

Saprophyte, a plant which lives on dead organic matter.

Scabrous, rough to the touch.

Scale, a thin structure, usually either a reduced leaf or an outgrowth of the epidermis; adj., scaly.

Scape, a leafless flower stalk arising from the ground; adj., scapose.

Scarious, thin, dry and membranaceous.

Scorpioid, with the axis of the inflorescence coiled at the tip like the tail of a scorpion.

Scurf, small branlike scales on the stem and leaves.

Secund, with the parts or flowers inserted on or turned to one side of an axis.

Seed, the mature fertilized ovule.

Sepal, a segment or division of the calyx.

Septicidal, said of a capsule that dehisces along the partitions dividing the cells.

Septum (pl. septa), a partition.

Sericeous, silky with straight soft hairs.

Serrate, bearing forward pointing teeth on the margin.

Sessile, without a stalk.

Setose, bristly, beset with bristles.

Sheath, a tubular structure, as the lower part of a grass leaf.

Siliceous, pertaining to silica.

Simple, of one piece or series, not compound.

Sinus, a cleft or recess between two lobes.

Sordid, dirty in color, an impure white.

Sorus (pl. sori), a cluster of sporangia in ferns.

Spadix, a spike with a fleshy axis, as in the Arum Family.

Spathe, a large showy bract enclosing or subtending a flower cluster.

Spike, a flower cluster in which the flowers are sessile on a common elongate axis; adj. spicate.

Spikelet, flower or flowers (florets) subtended by a pair of glumes in the flower clusters of grasses.

Spiral, placed as though wound around an axis.

Sporangium (pl. sporangia), the small receptacle or case containing spores.

Sporocarp, a body produced by certain fern allies which contains clusters of sporangia.

Spur, a hollow and slender extension of some part of a flower.

Stamen, the male organ of a flower, consisting of filament, anther, and pollen.

Staminate, furnished with a stamen or stamens.

Staminodium (pl. staminodia), a sterile or abortive stamen.

Stellate, starlike, where several similar parts spread from a common center as in a star.

Stigma, the part of the pistil which receives the pollen.

Stipe, a stalk in certain structures, the stalk of a fern frond; adj., stipitate.

Stipules, the appendages at the base of the petiole in the leaves of some plants; adj., stipular.

Stolon, a runner or creeping stem tending to root.

Stoloniferous, producing stolons.

Striate, striped or marked with fine longitudinal parallel furrows and ridges.

Strict, close, narrow, and straight.

Strigose, bearing appressed straight stiff hairs or bristles.

Style, the slender part of a pistil between the ovary and stigma.

Sub-, as a prefix, indicating about, nearly, somewhat.

Subtend, to extend under.

Subulate, tapering to a fine sharp point, awlshaped.

Suffrutescent, somewhat shrubby.

Sulcate, grooved or furrowed.

Superior, above some other organ; superior ovary, when all the floral parts are on the receptacle below or free from the ovary.

Suture, the line along which contiguous parts are joined.

Symmetric, having the same number of parts in each circle of the flower.

Tap root, a root with a stout tapering body.

Tendril, a slender structure derived from stems or leaves by means of which a plant may make itself secure.

Terete, cylindric and usually tapering.

Terrestrial, designating plants growing in dry ground.

Throat, the expanded part of a gamopetalous corolla between the limb and tube.

Tomentose, clothed with dense matted woolly hairs.

Toothed, furnished with teeth or short sharp projections on the margin, the teeth pointing straight out.

Trichome, any hairlike outgrowth of the epidermis.

Trifid, three cleft.

Truncate, as if cut off at the end.

Tube, the lower narrowed part of a calyx or corolla in which the parts are united.

Tuber, a short thickened underground stem.

Tubercle, a little tuber or warty swelling; adj., tuberculate.

Tuberous, having or resembling a tuber.

Turbinate, shaped like a top.

Turgid, swollen, but not with air.

Turion, vigorous young shoot springing out of the ground.

-ulate, -ulose, -ulous, suffixes of adjectives indicating the diminutive.

Umbel, an inflorescence in which the branches arise from the same point like the ribs of an umbrella; adj., umbellate.

Umbellet, a secondary umbel.

Uncinate, hooked.

Undulate, wavy.

Unisexual, having only stamens or pistils.

Utricle, a small, thin-walled, 1-seeded fruit.

Valve, one of the pieces into which a dehiscent capsule splits; a segment of the perianth.

Venation, the manner of veining.

Ventral, pertaining to the inner side of a structure.

Ventricose, swollen or inflated on one side.

Vernation, the arrangement of leaves in a bud.

Verticillate, having a circular arrangement of similar parts around an axis.

Vestiture, the hairs clothing a structure.

Villous, bearing long weak hairs.

Virgate, wand-shaped; long, straight, and slender.

Viscid, having a sticky surface.

Viviparous, sprouting or germinating while attached to the parent plant.

Web, tangled or woolly hairs at the base of the lemmas in some species of *Poa.*

Whorl, the arrangement of similar parts in a circle around an axis; adj., whorled.

Wing, a membranaceous expansion attached to an organ; one of the two lateral petals in a pea flower.

Xerophilous, growing in arid places.

Index

SCIENTIFIC AND COMMON NAMES IN THE ENUMERATION OF THE FLORA

(Specific and varietal names are indexed only in genera with ten or more names. Synonyms are given in italics. No new scientific botanical names are published in MARIN FLORA.)

[307]

Abbotts Lagoon, 5-A
Almonte (4), 8-G
Angel Island, 9-H
Aurora School (29), 2-C
Bald Hill (39), 7-F
Bear Valley, 6-C to 7-C
Belvedere, 9-H
Big Carson Canyon (9), 7-E
Big Rock Ridge, 5-F to 6-F
Black Canyon (19), 7-G
Black Point, 5-G
Bolinas, 8-E
Bolinas Lagoon, 8-E
Bolinas Ridge, 6-D to 8-E
Burdell, 4-F
Camp Taylor (28), 6-D
Carson Ridge (7), 7-F
Chileno Valley, 3-D to 3-E
China Camp, 6-H
Corte Madera, 8-G
Deer Park (32), 7-F
Dillons Beach, 2-A
Drakes Bay, 6-B
Drakes Estero, 5-B to 6-B
Elk Valley, 9-G
Escalle (16), 8-G
Estero Americano, 1-A to 1-B
Fairfax, 7-F
Fairfax Hills (22), 7-F
Forest Knolls (26), 6-E
Frank Valley (6), 8-F
Gallinas Valley, 6-F to 6-G
Greenbrae (17), 7-G
Hamilton Field, 5-G
Hicks Valley, 4-D
Ignacio, 5-G
Inverness (38), 5-C
Inverness Ridge, 5-B to 6-C
Kentfield (15), 7-G
Lagunitas (27), 6-E
Lagunitas Creek, 5-C to 7-F
Larkspur (14), 8-G
Ledum Swamp (35), 5-B
Liberty Spring (13), 7-F
Lily Lake (30), 7-F
Little Carson Canyon (8), 7-E
Lucas Valley, 6-E
Manzanita (3), 8-G

Marshalls, 4-B
McClure Beach (36), 3-A
McKennans Landing (31), 8-E
McNears Landing, 6-H
Mill Valley, 8-G
Mount Tamalpais, 8-F
Mount Vision (34), 5-B
Mud Lake (33), 7-D
Muir Beach, 9-F
Muir Woods, 8-F
Nicasio, 5-E
Novato, 5-F
Olema, 6-D
Olema Marshes (21), 5-C
Papermill Creek, 5-D to 6-D
Pine Mt. (10), 7-E
Point Bonita (1), 9-G
Point Reyes, 7-A
Point Reyes Station, 5-C
Radio Station, 5-A
Richardson Bay, 8-G to 9-H
Ridgecrest Road (12), 7-F to 8-F
Rodeo Lagoon, 9-G
Ross, 7-G
Salmon Creek, 4-D
San Andreas Fault, 2-A to 8-E
San Anselmo, 7-F
San Anselmo Canyon (23), 7-F
San Antonio (20), 3-F
San Antonio Creek, 3-E to 3-F
San Geronimo (24), 6-E
San Geronimo Ridge (11), 6-E
San Quentin, 7-H
San Rafael, 7-G
San Rafael Hills (18), 6-G
Santa Venetia, 6-G
Sausalito, 9-H
Shell Beach (37), 4-B
Stinson Beach, 8-E
Tamalpais Valley (5), 8-G
Tennessee Cove, 9-G
Tiburon (2), 8-H
Tiburon Peninsula, 8-H
Tocaloma, 6-D
Tomales, 2-B
Tomales Bay, 3-A to 5-C
Tomales Point, 3-A
White Hill (25), 7-F

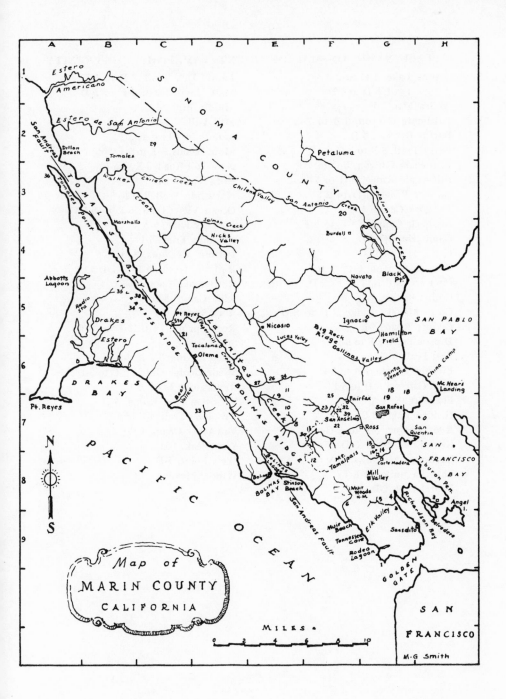

Map of
MARIN COUNTY
CALIFORNIA

PLACE NAMES ON MAP OF MOUNT TAMALPAIS AND VICINITY

Alpine Lake, 1-C to 2-A
Artura Trail, 3-D to 4-D
Azalea Flat, 5-E
Baltimore Canyon, 6-B to 7-C
Barths Retreat, 3-D
Berry Trail, 3-B to 3-C
Blithedale Canyon, 6-C to 7-E
Blithedale Ridge, 6-C to 7-D
Bolinas Ridge, 1-D to 2-E
Bootjack Camp, 4-E
Bootjack Canyon, 4-E to 5-E
Camp Handy, 3-B
Camp Hogan, 3-C
Cascade Canyon, 6-D to 6-E
Cataract Gulch, 1-C to 2-D
Collier Spring, 4-C
Corte Madera Ridge, 7-C to 8-D
County Road, 3-F
Dipsea Trail, 2-F to 5-F
East Peak, 5-C
Eldridge Grade, 4-A to 5-C
Fern Canyon, 5-D to 5-E
Fish Grade, 4-A
Hidden Lake, 3-C
Kent Ravine, 5-F to 5-G
Kent Trail, 2-A to 3-D
Lagunitas Meadows, 3-A
Lake Lagunitas, 4-B
Laurel Dell, 2-D

Laurel Dell Trail, 2-D to 3-E
Matt Davis Trail, 4-E to 5-D
Middle Peak, 4-D
Mill Valley, 7-E
Mountain Theater, 3-E
Muir Woods, 5-E to 5-F
New Toll Road, 3-E to 5-C
Nora Trail, 4-D
Northside Trail, 3-D to 5-C
Ocean View Trail, 5-E to 6-F
Phoenix Lake, 5-A
Pipeline Trail, 5-E
Potrero Meadows, 3-D
Railroad Grade, 4-D to 7-D
Rattlesnake Camp, 4-E
Rifle Camp, 3-D
Rock Spring, 3-E
Rock Spring Trail, 3-D to 4-D
Stage Road, 4-E
Steep Ravine, 3-F to 4-F
Swede George Gulch, 2-C to 3-D
Tenderfoot Trail, 6-E
Throckmorton Ridge, 5-D
Troop 80 Trail, 4-E to 5-E
Van Wyck Camp, 4-E
West Peak, 4-D
West Point, 4-D
Willow Meadow, 3-C

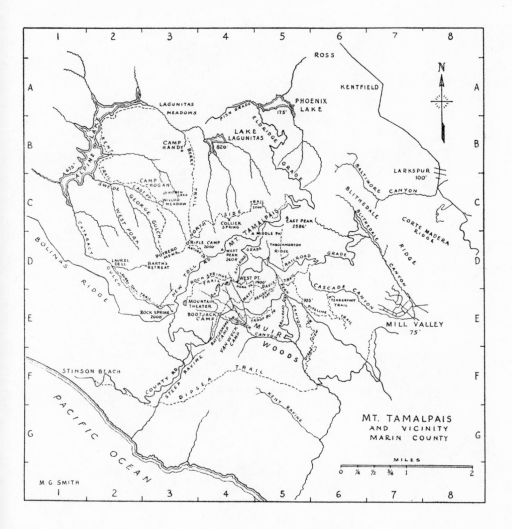

MARIN FLORA
Supplement (1969)
by John Thomas Howell

POLYPODIACEAE. Fern Family

2. Athyrium (p. 50)

1. **A. Filix-femina** (L.) Roth var. **cyclosorum** (Ledeb.) T. Moore. This is the correct varietal name for the common lady fern in Marin County (C. V. Morton, Amer. Fern Journ. 40: 241. 1950). The form with strongly revolute pinnules found in exposed places on Point Reyes Peninsula may be called var. *cyclosorum* fma. *strictum* (Gilbert) J. T. Howell.

3. Dryopteris (p. 50)

1. **D. arguta** (Kaulf.) Maxon. The author of this name has been corrected by C. V. Morton (Amer. Fern Journ. 58: 182. 1968).

2. **D. austriaca** (Jacq.) Woynar. This name for the spreading wood fern replaces *D. dilatata* (Hoffm.) Gray.

4. Polystichum (p. 50)

3. **P. Dudleyi** Maxon. The Dudley shield fern has a present southern distributional limit in the Santa Lucia Range in San Luis Obispo County *(Hoover 6672, 8807)*.

6. Polypodium (p. 51)

2a. **P. californicum** Kaulf. fma. **Parsonsiae** Morton. In typical *P. californicum* the frond segments are usually not even lobed, but in fma. *Parsonsiae* the segments are pinnately divided nearly to the midrib. This form was discovered by Mary Elizabeth Parsons in 1895 in a canyon near Kentfield (C. V. Morton, Amer. Fern Journ. 51: 75. 1961). A second form in which the segments are regularly and shallowly lobed has been found in Ross by Dr. F. W. Coe and is referable to fma. *Branscombii* Morton, originally described from Humboldt County.

9. Adiantum (p. 52)

3. **A.** × **Tracyi** C. C. Hall. *A. Jordanii* K. Müll. × *A. pedatum* L. var. *aleuticum* Rupr. This hybrid maidenhair, of occasional occurrence in the North Coast Ranges, was found once in Bear Valley, Marin County (Madroño 13: 196. 1956).

11. Cheilanthes (p. 52)

2a. **C. Carlotta-Halliae** Wagner & Gilbert. This is the fern described on page 53 of Marin Flora as a form of *C. siliquosa* Maxon in which the marginal sori are not continuous. In Marin County it is known from areas of serpentine on Tiburon Peninsula and Mount Tamalpais (Bootjack is the type locality). This attractive lace fern, which ranges south to San Luis Obispo County, probably originated as a hybrid between *C. siliquosa* and *C. californica* (Hook.) Mett.—*Aspidotis Carlotta-Halliae* (Wagner & Gilbert) Lellinger; *C. siliquosa* fma. *Carlotta-Halliae* (Wagner & Gilbert) Hoover.

5. **C. Covillei** Maxon. Crevices of rocks between Bootjack and Rock Spring, Mount Tamalpais, *Howell 33629, 34598*. The closely related *C. intertexta* (Maxon) Maxon also grows in the same general area but may be distinguished by long-ciliate scales on the lower side of the fronds.

ISOETACEAE. Quillwort Family
Isoetes (p. 53)

3. **I. Bolanderi** Engelm. Submerged or emergent aquatic in Lake Lagunitas, *H. E. Parks in 1925, Howell in 1926*. The identification of this plant which occurs here

so far below its usual montane habitat has been confirmed by both T. C. Palmer and C. F. Reed in the Herbarium of the University of California. In MARIN FLORA, this occasional quillwort of the Tamalpais ponds and reservoirs was considered a more aquatic aspect of *I. Howellii* Engelm., which on further study it may prove to be.

EQUISETACEAE. HORSETAIL FAMILY
Equisetum (p. 54)

3. **E. laevigatum** A. Br. This is the accepted name for the scouring rush formerly called *E. kansanum* Schaffn.

4. **E. hyemale** L. var. **affine** (Engelm.) A. A. Eat. According to specimens annotated by R. L. Hauke, this is the correct name for the giant scouring rush with perennial stems in Marin County that was called var. *californicum* Milde.

5. **E. × Ferrissii** Clute em. Hauke. *E. hyemale* var. *affine* × *E. laevigatum*. The coarse scouring rushes from Point Reyes Peninsula and Bolinas that were called *E. hyemale* var. *robustum* (A. Br.) A. A. Eat. have been referred by Hauke to this hybrid.

TAXACEAE. YEW FAMILY
2. Taxus (p. 56)

1. **T. brevifolia** Nutt. In 1958 a colony of the western yew was discovered just north of Samuel P. Taylor State Park by Robert H. Menzies and James Roof. The only other stations known in Marin County are on Inverness Ridge near Bear Valley.

TAXODIACEAE. REDWOOD FAMILY
Sequoia (p. 56)

1. **S. sempervirens** (Lamb.) Endl. A long-overlooked grove of second-growth redwoods on Pine Gulch Creek west of the Bolinas-Olema road has been rediscovered and described by the geologist Alan J. Galloway (Leafl. West. Bot. 9:66, 67. 1959). Although these redwoods may not occur west of the broad San Andreas fault zone, they do occur within the zone and are growing "on the younger sediments that are found west of the fault, not on Franciscan rocks that are found east of the fault." In MARIN FLORA the relation of redwood distribution to the geologic history of the region and its rocks is referred to on pages 7 and 56.

PINACEAE. PINE FAMILY
1. Pinus (p. 57)

a. Leaves ("needles") 2 in a bundle...b
a. Leaves 3 or 5 in a bundle..c
b. Cones strongly asymmetrical, the scales on one side of the cone prominently
    thickened.............................. ....1. *P. muricata*
b. Cones symmetrical, all of the scales alike and thin...........................
    ....1a. *P. muricata* fma. *remorata*
c. Needles 5 in a bundle. Cones symmetrical or nearly so..........4. *P. Torreyana*
c. Needles 3 in a bundle ....d
d. Cones symmetrical or nearly so, very large and heavy, the scales produced into
    a sharp hook..5. *P. Coulteri*
d. Cones asymmetrical, the scales not produced into a sharp hook..............e
e. Cones broadly ovoid to globose.................................2. *P. radiata*
e. Cones more slender, elongate-ovate or oblong.................3. *P. attenuata*

1. **P. muricata** D. Don. The bishop pine has been planted and has become naturalized on the south side of Mount Tamalpais near forestation plantations: at Double Bow Knot on old railroad grade and along fire-break below West Point. The new generation of trees is cone-bearing. The natural occurrence of the tree in Marin County is outlined on page 57.

1a. **P. muricata** fma. **remorata** (H. L. Mason) Hoover. A single tree with the symmetrical cone of this pine was found by Anne Leary and Barbara Sherfey along the road to McClure Beach in 1960. Apparently this tree has been cut down, the only known Marin County specimen of this form which is more usual southward, especially on Santa Cruz Island—hence the common name, Santa Cruz Island pine.—*P. remorata* H. L. Mason.

2. **P. radiata** D. Don. The Monterey pine is commonly reproducing itself in Marin County in the vicinity of mature planted trees: Sausalito; Tiburon; Mount Tamalpais (Double Bow Knot, West Point); Stinson Beach; Marshall. No cone-bearing seedlings have been seen.

3. **P. attenuata** Lemmon. The knobcone pines planted on the north side of Mount Tamalpais in the chaparral between Laurel Dell and Potrero Meadow were killed in the fire of 1945 (cf. p. 22) but a new cone-bearing generation now thrives there.

4. **P. Torreyana** Parry. The Torrey pine, "the rarest and most restricted in its distribution of all American pines," is commonly spontaneous in the vicinity of planted trees in Stinson Beach. Seedlings two or three years old were collected by Arthur Menzies in February, 1969.

5. **P. Coulteri** D. Don. The Coulter or big cone pine is occasionally spontaneous near trees planted in the chaparral on Mount Tamalpais. A notably large cone of this species was found under planted trees near Lake Lagunitas by Gregory Resnick (*No. 101*) in 1968. It measures 40 cm. (16 inches) in length, 2 to 4 cm. more than the maximum length usually given.

2. Pseudotsuga (p. 57)

1. **P. Menziesii** (Mirbel) Franco. The brilliant bibliographic research by the Portuguese botanist João do Amaral Franco has shown this to be the correct name of the Douglas fir.

CUPRESSACEAE. CYPRESS FAMILY

Cupressus (p. 57)

2. **C. macrocarpa** Hartw. Monterey cypress, like the Monterey pine, is commonly reproducing itself near cultivated trees, as in Sausalito, near Marshall, and elsewhere. It may be distinguished from the indigenous Sargent cypress (*C. Sargentii* Jeps.) by its greener leaves without dorsal glandular pits and by its smooth seeds without a glaucous cast.

POTAMOGETONACEAE. PONDWEED FAMILY

1. Potamogeton (p. 59)

3. **P. pusillus** L. This is the name now accepted for the small pondweed identified as *P. panormitanus* Biv.

ALISMATACEAE. WATER-PLANTAIN FAMILY
1. Sagittaria (p. 61)

2. **S. cuneata** Sheldon. Plants of this arrowhead from Chileno Valley have been redetermined by Peter Rubtzoff (Leafl. West. Bot. 9: 77, 78. 1960). In it the terminal lobe of the 3-lobed leaves is much larger than the basal lobes, while in *S. latifolia* Willd. all the lobes are about equal.

2. Alisma (p. 61)

2. **A. lanceolatum** Withering. This Old World water-plantain, which has been collected by Peter Rubtzoff at a number of stations in Sonoma County, was found by Lilian McHoul in 1965 along Corte Madera Creek at the College of Marin in Kentfield. Its petals are decidedly pink, whereas those of the common *A. triviale* Pursh (the presently accepted name of the American water-plantain) are white, rarely faintly pinkish.

4. Echinodorus

1. **E. Berteroi** (Sprengel) Fassett. Clumps of this burhead were found in the shallow water of a small reservoir between Salmon Creek and Chileno Valley. The genus resembles *Sagittaria* (p. 61) in the spiral arrangement of the carpels, but in *Echinodorus* all the flowers in the inflorescence are perfect whereas in *Sagittaria* the upper flowers are staminate.—*E. cordifolius* of authors.

HYDROCHARITACEAE. FROGBIT FAMILY
Elodea (as *Anacharis*, p. 61)

1. **E. canadensis** Michx. The proper generic name for the waterweeds is *Elodea* not *Anacharis*.

GRAMINEAE. GRASS FAMILY
1. Bromus (p. 65)

4a. **B. stamineus** E. Desv. This Chilean brome is now widespread in Marin County: Tiburon Peninsula, Mount Tamalpais (East Peak), Stinson Beach, Olema.

4b. **B. Willdenovii** Kunth. Commonly called rescue grass, this coarse South American brome occurs along the road in Frank Valley. Both this and *B. stamineus* may be distinguished from other perennial bromes in Marin County by the 9- to 13-nerved, strongly keeled lemmas which in *B. stamineus* are long-awned but in *B. Willdenovii* are nearly or quite awnless.—*B. catharticus* Vahl.

8. **B. mollis** L. fma. **leiostachys** (Hartman) Fernald. The grass called *B. racemosus* on page 67 of MARIN FLORA has been shown to be a glabrous form of soft chess by Peter Rubtzoff (Leafl. West. Bot. 9: 67, 68. 1959).

12. **B. diandrus** Roth. This is the now-accepted name of the ripgut grass in California, not *B. rigidus* Roth.

12a. **B. sterilis** L. This annual brome, which resembles *B. diandrus* but is somewhat smaller in its spikelets, has been found on the ridge between Mill Valley and Muir Woods and at Shell Beach near Inverness.

13a. **B. tectorum** L. var. **glabratus** Spenner. The form of cheat grass with glabrous spikelets has been found near Mountain Theater on Mount Tamalpais.

3. Festuca (p. 68)

1a. **F. arundinacea** Schreb. Because the alta fescue has been widely planted in pastures, it is to be expected in many places in meadows, along roads, and in waste ground. It has been found on Tiburon Peninsula, on Mount Tamalpais (near Bootjack), near Olema, and on Point Reyes Peninsula near Tomales Bay State Park.

1b. **F. pratensis** Huds. The meadow fescue has also been planted as a pasture grass but the only naturalized occurrence known in Marin County is near Olema (*David Morgan in 1956*).—*F. elatior* L. in part.

Both *F. arundinacea* and *F. pratensis* are introduced perennials from Europe, and though closely related, they may be distinguished by their lower leaf-sheaths which remain membranous in *F. arundinacea* but which tend to become fibrous in *F. pratensis*. These grasses resemble *F. californica* Vasey in their awnless or nearly awnless lemmas, but they do not have the handsome tufted habit of the native plant nor its villous-pubescent leaf-collar.

4. Catapodium (as *Scleropoa*, p. 70)

1. **C. rigidum** (L.) C. E. Hubbard. A sidewalk grass rare in Marin County should now be known by this name.

6. Glyceria (p. 70)

2, 3. **G. leptostachya** Buckl. and **G. occidentalis** (Piper) J. C. Nels. These two species of mannagrass, formerly believed to reach a southern distributional limit in Marin County, have been reported from the San Francisco Peninsula, the former from San Francisco, the latter from San Mateo County.

Southern distributional records for *Glyceria elata* (Nash) Hitchc. by Rubtzoff (Leafl. West. Bot. 9: 165, 166. 1961) are from Lake and Sonoma counties, not "Sonoma and Marin" counties as given in Munz Supplement to A California Flora (p. 187).

9a. Eragrostis

1. **E. diffusa** Buckl. Although the diffuse lovegrass is native in parts of California to the south, in Marin County it has been seen only as a garden or roadside weed in Ross and near the town of Point Reyes. It is an annual grass that may be distinguished from the annual species of *Poa* (p. 71) by its 3-nerved glabrous lemmas, as well as by its distinctive appearance. Because the glumes are papery in this lovegrass, it may be keyed out (p. 62) to *Melica,* but all melicas are perennial and their lemmas (as in *Poa*) have more than 3 nerves.

12a. Cortaderia

1. **C. rudiuscula** Stapf. The noxiously invasive weedy pampasgrass of coastal California would appear to be this rosy-plumed Argentinian species rather than *C. Selloana* (Schult.) Aschers. & Graebn., the cultivated ornamental which may not become naturalized. In Marin County, *C. rudiuscula* is weedy on Tiburon Peninsula, in San Rafael Hills, at Stinson Beach, and perhaps elsewhere. It may be distinguished by its tussocky habit and plume-like inflorescences.

13. Melica (p. 73)

3. **M. Geyeri** Munro var. **aristulata** J. T. Howell. This form of *M. Geyeri* was reported only from Marin County until it was collected by Beecher Crampton (*No. 1964*) in Tehama County west of Paskenta. It is to be expected elsewhere in the North Coast Ranges.

15a. Aegilops

1. **A. triuncialis** L. Goatgrass was found near the summit of Mount Tamalpais in 1953 but has not been seen since. A bad weed on rangelands, it differs from *Triticum* (p. 74) in its strongly nerved 3-awned lower glume which in *Triticum* is scarcely nerved and either awnless or 1-awned.

17. Elymus (p. 74)

4. **E. × vancouverensis** Vasey emend. Bowden. The parents of this coastal wild-rye are *E. mollis* Trin. and *E. triticoides* Buckl.

5a. **E. pacificus** Gould. The rare maritime dune grass, *Agropyron arenicola* Davy, originally described from Point Reyes Peninsula, is now referred to the genus *Elymus*. It may be distinguished from the other wild-ryes in the uniformly solitary spikelet at each node of the rachis.

6. **E. Caput-Medusae** L. Medusa-head grass, a bad weed on rangelands, was found in Chileno Valley in 1959. It is a barley-like annual with long-awned lemmas.

20. Hordeum (p. 76)

4. **H. geniculatum** All. This name replaces *H. Hystrix* Roth.

6. **H. glaucum** Steud. This name replaces *H. Stebbinsii* Covas.

25. Koeleria (p. 78)

1. **K. macrantha** (Ledeb.) Schultes. According to Werner Greuter (Candollea 23: 85. 1968), this is the correct designation for the junegrass that is so widespread in the Northern Hemisphere. The plant reported from Tiburon Peninsula as *Poa Douglasii* Nees (p. 71) was a misidentified junegrass.

25a. Schismus

1. **S. barbatus** (L.) Thell. This Old World annual grass that is now widespread in the drier parts of southern California occurred as a waif in San Rafael (*True 4093*). In Marin County the genus *Schismus* with about five flowers in each spikelet is readily separable from *Aira* (p. 79) with only two flowers in each spikelet.

28. Aira (p. 79)

1. **A. elegans** Willd. This name replaces *A. capillaris* Host.

29. Avena (p. 79)

2a. **A. fatua** L. var. **glabrata** Peterm. Fallow field on Point Reyes Peninsula near Bolinas (*True in 1969*). This smooth oat differs from typical *A. fatua* in its glabrous or subglabrous lemma and from *A. sativa* L., the cultivated oat, in its bent and twisted awns.

33. Calamagrostis (p. 80)

4. **C. ophitidis** (J. T. Howell) Nygren. The serpentine reedgrass is now regarded as a species quite distinct from *C. purpurascens* R. Br.

34. Ammophila (p. 81)

2. **A. breviligulata** Fernald. The American beachgrass was discovered by Douglas Ripley (*No. 947*) on dunes at Swimmers Beach on the southeast side of Angel Island. Native in eastern North America, it may be differentiated from the more

commonly naturalized European beachgrass, *A. arenaria* (L.) Link, by its short (1–3 mm. *versus* 10–30 mm.) ligules.

35. Agrostis (p. 81)

1. **A. avenacea** Gmel. This name replaces *A. retrofracta* Willd.

4. **A. stolonifera** L. var. **major** (Gaud.) Farwell. For the widespread redtop, this name replaces *A. alba* L.

9a. **A. Blasdalei** Hitchc. var. **marinensis** Crampton. This form with florets somewhat larger than in the typical form was described from rocks "1.25 mi. east of Dillon Beach."

12. **A. densiflora** Vasey. Kenton Chambers has shown that this name should replace *A. californica* Trin. (Madroño 18: 251. 1966).

12a. **A. clivicola** Crampton var. **punta-reyesensis** Crampton. This recently recognized bentgrass from Point Reyes Peninsula and Dillon Beach may be distinguished from *A. densiflora* by its somewhat larger florets.

The following synopsis of the several tufted maritime bentgrasses is adapted from Crampton's treatment (Brittonia 19: 174–177. 1967):

a. Leaf-blades involute, filiform; callus-hairs lacking; anthers 1–2 mm. long.....b

a. Leaf-blades flat; callus-hairs present; anthers 0.5–1 mm. long...............c

b. Lemmas 2 mm. long or less; Marin County to Mendocino County
....9. *A. Blasdalei* var. *Blasdalei*

b. Lemmas 2.5–3 mm. long; near Dillon Beach....9a. *A. Blasdalei* var. *marinensis*

c. Lemmas 2 mm. long or less; south to San Luis Obispo County, north to "about Lincoln County, Oregon" ....12. *A. densiflora*

c. Lemmas 2.5–3 mm. long ....d

d. Culms erect; Marin and Sonoma counties..12a. *A. clivicola* var. *punta-reyesensis*

d. Culms spreading or prostrate; Sonoma and Mendocino counties
....12b. *A. clivicola* var. *clivicola*

36. Alopecurus (p. 83)

1. **A. aequalis** Sobol. var. **sonomensis** Rubtzoff. This name replaces *A. geniculatus* for the Marin County plant. (Cf. Leafl. West. Bot. 9: 170–172. 1961.)

37. Polypogon (p. 83)

1a. **P. australis** Brogn. The rather widespread occurrence of this South American grass in California has been described by Peter Rubtzoff (Leafl. West. Bot. 9: 166–169. 1961). In Marin County it has been found only at Inverness (*Morgan, Howell*). It is closely related to *P. interruptus* H.B.K. but may be distinguished by its shorter ligule (about as long as wide or shorter in *P. australis* and to twice as long as wide in *P. interruptus*). Both species are natives of the New World, although on page 84 *P. interruptus* is described as an introduction "from Europe."

38. Phleum (p. 84)

1. **P. alpinum** L. As long ago as 1892 this timothy was reported from San Francisco, a record overlooked when it was stated that the species reached a southern coastal distributional limit in Marin County.

40a. Oryzopsis

1. **O. miliacea** (L.) Benth. & Hook. Commonly called smilo grass, this widely distributed Mediterranean species was found in a weedy area near Novato in 1968 by Peter Rubtzoff (*No. 5970*). From *Agrostis* (p. 81), *Oryzopsis* differs in its indurate lemma and from *Stipa* (p. 84) it differs in the deciduous awn on its lemma.

43. Beckmannia (p. 85)

1. **B. syzigachne** (Steud.) Fernald. The American sloughgrass has been collected in the Santa Cruz Mountains.

46a. Ehrharta

1. **E. calycina** Smith. This attractive grass, introduced for forage from South Africa, was collected by David Morgan in 1960 near McClure Ranch on Point Reyes Peninsula. *Ehrharta* differs from *Hierochloe* (p. 85) in its sterile lower florets and from *Anthoxanthum* (p. 85) in its open panicle and awnless lemmas.

48. Digitaria (p. 86)

2. **D. Ischaemum** (Schreb.) Schreb. Smooth crabgrass has been found as a garden weed in San Rafael by Gordon True. It may be distinguished from *D. sanguinalis* (L.) Scop. by its hairless leaf sheaths.

53. Setaria (p. 87)

2. **S. lutescens** (Weigel) F. T. Hubbard. Yellow bristlegrass has been found as a roadside and garden weed on Mount Tamalpais, in San Rafael, and elsewhere in the county.—*S. glauca* of authors.

3. **S. viridis** (L.) Beauv. Green bristlegrass grew with yellow bristlegrass as a garden weed in San Rafael (*Howell 30110*). It has also been found in waste ground at head of Bolinas Lagoon. Both grasses are native to Europe.

4. **S. Faberi** W. Herrmann. In Marin County found only by David Morgan on the Redwood Highway at the Sonoma County line. It is a native of China.

The following key will distinguish the bristlegrasses in Marin County:

a. Plants perennial . . . . 1. *S. geniculata*
a. Plants annual . . . . b
b. Bristles below each spikelet more than 5 . . . . 2. *S. lutescens*
b. Bristles below each spikelet 1–3 c
c. Blades glabrous; fertile lemma finely papillose . . . . 3. *S. viridis*
c. Blades usually hairy; fertile lemma finely rugulose . . . . 4. *S. Faberi*

53a. Pennisetum

1. **P. setaceum** (Forsk.) Chiov. Fountain grass, a beautiful garden ornament from Africa, occasionally escapes to roadsides and waste ground, as in San Rafael and on Tiburon Peninsula.

2. **P. clandestinum** Hochst. Kikuyu grass, a lawn grass or ground cover (that should never be planted because it can become a smothering weed) has escaped in San Rafael and Tiburon. A native of Africa.

In *Setaria* (p. 65), the bristles remain attached to the inflorescence; in *Pennisetum* the bristles are shed with the spikelets. *Pennisetum setaceum* is a tufted plant 1 to 3 feet tall with showy inflorescences; *P. clandestinum* is low-growing with creeping stems and with inconspicuous inflorescences partly enclosed by leaf sheaths.

55. Sorghum (p. 87)

2. **S. bicolor** (L.) Moench. This name replaces *S. vulgare* Pers.

CYPERACEAE. SEDGE FAMILY
2. Eleocharis (p. 88)

1. **E. quinqueflora** (F. X. Hartmann) O. Schwarz var. **Suksdorfiana** (Beauverd) J. T. Howell. This name replaces *E. pauciflora* (Lightf.) Link for the spike-rush that is rare in swales on Point Reyes Peninsula. In spite of the specific epithet, the spikes in the variety may have as many as 12 flowers.

3a. **E. Parishii** Britt. Parish's slender-stemmed spike-rush, not before reported from Marin County, is not uncommon in the perennial seepages on the slopes below Old St. Hilary's Church in Tiburon (*Howell 41409*). It differs from *E. acicularis* (L.) R. & S. in its taller culms and its smooth or finely reticulated (not striate) achenes.

In his Supplement to A California Flora (p. 183), Dr. Munz cites Marin County for both *Eleocharis acicularis* (L.) R. & S. var. *radicans* (Poir.) Britt. and *E. parvula* (R. & S.) Link. Although Marin County is within the range of these spike-rushes, no specimens from the county have been seen.

3. Scirpus (p. 89)

5. **S. californicus** (C. A. Mey.) Steud. The northern distributional limit of the California tule has been extended to Napa and Sonoma counties.

5. Carex (p. 91)

5, 26, and 35. **C. vicaria** Bailey, **C. lanuginosa** Michx., and **C. rostrata** Stokes. These species, which were reported as reaching a southern distributional limit in Marin County, have been found in the Santa Cruz Mountains, and *C. lanuginosa* occurs also in San Luis Obispo County and on Mount Diablo.

10. **C. montereyensis** Mkze. This sedge reaches a northern distributional limit in Sonoma County.

23. **C. mendocinensis** Olney. The names *C. debiliformis* Mkze. and *C. mendocinensis* Olney have been shown to apply to the same sedge (Leafl. West. Bot. 6: 157–161. 1951). It is now known from Monterey and San Luis Obispo counties where it was found in groves of *Cupressus Sargentii* by Clare B. Hardham; and in Marin County its range may be extended to the seepages below Old St. Hilary's. The plant in MARIN FLORA called *C. mendocinensis* (p. 95) is the rare hybrid, *C. gynodynama* Olney × *C. mendocinensis*.

25. **C. luzulina** Olney. An outlying occurrence of this sedge is in the cypress grove on Cypress Mountain, San Luis Obispo County, where it was found by Mrs. Hardham (*No. 4770, 5660*).

LEMNACEAE. DUCKWEED FAMILY
3. Wolffiella

1. **W. lingulata** (Hegelm.) Hegelm. This interesting duckweed has been collected in a pond about 2 miles south of Olema, *Smith & Chisaki 3132*. It may be

distinguished from the rhizophorous duckweeds, *Spirodela* and *Lemna* (p. 97), by its oblong rootless fronds.

LILIACEAE. Lily Family

5. Allium (p. 103)

1*a*. **A. Ampeloprasum** L. The wild leek from the Old World with its 2-parted bulb was found as a roadside escape in Tiburon by Javier Peñalosa (*No. 1951*). It differs from all other onions in Marin County in its robust habit (over 3 feet tall) and its dense umbel of very numerous pink flowers.

6*a*. **A. neapolitanum** Cyr. This attractive European onion, that becomes weedy in gardens and naturalized in fields, grew spontaneously in a garden in Larkspur (*Blanche Clear in 1958*) and in a vacant lot in San Rafael (*Howell in 1969*). It differs from *A. triquetrum* in its subterete scape and in its obtuse perianth-segments that become thin-papery in age.

6. Muilla (p. 103)

1. **M. maritima** (Torr.) Wats. This little lily, so rare in Marin County, is found northward through the Coast Ranges at least to Glenn County.

6*a*. Nothoscordum

1. **N. inodorum** (Ait.) Nicholson. This garden plant, that can become established both by seeds and bulb-offsets, occurred as a weed in a nursery near Mill Valley (*Garvey in 1961*). *Nothoscordum* differs from *Allium* and *Muilla* (p. 103) in its short but definite perianth-tube and from *Brodiaea* (p. 108) in its unjointed flower stalks. According to K. Krause, it is a native of subtropical America.

8. Lilium (p. 105)

2. **L. occidentale** Purdy. According to Peter Rubtzoff (Wasmann Journ. Biol. 11: 165. 1953), the lily at Ledum Swamp on Point Reyes Peninsula is the western lily, which here reaches its southernmost distributional limit.

15. Maianthemum (p. 107)

1. **M. kamtschaticum** (Cham.) Nakai. The Pacific May-lily, that has a distribution through the lands bordering the northern Pacific Ocean from Japan to California, is to be known by this name instead of *M. dilatatum* (Wood) Nels. & Macbr. (Cf. Ingram, Baileya 14: 54–59. 1966. Ingram's common name, Oregon coltsfoot, must be in error since coltsfoot applies to *Petasites* in the Pacific States or to the related *Tussilago* elsewhere.) In California, Pacific May-lily reaches its southernmost stations in San Mateo County, not in Sausalito.

IRIDACEAE. Iris Family

2. Iris (p. 108)

1. **I. longipetala** Herbert. This iris has been found by Peter Rubtzoff at two stations in Sonoma County, the presently known northern distributional limit.

2. **I. Douglasiana** Herbert. According to Dr. L. W. Lenz (Aliso 4: 1–72. 1958), two hybrids with the Douglas iris as one parent occur in Marin County: *I. Douglasiana* × *I. Fernaldii* Foster (Corte Madera Ridge; southeast of Fairfax), and *I. Douglasiana* × *I. macrosiphon* Torr. (Inverness Ridge). The first is particularly

interesting because one of the parent-suspects, *I. Fernaldii* with a geographic distribution from Lake County to Santa Cruz County, has never been reported from Marin County.

3. Chasmanthe

1. **C. aethiopica** (L.) N. E. Br. Perhaps originating from garden refuse but persisting adventitiously, this South African plant has been found along the Stinson Beach road, in Tiburon near Old St. Hilary's, in Mill Valley, and in San Rafael. It may be distinguished from *Iris* and *Sisyrinchium* (p. 108) by its *Gladiolus*-like habit and flowers. Its perianth is decidedly oblique with the uppermost perianth-segment longest and somewhat hooded.

4. Crocosmia

1. **C. × crocosmiflora** (Lemoine) N. E. Br. *C. aurea* Planch. × *C. Pottsii* (Baker) N. E. Br. This garden plant of hybrid origin commonly known as montbretia persists along trails on the south side of Mount Tamalpais where it may have been planted by an over-enthusiastic nature lover who would improve upon the flora. It is often relictual in old gardens and may become established from garden refuse. *Crocosmia* resembles *Chasmanthe* but the two may be separated by the lower part of the perianth tube which is tubular in *Crocosmia* but filiform-constricted in *Chasmanthe*.

ORCHIDACEAE. ORCHID FAMILY
1. Calypso (p. 109)

1. **C. bulbosa** (L.) Oakes. In 1963, this orchid was discovered in the Santa Cruz Mountains near Big Basin Redwoods State Park, a range extension of about 50 miles southward from Marin County.

3. Epipactis (p. 109)

2. **E. Helleborine** (L.) Crantz. The European helleborine is weedily spontaneous in Marin gardens where it was never planted: Mill Valley (*Helen Frank in 1956. Frances Fullerton in 1961*); San Rafael (*Gordon True in 1967*).

6. Goodyera (p. 110)

1. **G. oblongifolia** Raf. Rattlesnake plantain has been detected as a rare plant in the Santa Cruz Mountains where it was found near Boulder Creek by Vesta Hesse in 1950.

7. Corallorhiza (p. 111)

2a. **C. maculata** Raf. fma. **immaculata** (Peck) J. T. Howell. The occasional form of spotted coralroot in which the lip is unspotted has been found by Barbara Menzies and Caroline Ramberg near the foot of Cataract Gulch; and Dr. D. G. Howard has reported it from the Ben Johnson Trail and the Fairfax Hills.

FAGACEAE. OAK FAMILY
3. Quercus (p. 114)

1a. **Q. × Chasei** McMinn, Babcock, & Righter. This name has been given to the hybrid between *Q. agrifolia* Née and *Q. Kelloggii* Newb. (p. 114).

8a. **Q. × subconvexa** Tucker. This is the hybrid between *Q. durata* Jeps. and *Q. Garryana* Dougl. described on page 115. According to Dr. Tucker (Madroño 12: 119, 120. 1953) it is known from Marin and Santa Clara counties.

8b. **Q.** × **Howellii** Tucker. The hybrid between *Q. dumosa* Nutt. and *Q. Garry-ana* (p. 116) appears to be known only from Fish Grade on the north side of Mount Tamalpais (Tucker, loc. cit., p. 125, 126).

URTICACEAE. Nettle Family
3. Soleirolia

1. **S. Soleirolii** (Req.) Dandy. Baby's tears, a groundcover that can become persistently and annoyingly weedy in moist shady gardens, has been seen on a moist bank in Fern Canyon on Mount Tamalpais far from any habitation. This Mediterranean plant differs from the other nettles in Marin County (pp. 116, 117) in its creeping rooting stems and alternate innocuously hairy leaves.—*Helxine Soleirolii* Req.

CANNABACEAE. Hemp Family
Cannabis

1. **C. sativa** L. Hemp or marijuana has at least twice been found growing casually in Marin County: in 1919 on Tiburon Peninsula and in 1962 near Mill Valley. Among herbaceous apetalous dicotyledons in Marin County (p. 40), it is most closely related to the nettles (*Urticaceae*), but it lacks stinging hairs and is readily distinguished by its alternate digitately compound leaves. It is native in southeastern Russia.

POLYGONACEAE. Buckwheat Family
1. Polygonum (p. 118)

3a. **P. capitatum** Buch.-Ham. An attractive groundcover from the Himalaya Mountains adventitiously inclined, has been observed as an escape in San Rafael and Mill Valley. From all knotweeds and smartweeds in Marin County it may be distinguished by its creeping habit and bright pink flowers in small subglobose heads that are usually solitary or paired.

10. **P. marinense** Mertens & Raven. This name replaces *P. Fowleri* Robins. (p. 120). Fowler's knotweed approaches California no nearer than the Puget Sound region in Washington, while the Marin County plant is a localized endemic with type locality at the head of Drake's Estero on Point Reyes Peninsula.

Mertens and Raven (Madroño 18: 89–91. 1965), following European botanists, recognize two species as constituting the variable dooryard knotweeds of Marin County:

12. **P. aviculare** L. A rare plant occurring on the edge of salt marshes at Almonte (between Marin City and Mill Valley) and at Greenbrae.

13. **P. arenastrum** Jord. The dooryard knotweed common throughout the county, occurring in hard ground about dwellings, along roads, and in waste places. In this species the perianth parts are united about half their length; in *P. aviculare* they are united only at the base.

2. Rumex (p. 120)

2b. **R. salicifolius** Weinm. fma. **ecallosus** J. T. Howell. According to Dr. K. Rechinger this aspect of the variable *R. salicifolius* complex is referable to his species *R. californicus*.

CHENOPODIACEAE. Goosefoot Family
3. Chenopodium (p. 124)

5. **C. macrospermum** Hook. f. var. **farinosum** (Wats.) J. T. Howell. The northern distributional limit is in Humboldt County, not in Marin County.

7. **C. album** L. Lambs-quarters is not common in Marin County, being known only from Tamalpais Valley and Ross. The other plants that were referred to it have been redetermined by Dr. Herbert A. Wahl as the two following.

7a. **C. Berlandieri** Moq. var. **Zschackei** (Murr) Murr. In northern Marin County on San Antonio Creek, but to be expected elsewhere in waste ground and on roadsides.

7b. **C. strictum** Roth var. **glaucophyllum** (Aellen) H. A. Wahl. Weedy in San Rafael and probably elsewhere.

The three species are closely related but may be distinguished by this key:

a. Seeds slightly roughened, foveolate-reticulate...............7a. *C. Berlandieri*
a. Seeds smooth or nearly so ....b
b. Fruit closely covered by the calyx lobes.........................7. *C. album*
b. Fruit exposed by the spreading calyx lobes....................7b. *C. strictum*

4. Atriplex (p. 125)

1b. **A. rosea** L. Locally common in waste ground bordering the estuary of San Rafael Creek in San Rafael. From *A. patula* L. and its var. *hastata* (L.) Gray, the only other annual kinds of *Atriplex* in Marin County, *A. rosea* may be readily separated by its permanently white-scurfy foliage. A native of the Old World, it is widespread and common southward in California.

AMARANTHACEAE. Amaranth Family
Amaranthus (p. 126)

2. **A. hybridus** L. Rare in Marin County (Fairfax).

2a. **A. Powellii** Wats. Occasional (Mill Valley, Novato, Bolinas Lagoon, Inverness). In *A. hybridus* the seeds are nearly round and scarcely 1 mm. in diameter; in *A. Powellii* the seeds are elliptic and 1.1 to 1.3 mm. long.

4. **A. blitoides** Wats. The indigenous North American plant is now recognized as distinct from the European *A. graecizans* L.

NYCTAGINACEAE. Four-o'clock Family
2. Mirabilis

1. **M. Jalapa** L. Marvel-of-Peru, an old-fashioned garden favorite, grew without cultivation in waste ground in Sausalito, San Rafael, and Tomales. The disposition of its showy flowers differs from that in *Abronia* (p. 127) in that there is only one flower to each involucre, not many in a head.

AIZOACEAE. Carpetweed Family
1a. Glinus

1. **G. lotoides** L. Abundant on the desiccated strand of Bon Tempe Reservoir in 1956. It is interesting that also present were two other annual carpetweeds, *Mollugo verticillata* L. and *Cypselea humifusa* Turp. (p. 127). The glinus, a native of Europe, was easily distinguished from both by its stellate-pubescent herbage.

3. Mesembryanthemum (p. 127)

3. M. floribundum Haw. Established in large patches on the Duxbury bluffs near Bolinas and perhaps elsewhere along the coast; also in Tiburon. This South African species is widely planted as a ground cover and may be recognized by its short cylindric leaves that can be almost smothered in a welter of pink bloom.

4. Tetragonia (p. 128)

1. T. tetragonioides (Pallas) Kuntze. This is the correct name for New Zealand spinach.

PORTULACACEAE. Purslane Family
3. Montia (p. 129)

5b. M. spathulata (Dougl.) Howell var. **rosulata** (Eastw.) J. T. Howell. This small miner's lettuce, that was believed to be one of Marin County's most localized endemics, has been found near Occidental, Sonoma County, according to R. C. Bacigalupi.

CARYOPHYLLACEAE. Pink Family
5a. Lychnis

1. L. Coronaria (L.) Desr. The European mulleinpink (also one of the diverse types called "dusty miller") grew adventitiously on the old railroad grade near Camp Alice Eastwood on Mount Tamalpais. It is a white-woolly plant and as such is readily separated from species of *Silene* (p. 133) with which it might be confused.

9a. Scleranthus

1. S. annuus L. Knawel, a native of Europe, has been detected as an inconspicuous annual on a serpentine flat near Forest Knolls by Beecher Crampton (*No. 4205*). It is related to *Paronychia* (p. 135) but its leaves are exstipulate and its calyx lobes not bristle-tipped.

PAPAVERACEAE. Poppy Family
4a. Stylomecon

1. S. heterophylla (Benth.) G. Taylor. The discovery of wind poppy on Angel Island by Douglas Ripley (*No. 1068*) adds this beautiful wild flower to the flora of Marin County. It resembles the native *Papaver* (p. 141) but differs essentially in its short slender style that bears a globose stigma.

CRUCIFERAE. Mustard Family
4. Brassica (p. 144)

1a. B. oleracea L. var. **acephala** DC. Common Kale or cow cabbage grew in "a small but vigorous colony" on Tiburon Peninsula according to Javier Peñalosa (Wasmann Journ. Biol. 21: 40. 1963), undoubtedly a fugitive from cultivation.

3a. B. juncea (L.) Coss. Indian mustard, a native of the Old World, was collected once in Blithedale Canyon, Mill Valley (*Howell 26619*). It differs from *B. Kaber* in its angled pods with membranous walls that are not constricted between seeds.

4a. Diplotaxis

1. D. muralis (L.) DC. Sand-rocket has been found as a garden weed in San Rafael (*True 4201*). It is an annual resembling *Brassica* (p. 144) but it differs in its beakless pod in which the small elliptical seeds are arranged in two rows.

5. Raphanus (p. 145)

1. **R. sativus** L. C. A. Panetsos and H. G. Baker (Genetica 38: 243–274. 1968) have shown that the displays of wild radish are not the simple result of the garden radish escaping from cultivation but rather the result of introgressive hybridization of the purple-flower garden plant (*R. sativus* L.) with the yellow-flower jointed charlock (*R. Raphanistrum* L.). The extensive and colorful floral effects of radish in the Coast Ranges confirm the conclusion of these workers that "introgression of *R. Raphanistrum* characters appears to have been a major factor in converting the erstwhile crop plant *R. sativus* into a highly successful weed in California."

10. Cardamine (p. 147)

2. **C. pensylvanica** Muhl. The Pennsylvania bitter-cress was found by Peter Raven as a streamside casual at Stinson Beach (*No. 9954*). A widespread North American plant that is rare in California, this bitter-cress differs from the relatively common *C. oligosperma* in its narrow almost filiform pods (0.5 mm. *versus* 1–1.5 mm.).

15. Lepidium (p. 148)

2. **L. strictum** (Wats.) Rattan. This name replaces *L. pubescens* Desv. which was originally applied by A. N. Desvaux to a different kind of plant.

3a. **L. latifolium** L. The tall stems of this European pepper-grass formed an herbaceous thicket in the old railroad yards in Tiburon, where it was discovered by Lilian McHoul in 1965. It is the only perennial pepper-grass in Marin County.

20. Thysanocarpus (p. 149)

3. **T. laciniatus** Nutt. var. **crenatus** Brew. With the discovery of this more austral fringe-pod in Tehama County (*Wagnon 12663* in 1953), Marin County stations are no longer the northernmost.

CRASSULACEAE. STONE-CROP FAMILY
2. Sedum (p. 150)

1a. **S. album** L. A garden plant with a propensity to become spontaneous, this Old World stone-crop grew weedily in Bolinas (*Howell 34605*).

3. Echeveria (p. 151)

1. **E. cymosa** Lemaire. This would seem to be the proper name for the Marin rock-lettuce, not *E. laxa* Lindl.—*Dudleya cymosa* (Lemaire) Britt. & Rose.

2. **E. farinosa** Lindl. The Marin sea-lettuce belongs to this variable species that is found as far north as Oregon. *Echeveria caespitosa* (Haw.) DC is not reported north of Monterey County.—*Dudleya farinosa* (Lindl.) Britt. & Rose.

ROSACEAE. ROSE FAMILY

At least three genera of woody plants with pomaceous fruits have escaped from cultivation in Marin County. These and the indigenous pomaceous plants may be distinguished by the following key.

a. Leaves evergreen (or half-evergreen in *Cotoneaster*)..........................b
a. Leaves deciduous...d
b. Stems more or less spinose; leaves obtuse or acute; styles 5, connate below by
   their inner faces ....*3b. Pyracantha*

b. Stems not spinose; leaves acute or subacute; styles 2 or 3, distinct............c

c. Leaves toothed..3. *Photinia* (p. 156)

c. Leaves entire...3*a*. *Cotoneaster*

d. Leaves roundish, toothed above the middle.............4. *Amelanchier* (p. 156)

d. Leaves ovate to obovate, at least some toothed to the base....................e

e. Stems spinose..5. *Crataegus* (p. 156)

e. Stems not spinose..5*a*. *Malus*

3*a*. Cotoneaster

1. **C. pannosa** Franchet. Naturalized in Stinson Beach and occasionally spontaneous in Sausalito, Tiburon, Marshall, and elsewhere. A native of China.

3*b*. Pyracantha

1. **P. angustifolia** (Franchet) Schneider. Occasional along fence-rows near Stinson Beach and probably spontaneous elsewhere. A native of China.

2. **P. Fortuneana** (Maxim.) Li. Rare as an escape from cultivation, as in Stinson Beach. A native of China.—*P. crenato-serrata* (Hance) Rehder.

The two species of fire-thorn are common in cultivation but can scarcely be considered naturalized. They are readily distinguished by their leaves: in *P. angustifolia* the leaves are narrowly oblong with margins entire or nearly so; in *P. Fortuneana* the leaves are obovate or obovate-oblong with margins toothed.

5*a*. Malus

1. **M. domestica** Borkh. Seedling apple may occur occasionally at lower altitudes in California and in Marin County was seen along the road in Frank Valley below Muir Woods. Introduced from the Old World.

Usually *M. sylvestris* is the name given the cultivated apple. According to Terpó (Fl. Europaea 2: 67. 1968), *M. sylvestris* Miller is a wild European species that may be cultivated as a rootstock.

6. Rubus (p. 156)

1. **R. discolor** Weihe & Nees. This is the currently acceptable name for the Himalaya-berry that has been so generally known as *R. procerus* P. J. Mueller.

16. Prunus (p. 161)

2*a*. **P. cerasifera** Ehrh. Occasionally the cherry-plum grows without cultivation, as on Tiburon Peninsula where it was found by Javier Peñalosa (*No. 2253*).

LEGUMINOSAE. Pea Family
1. Acacia (p. 163)

4. **A. longifolia** (Andrews) Willd. Golden wattle has become naturalized at several places along the California coast and now should be added to the list of plants adventive in Marin County: Tiburon Peninsula (*Peñalosa 2271*); Fort Cronkite (*Kawahara 508*).

2. Albizia (p. 163, as "Albizzia")

1. **A. distachya** (Vent.) Macbr. This name replaces *A. lophantha* (Willd.) Benth.

2a. Cassia

1. **C. tomentosa** L.f. The tomentose senna, a native of Mexico, is spontaneous in Bolinas and perhaps in other places near the coast where it is cultivated as a garden shrub or windbreak. Belonging to the Senna Subfamily of the Pea Family, it may be distinguished from all other members of that family in Marin County by its slightly irregular corolla in which all the petals are distinct (not mimosoid, as in *Acacia*, and not papilionaceous as in *Lupinus*, etc.).

5. Lupinus (p. 163)

1. **L. grandifolius** Lindl. This beautiful lupine from coastal central California may have been that variant in the widespread *L. polyphyllus* complex that was involved in the origin of the garden plants known as the Russell lupines. The other parent used in the hybridization was probably the Mexican species *L. Hartwegii* Lindl. (Cf. Journ. Roy. Hort. Soc. 91: 113. 1966.) Usually the flowers of *L. grandifolius* are a beautiful violet, but in 1969 Barbara Menzies found a pinkish-buff form near the Bolinas-Olema road.

7. **L. propinquus** Greene. For those who may wish to separate the violet-flowered shrubby lupine from the yellow-flowered *L. arboreus* Sims, the one with violet flowers should be called *L. propinquus* Greene, not *L. rivularis* Dougl.

14. **L. polycarpus** Greene. This name must replace *L. micranthus* Dougl. since there is an earlier European *L. micranthus* Guss.

9. Medicago (p. 167)

4. **M. polymorpha** L. Bur clover, whether it has longer prickles (*M. hispida* Gaertn.) or shorter prickles (*M. apiculata* Willd.), is regarded in European floras as *M. polymorpha* var. *polymorpha*. The less common variant, in which the fruits are without prickles [4b. *M. hispida* var. *confinis* (Koch) Burnat], is called *M. polymorpha* var. *brevispina* (Benth.) Heyn.

11. Trifolium (p. 168)

21a. **T. subterraneum** L. Subterranean clover was found by Jeanette Coyle (*No. 275*) along the trail at Audubon Canyon Ranch in 1967. It is abundant in former pastures in Bear Valley, Point Reyes National Seashore. From all other Californian clovers this Old World species differs in its bur-like fruiting inflorescence which develops from a mixture of fertile and sterile flowers.

23a. **T. hirtum** All. Rose clover is now one of the common introductions from Europe and can be expected through most of California's hills and valleys. In Marin County it has been detected on hills near Marshall and along the highway near Nicasio Reservoir. It is like *T. Macraei* H. & A. in the enlarged stipules that closely subtend the heads but the two may be readily separated by the showier bright pink flowers of the rose clover in which the very short calyx tube is about one-fifth the length of the linear calyx lobes.

25a. **T. incarnatum** L. Commonly introduced in pastures, crimson clover occasionally becomes established, as in the hills west of Ignacio (*G. T. Robbins 3618*). No other Californian clover has a corolla as colorful as in this showy European species.

12. Lotus (p. 172)

13. **L. Biolettii** Greene. According to R. F. Hoover (Leafl. West. Bot. 10: 346, 347. 1966), *L. junceus* (Benth.) Greene is restricted to coastal slopes from Monterey County southward and *L. Biolettii* (with Mount Tamalpais type locality) from Monterey County northward. A variant with pods "coiled at least into a complete circle," *L. Biolettii* var. *spiralis* Hoover, also grows on Mount Tamalpais.

14a. **L. scoparius** (Nutt.) Ottley fma. **prostratus** Hoover. This is the mat-forming plant of coastal sand hills and dunes.

18. Lathyrus (p. 178)

2a. **L. sphaericus** Retz. The 2-foliolate leaves with or without tendrils and the small solitary short-stalked terra cotta flowers of this Old World annual pea set it off from all sweet peas in California. It was found by Barbara Menzies in May, 1969, north of Bolinas on the road to Olema.

3. **L. vestitus** Nutt. This is the common and variable wild sweet pea in Marin County. According to C. L. Hitchcock (Univ. Wash. Publ. Biol. 15: 17–19. 1952), it is represented by three subspecies which may be separated as follows:

a. Plants glabrous. Stems scandent, more than 1.5 feet tall.

....*3c. L. vestitus* ssp. *Bolanderi*
a. Plants more or less pubescent...b
b. Stems trailing or erect, scarcely scandent, generally less than 1.5 feet tall
....*3a. L. vestitus* ssp. *vestitus*
b. Stems scandent, more than 1.5 feet tall...........*3b. L. vestitus* ssp. *puberulus*

3a. **L. vestitus** Nutt. ssp. **vestitus**. Occasional from Mount Tamalpais and Bolinas Ridge to Point Reyes Peninsula (Inverness, Point Reyes).

3b. **L. vestitus** ssp. **puberulus** (White) C. L. Hitchcock. The commonest variant, growing in the chaparral and brush and in open woods: Mount Tamalpais (Blithedale Canyon) and San Rafael to the Carson country and Point Reyes Peninsula. The sweet pea described from Olema as *L. polyphyllus* Nutt. var. *insecundus* Jeps. belongs here.

3c. **L. vestitus** ssp. **Bolanderi** (Wats.) C. L. Hitchcock. Bolander's sweet pea which was formerly considered the most common is the rarest, it being known from the following collections: Angel Island (*Eastwood in 1925*), Tiburon Peninsula (*Peñalosa in 1961*), Mount Tamalpais along the old railroad grade above Muir Woods (*Eastwood in 1924*), and Arroyo Hondo near Bolinas (*True 4768*).

4. **L. Jepsoni** Greene ssp. **californicus** (Wats.) C. L. Hitchcock. This name replaces *L. Watsoni* White according to Hitchcock's revision. In the Fairfax Hills a hybrid between the California sweet pea and *L. vestitus* ssp. *puberulus* has the flower of the former and the unwinged stem of the latter.

GERANIACEAE. Geranium Family
1. Pelargonium (p. 180)

1. **P. grossularioides** (L.) Ait. This is the correct name for the small-flowered weedy perennial from South Africa.

2. Geranium (p. 180)

6. **G. Solanderi** Carolin. In a paper on the "Genus Geranium in the south western Pacific area" (Proc. Linn. Soc. New South Wales 89: 326–361. 1964), R. C. Carolin

explains why the long-used name *G. pilosum* Sol. ex Forst.f. (not Cav.) is not tenable.

7. **G. potentilloides** L'Hér. This is the name Carolin accepts for the perennial Australian cranesbill that has been found so rarely near Olema: by Alice Eastwood and others in 1898 and identified by her as *G. sibiricum* L. (Erythea 6: 117. 1898); and recently in 1965 by Lilian McHoul and identified by Margaret Bergseng as *G. microphyllum* Hook.f. (Madroño 18: 213. 1966). In 1969 it was found to be common near Bear Valley *(McHoul; Howell)*. In this geranium the peduncles are 1-flowered, while in *G. Solanderi* and *G. retrosum* L'Hér. the peduncles are usually 2-flowered.

8. **G. anemonifolium** L'Hér. This interesting plant, native in the Canary Islands and Madeira, was found as an escape from cultivation in Cascade Canyon above Mill Valley *(Howell 26630)*. It is a rather robust perennial with showy purplish flowers in a markedly glandular-hairy inflorescence.

OXALIDACEAE. Oxalis Family
1. Oxalis (p. 182)

2. **O. latifolia** H.B.K. Duncan Porter (Wasmann Jour. Biol. 26: 7. 1968) has shown this to be the correct name for the garden plant that has been called *O. Martiana* Zucc. in California.

4. **O. Pes-caprae** L. The Bermuda buttercup should be known by this name, not *O. cernua* Thunb.

LINACEAE. Flax Family
1. Linum (p. 183)

2. **L. bienne** Miller. This name for the pretty small-flowered Old World flax antedates *L. angustifolium* Hudson.

CALLITRICHACEAE. Water Starwort Family
1. Callitriche (p. 185)

3. **C. trochlearis** Fassett. The collection from east of Aurora School in northern Marin County called *C. stenocarpa (Howell 21706A)* has been redetermined by Peter Rubtzoff who would also refer here the collection from Potrero Meadows *(Howell 17889)* that was called *C. Bolanderi*.

4. **C. verna** L. This name replaces *C. palustris* L.

5. **C. heterophylla** Pursh. This name replaces *C. Bolanderi* Hegelm.

LIMNANTHACEAE. Meadow Foam Family
1. Limnanthes (p. 186)

In a monographic study of the genus *Limnanthes* (Univ. Calif. Publ. Bot. 25: 455–512. 1952), C. T. Mason describes four color forms of *L. Douglasii* R. Br. Three of these [all except var. *rosea* (Benth.) C. T. Mason which grows in interior central California] are now known from Marin County.

1*a*. **L. Douglasii** R. Br. var. **Douglasii**. Petals yellow with conspicuous white tips. This is the commonest and most widespread form, coloring meadows and springy slopes from Mount Tamalpais northward to the Sonoma County line and beyond.

1*b*. **L. Douglasii** var. **sulphurea** C. T. Mason. Petals entirely yellow. This attractive form of the species grows in boggy places along the road to the lighthouse on Point Reyes Peninsula, the type locality.

1c. **L. Douglasii** var. **nivea** C. T. Mason. Petals white when fresh, the lower part drying pale lemon-yellow. This variant was found growing with var. *Douglasii* in a marsh near the Sonoma County line on the road from Petaluma to Tomales by Arthur and Barbara Menzies. Although it has been known from several stations to the north in Sonoma County, this is the first record of var. *nivea* in Marin County.

MALVACEAE. MALLOW FAMILY
2a. Althaea

1. **A. rosea** Cav. An occasional fugitive from cultivation in California, hollyhock grows spontaneously in Tiburon (*Peñalosa 1310*) and may be expected elsewhere. This native of China, so long a garden favorite, is closely related to *Malva* (p. 191) and *Lavatera* (p. 192) from which it may be distinguished by its characteristic habit and more numerous bractlets.

3. Sidalcea (p. 192)

3. **S. rhizomata** Jeps. The marsh checker-bloom, formerly believed to be restricted to Point Reyes Peninsula (cf. p. 27), has been collected on the Mendocino County coast.

5. **S. Hickmanii** Greene ssp. **viridis** C. L. Hitchcock. One of California's rarest plants, this mallow is known only from Carson Ridge.

ELATINACEAE. WATERWORT FAMILY
1. Elatine (p. 194)

1. **E. heterandra** H. L. Mason. The waterwort from the laguna in Chileno Valley has been determined by Peter Rubtzoff as this species which was described in 1956.

CISTACEAE. ROCK-ROSE FAMILY
2. Cistus (p. 194)

1. **C. incanus** L. ssp. **corsicus** (Loisel.) Heywood. According to E. F. Warburg (Fl. Europaea 2: 283. 1968), this is the name preferred for the rock-rose that has become naturalized in Marin County. In 1962 R. M. Brown found it on Mount Tamalpais along the fire road on Corte Madera Ridge.

VIOLACEAE. VIOLET FAMILY
1. Viola (p. 194)

6a. **V. tricolor** L. Wild pansy was commonly spontaneous as a garden weed in Bolinas and is to be expected wherever the small-flowered European native is planted. It differs from all native species in its large leaflike stipules.

MYRTACEAE. MYRTLE FAMILY
2. Eugenia

1. **E. apiculata** DC. Spontaneous in the Bolinas garden of Mrs. James Jenkins. This Chilean shrub, which may be readily distinguished by its petaliferous flowers and fleshy fruit from *Eucalyptus* (p. 196), has more recently been called *Myrceugenella apiculata* (DC.) Krausel. For use in our local Californian floras the older well-known generic name seems preferable.

ONAGRACEAE. Evening Primrose Family
1. Ludwigia (p. 197)

1. **L. uruguayensis** (Camb.) Hara. Since the genus *Ludwigia* is now enlarged to include *Jussiaea* (p. 197), this rank South American species has been renamed. It was first made known in California from Tiburon but has since been found at several places in southern California.—*Jussiaea uruguayensis* Camb.

2. **L. palustris** (L.) Ell. var. **pacifica** Fernald & Griscom. The range of this species extends southward through the Coast Ranges to southern California.

3. Zauschneria (p. 197)

1. **Z. californica** Presl ssp. **mexicana** (Presl) Raven. Typical *Z. californica* Presl, characterized by linear leaves, is not found north of Monterey County, while the more variable ssp. *mexicana* with broader leaves is found from the North Coast Ranges to Baja California.

4. Epilobium (p. 198)

1. **E. Halleanum** Hausskn. This willow-herb has been found in the Santa Cruz Mountains by John H. Thomas, a southern distributional limit in the Coast Ranges.

5a. **E. foliosum** (T. & G.) Suksdorf. The small-flowered annual willow-herb that was originally described from Mount Tamalpais as *E. minutum* Lindl. var *Biolettii* Greene is now accepted as a distinct species.

6. Clarkia (p. 199)

The genus *Clarkia* is enlarged to include *Godetia* (p. 200) according to the monographic work of Harlan and Margaret Lewis (Univ. Calif. Publ. Bot. 20: 241–392. 1955). The following are their names for the clarkias of Marin County:

1. **C. unguiculata** Lindl.—*C. elegans* Dougl. (not Poiret).

2. **C. concinna** (F. & M.) Greene.

3. **C. purpurea** (Curtis) Nels. & Macbr. This variable species is represented by three subspecies:

3a. **C. purpurea** ssp. **purpurea.** Collections are cited by Lewis and Lewis from Sausalito, Tiburon, and Tomales Point.—*Godetia purpurea* (Curtis) G. Don.

3b. **C. purpurea** ssp. **viminea** (Dougl.) Lewis & Lewis. The only Marin County collection cited is from Olema.—*Godetia viminea* (Dougl.) Spach.

3c. **C. purpurea** ssp. **quadrivulnera** (Dougl.) Lewis & Lewis. The commonest form of the species in Marin County but no collections are cited from there. It has been seen from Frank Valley, San Rafael Hills, Forest Knolls, and the Carson country.—*Godetia quadrivulnera* (Dougl.) Spach; *G. purpurea* var. *parviflora* (Wats.) C. L. Hitchc.

4. **C. Davyi** (Jeps.) Lewis & Lewis. Point Reyes is the type locality of this interesting maritime plant.—*Godetia quadrivulnera* var. *Davyi* Jeps.

5. **C. amoena** (Lehm.) Nels. & Macbr. One of the most variable species in the genus as well as one of the most beautiful, it is represented by two subspecies:

5a. **C. amoena** ssp. **amoena.** Collections are cited from Olema, Point Reyes near the lighthouse, and Tomales.—*Godetia amoena* (Lehm.) G. Don.

5b. **C. amoena** ssp. **Huntiana** (Jeps.) Lewis & Lewis. A collection is cited from Tocaloma; others are from Lagunitas and Nicasio.—*Godetia amoena* fma. *Huntiana* Jeps.

6. **C. rubicunda** (Lindl.) Lewis & Lewis. This species, which is also represented by two subspecies in Marin County by Lewis and Lewis, was not distinguished in MARIN FLORA from the large-flowered plants called farewell-to-spring. The species is not reported north of Marin County.

6a. **C. rubicunda** ssp. **rubicunda.** Stinson Beach is the only station cited in Marin County.—*Godetia rubicunda* Lindl.

6b. **C. rubicunda** ssp. **Blasdalei** (Jeps.) Lewis & Lewis. This farewell-to-spring is cited in Marin County only from Tiburon.—*Godetia Blasdalei* Jeps.

7. **C. gracilis** (Piper) Nels. & Macbr. Lewis and Lewis report two subspecies for Marin County:

7a. **C. gracilis** ssp. **gracilis.** Occasional from Mount Tamalpais to the Carson country and San Rafael Hills.—*Godetia gracilis* Piper; *G. lassenensis* Eastw. var. *concolor* (Jeps.) J. T. Howell.

7b. **C. gracilis** ssp. **sonomensis** (C. L. Hitchc.) Lewis & Lewis. Marin County is mapped as the southern distributional limit of this subspecies (op. cit., figure 6) but no collection is cited from the county.—*Godetia lassenensis* Eastw. var. *sonomensis* (C. L. Hitchc.) J. T. Howell.

The twelve taxonomic units in *Clarkia* recognized by Lewis and Lewis in Marin County may be distinguished as follows:

a. Petals distinctly clawed, the claws equaling the blade or a little shorter.......b
a. Petals scarcely, if at all, clawed...c
b. Petals entire; stamens 8.....................................1. *C. unguiculata*
b. Petals 3-lobed; stamens 4.....................................2. *C. concinna*
c. Inflorescence erect in bud...d
c. Inflorescence nodding in bud..k
d. Capsule 8-ribbed (or 8-grooved) when immature............................e
d. Capsule 4-ribbed (or 4-grooved) when immature............................h
e. Stems prostrate or decumbent; leaves oblanceolate to elliptic or obovate, obtuse
....4. *C. Davyi*
e. Stems decumbent or erect; leaves linear-lanceolate to linear-oblanceolate, rarely wider, acute...f
f. Leaves lanceolate to ovate; flowers crowded at end of stem
....3a. *C. purpurea* ssp. *purpurea*
f. Leaves linear or linear-lanceolate; flowers scattered or rarely congested.......g
g. Petals 1.5 cm. long or more......................3b. *C. purpurea* ssp. *viminea*
g. Petals less than 1.5 cm. long.............3c. *C. purpurea* ssp. *quadrivulnera*
h. Petals usually pencilled or blotched with red in the center, not bright red at base i
h. Petals bright red only at base...j
i. Plants stout; flowers tending to be congested........5a. *C. amoena* ssp. *amoena*
i. Plants slender; flowers scattered................5b. *C. amoena* ssp. *Huntiana*
j. Plants slender; flowers scattered..............—6a. *C. rubicunda* ssp. *rubicunda*
j. Plants stout; flowers congested..................6b. *C. rubicunda* ssp. *Blasdalei*

k. Petals 2 cm. long or less, without a red spot..........*7a. C. gracilis* ssp. *gracilis*
k. Petals more than 2 cm. long, with a central red spot

....*7b. C. gracilis* ssp. *sonomensis*

8. Oenothera (p. 201)

1. **O. Hookeri** T. & G. The showy evening primrose is represented by two sub-species: ssp. *Hookeri* from Stinson Beach and Olema with petals more than 2.5 cm. long; and ssp. *Wolfii* Munz on the road east of Aurora School with petals 2 to 2.5 cm. long. Marin County is the southern distributional limit for ssp. *Wolfii*.

The remaining members of the genus *Oenothera* in Marin County have recently been treated by Peter H. Raven in "A revision of the genus Camissonia" (Contrib. U.S. Nat. Herb. 37: 161–396. 1969). Although the conservative retention of the genus *Oenothera* is adhered to in the following floristic account, it borrows from Raven's monograph for pertinent taxonomic and phytogeographic data.

2. **O. strigulosa** F. & M.—*O. contorta* Dougl. var. *strigulosa* (F. & M.) Munz; *Camissonia strigulosa* (F. & M.) Raven.

3. **O. micrantha** Hornem. The species is found northward to the Sonoma County coast, the northern limit of distribution and probably the region where first collected.—*Camissonia micrantha* (Hornem.) Raven.

3a. **O. hirtella** Greene. A collection of this widespread Californian plant is cited by Raven from Mill Valley, *Bioletti* in 1892. It differs from *O. micrantha* in its broader upper leaves and more slender capsule.—*O. micrantha* var. *hirtella* (Greene) Jeps.; *Camissonia hirtella* (Greene) Raven.

4. **O. cheiranthifolia** Hornem.—*Camissonia cheiranthifolia* (Hornem.) Raimann.

5. **O. ovata** Nutt.—*Camissonia ovata* (Nutt.) Raven.

5a. **O. graciliflora** H. & A. The only known collection of this species from Marin County was made on Mount Tamalpais, *Wood in 1915*. It differs from *O. ovata* in its annual root and short angled capsule.—*Camissonia graciliflora* (H. & A.) Raven.

HALORAGIDACEAE. Water-milfoil Family
3. Gunnera

1. **G. tinctoria** (Molina) Mirbel. Naturalized in the wet sloping meadow above Heart's Desire Beach in Tomales Bay State Park (*Howell 28293*). This terrestrial Andean herb with its gigantic roundish leaves is so unlike the delicate aquatic milfoils and mare's tails (p. 201), one may well marvel at their family relationship!—*G. chilensis* Lamk.

UMBELLIFERAE. Parsley Family
8. Torilis (p. 206)

2. **T. heterophylla** Guss. Garden weed in Larkspur (*Howell 42251*); to be expected on grassy or brushy slopes. This European hedge-parsley differs from the more common *T. nodosa* (L.) Gaertn. in its long-pedunculate umbels and from *Anthriscus*, the bur-chervil, in its more coarsely divided leaves, 2- or 3-rayed umbels, and nearly beakless fruits.

11. Anthriscus (p. 206)

1. **A. Caucalis** Bieberstein. The bur-chervil is given this relatively simple name in Flora Europaea (2:326. 1968), instead of *A. neglecta* Boiss. & Reut. var. *Scandix* (Scop.) Hyl. or *A. scandicina* (Weber) Mansfeld.

13a. Petroselinum

1. **P. crispum** (Miller) A. W. Hill. Common parsley may persist and become weedy in gardens or escape to waste ground, as in Blithedale Canyon above Mill Valley (*Raven 2893*). Though it is generally easily recognized, botanically it may be separated from *Apium* (p. 206) by its pedunculate umbels and yellow petals. It is native in Europe.

13b. Coriandrum

1. **C. sativum** L. An old herbarium specimen records the occurrence of the European herb, coriander, in Marin County (Lagunitas, *J. P. Moore in 1878*). The globose fruits, which are used for seasoning, do not separate into two parts (mericarps). The leaves are diverse, the lowest with broad toothed segments, the uppermost with narrow entire segments. The petals are white and markedly unequal.

ERICACEAE. Heather Family

2. Arctostaphylos (p. 212)

1. **A. Uva-ursi** (L.) Spreng. Kinnikinnick has been reported from several stations south of the Golden Gate, while a remarkable occurrence in the Sierra Nevada above Convict Lake is also far south of Marin County.

4. **A. Manzanita** Parry. The Parry manzanita grows in the hills north of Mount Diablo near Somersville and has been found near El Sobrante by Walter and Irja Knight. Contra Costa County marks the southern distributional limit for the species.

A single shrub of this species grows on the south side of Mount Tamalpais along the old railroad grade. It was erroneously reported as *A. glauca* Lindl., a South Coast Range species reaching a northern limit on Mount Diablo.

10. **A. sensitiva** Jeps. As a genus, *Schizococcus* (p. 214) is too closely related to *Arctostaphylos* to be maintained. As an indication of the close genetic relationship of the two groups, a hybrid between *A. sensitiva* and *A. Cushingiana* Eastw. has been found on Throckmorton Ridge on Mount Tamalpais (cf. Leafl. West. Bot. 7: 265. 1955.)—*Schizococcus sensitivus* (Jeps.) Eastw.

4. Ledum (p. 215)

1. **L. glandulosum** Nutt. ssp. **columbianum** (Piper) C. L. Hitchcock var. **australe** C. L. Hitchcock. A collection made in July, 1903, by A. D. E. Elmer (*No. 4944*) at "Pt. Reyes Post Office" is the type of this variety which is found from Monterey County north to Clatsop County, Oregon. Old maps show Point Reyes [Post Office] to have been near the present-day radio stations.

7. Vaccinium (p. 215)

2. **V. arbuscula** (Gray) Merriam. This low-growing deciduous huckleberry, also found on Point Reyes Peninsula by Elmer in 1903, has not been detected there since then. It is probably to be sought in brushy places bordering swales, although on San Bruno Mountain in San Mateo County (its southern distributional limit, where it was discovered by James Roof), it is restricted to rocky outcrops.

10. Hemitomes

1. **H. congestum** Gray. This rare Pacific coast saprophyte, which Jepson called gnome plant, is properly named *Hemitomes* rather than *Newberrya* (p. 216). Many

years ago it was thought that, as a plant name, *Hemitomes* Gray (1857) was too like the older *Hemitomus* L'Héritier (1804) to be tenable; but now the two words are accepted as different. Heretofore known in Marin County only from the east slope of Mount Tamalpais, *Hemitomes* was found in June, 1969, by Mrs. Amalie Eisig in Samuel P. Taylor State Park.

PRIMULACEAE. Primrose Family
2. Anagallis (p. 216)

2. **A. minima** (L.) Krause. Since the genus *Centunculus* (p. 216) is better considered a part of *Anagallis,* inconspicuous chaffweed moves in with the pimpernels with the botanical designation given.

5. Dodecatheon (p. 217)

1. **D. Hendersonii** Gray ssp. **cruciatum** (Greene) H. J. Thompson. All Marin County flowers that have been examined are 4-merous, the mark of this subspecies. In typical *D. Hendersonii,* which occurs in Sonoma County, the flower parts are in 5's.

PLUMBAGINACEAE. Thrift Family
2. Limonium (p. 217)

1. **L. californicum** (Boiss.) Heller. The wild statice of Marin salt marshes is better treated as a species, rather than as *L. commune* S. F. Gray var. *californicum* (Boiss.) Greene.

OLEACEAE. Olive Family
1. Fraxinus (p. 217)

1. **F. latifolia** Benth. This name is older than *F. oregona* Nutt. and so must be used for the Oregon ash.

GENTIANACEAE. Gentian Family
2. Cicendia

1. **C. quadrangularis** (Lamk.) Griseb. Because the generic name *Microcala* (p. 218) is not legally acceptable, the one given here must be used.

CONVOLVULACEAE. Morning-glory Family
1. Dichondra (p. 219)

1. **D. Donnelliana** Tharp & Johnston. The wild dichondra of the California Coast Ranges has been interpreted as indigenous and has been named and described from a collection made on Bolinas Ridge by H. L. Mason

4. Cuscuta (p. 221)

2. **C. Ceanothi** Behr. This earlier name replaces *C. subinclusa* D. & H.

POLEMONIACEAE. Gilia Family
4. Gilia (p. 223)

3a. **G. clivorum** (Jeps.) V. Grant. This is the plant described in Marin Flora as the small-flowered form of *G. millefoliata* F. & M. under the possible name of *G. multicaulis* Benth. var. *clivorum* Jeps. According to Grant the plant ranges from Lake County to southern California.

5. Linanthus (p. 224)

1. **L. pygmaeus** (Brand) J. T. Howell .Mainland plants have paler, somewhat larger corollas and have been called ssp. *continentalis* Raven. *Linanthus pygmaeus* was originally described from Guadalupe Island, Baja California.

3. **L. grandiflorus** (Benth.) Greene. The northern distributional limit is in Sonoma County, not Marin County.

HYDROPHYLLACEAE. PHACELIA FAMILY
4. Phacelia (p. 227)

1. **P. californica** Cham. The form of this species with more hispid pubescence has twice been named in Marin County: in 1902 from "Sequoia Canyon" (*i.e.* the canyon in which Muir Woods is situated) as *P. Biolettii* Greene; and in 1913 from Mount Tamalpais as *P. magellanica* (Lamk.) Cov. ssp. *barbata* Brand fma. *Jepsonii* Brand.

4a. **P. tanacetifolia** Benth. Although this species is widespread in the hills and valleys of California, it is doubtful whether it is indigenous in Marin County. The specimen seen from a roadcut near Alto probably represents a fugitive from cultivation. It differs from *P. distans* Benth. in its more colorful lavender-blue flowers with long-exserted stamens.

BORAGINACEAE. BORAGE FAMILY
4. Allocarya (p. 231)

1a. **A. Chorisiana** (Cham.) Greene. Choris' white-flowered forget-me-not was discovered near Bolinas by Anne Leary and Barbara Sherfey in 1960, the first specimens to be reported north of the Golden Gate. This species was first collected in San Francisco in 1816 by Chamisso who named it for Login Andreevich Choris, the artist who together with Chamisso and Eschscholtz was on Kotzebue's first expedition. Besides differences in the fruits, *A. Chorisiana* may be separated from *A. undulata* Piper by its larger, often showy corollas.

4. **A. stipitata** Greene. Occurring locally in a depression on the grassy summit ridge of Tiburon Peninsula near Ring Hill where it was discovered in 1969 by Mrs. William Cuthbertson and Mrs. George Ellman. This species, which is characteristic of the flora of the rain pools in the Central Valley of California, differs from all other allocaryas in Marin County in the basal attachment of the nutlets. It is like *A. Chorisiana* in the showy white corolla lobes.

5. Amsinckia (p. 231)

1a. **A. lunaris** Macbr. This fiddleneck which occurs rather rarely in the Coast Ranges of central California is known from at least two collections in Marin County: San Geronimo, *Clara Tose in 1927* (determined by W. Suksdorf); head of Nicasio Creek, *Eastwood & Howell 5497* (determined by F. Chisaki). In this species the corolla is somewhat irregular and bilaterally symmetrical, while in *A. intermedia* F. & M. the corolla is radially symmetrical.

2. **A. Menziesii** (Lehm.) Nels. & Macbr. This older specific name replaces *A. Helleri* Brand.

7. Myosotis (p. 232)

1. **M. latifolia** Poir. This is the correct name for the naturalized forget-me-not, replacing *M. sylvatica* Ehrh.

2. **M. versicolor** (Pers.) Sm. The forget-me-not with tiny corollas (2 mm. across) that change from yellow to blue and violet in age has been found in two places in low wet ground: pond south of Olema (*Alix Wennekens*); roadside near Woodacre (*Arthur Weston*). The plant is native in Europe.

8. Echium (p. 232)

2. **E. plantagineum** L. An attractive addition to a weedy roadside near Point Reyes Station, this herbaceous species from southern Europe is readily separated from its shrubby naturalized relative, *E. fastuosum* Ait. In *Echium* the flowers are borne in bracteate spikes and the slightly irregular tubular corollas are crestless; in *Cynoglossum* and *Myosotis,* two other genera with blue corollas, the flowers are borne in bractless racemes and the regular salverform corollas are crested in the throat.

9. Borago

1. **B. officinalis** L. Borage has been found as a garden weed in Larkspur and in Tiburon. It is a native of the Old World where in some parts it is grown as a culinary herb. It differs from other members of *Boraginaceae* in Marin County in its flat blue corolla that is a half inch or more across.

LABIATAE. Mint Family
5. Lamium (p. 234)

2. **L. purpureum** L. The red henbit, a rare European weed in the San Francisco Bay area, was found by Lilian McHoul in 1967 at the Bolema Club between Olema and Bolinas. It may be separated from giraffe's head (*L. amplexicaule* L.) by the petiolate leaves of the inflorescence.

5a. Pogogyne

1. **P. serpylloides** (Torr.) Gray. This delicate annual mint, which generally occurs along the edge of brushy thickets, was collected in Tomasini Canyon just northeast of Point Reyes Station in May, 1961 (*Bacigalupi & Heckard 7721*). It may be separated from *Lamium* (pp. 233, 234) by its much smaller corollas (3–5 mm. instead of 10–15 mm.) and by only 2 (instead of 4) fertile stamens. It is widespread in the Coast Ranges and may be expected elsewhere in the hills of northern Marin County.

6. Stachys (p. 234)

3a. **S. arvensis** L. The field hedge-nettle, which grew as a garden weed in Ross (*Howell 11809*), may be readily distinguished from other Californian species by its annual root and small pink corollas that are scarcely longer than the calyx. It is native in Europe.

7. Salvia (p. 235)

3a. **S. microphylla** H. B. K. This shrubby red-flowered sage occurred in a large naturalized patch along the Waldo Grade in the Sausalito Hills (*H. Leach in 1957*). Formerly known as *S. Grahamii* Benth., it is an old-fashioned garden favorite from Mexico that occasionally escapes from cultivation.

10. Satureja (p. 235)

1. **S. Chamissonis** (Benth.) Briq. According to Epling and Játiva (Brittonia 18: 245. 1966), Chamisso's name, not that of Douglas, is to be associated with the yerba buena.

12. Monardella (p. 236)

1b. **M. villosa** Benth. var. **subglabra** Hoover. Among the variations in coyote mint in Marin County, the nearly glabrous type on Tiburon Peninsula would be treated as a variety by Robert F. Hoover and as a species by Clare B. Hardham. Mrs. Hardham would also accord var. *franciscana* (Elmer) Jeps. specific status (*M. franciscana* Elmer).

SOLANACEAE. NIGHTSHADE FAMILY
1. Solanum (p. 237)

2a. **S. laciniatum** Ait. A robust handsome shrub with large violet corollas, this nightshade has been noted as a casual escape from cultivation in Steep Ravine on Mount Tamalpais and in Bolinas. A native of New Zealand, it has been confused with the Australian *S. aviculare* Forst. f., which is not known to occur in California.

3. **S. nodiflorum** Jacq. As noted on page 294, this is the correct name for the widespread plant that has been called *S. nigrum* L. in Marin County.

1a. Lycopersicon

1. **L. esculentum** Mill. The tomato was found in waste ground in Tiburon and may occur as a casual escape elsewhere. It is a native of South America. From the species of *Solanum* in Marin County it is readily distinguished by its compound leaves and yellow corollas.

2. Salpichroa (p. 237)

1. **S. origanifolia** (Lamk.) Baill. This name replaces *S. rhomboidea* (Gill. & Hook.) Miers for the lily-of-the-valley vine.

3. Physalis (p. 238)

1. **P. philadelphica** Lamk. U. T. Waterfall now uses this prior name for the tomatillo instead of *P. ixocarpa* Brot.

2. **P. pubescens** L. This ground-cherry, which is native in eastern and southern parts of America, has been found twice: Stinson Beach, *Pollard in 1953* (determined by U. T. Waterfall as var. *grisea* Waterfall); head of Bolinas Lagoon, *McHoul in 1968*. It differs from *P. philadelphica* in its more pubescent herbage, more sharply angled fruiting calyx, and smaller corollas (less than 1 cm. in diameter).

4. Nicotiana (p. 238)

3. **N. glauca** Grah. A weed in waste ground on Tiburon Peninsula, the South American tree tobacco may be expected in the drier eastern parts of Marin County. It differs from all other tobaccos in California in its shrubby habit and yellow corollas.

SCROPHULARIACEAE. FIGWORT FAMILY
3. Antirrhinum (p. 240)

2. **A. Kelloggii** Greene. It is now agreed that this is the name of the pretty wild snapdragon formerly called *A. Hookerianum* Millspaugh.

4. Collinsia (p. 240)

3. **C. bartsiaefolia** Benth. var. **hirsuta** (Kell.) Penn. According to R. C. Bacigalupi this is the name of the plant collected by Alice Eastwood on Point Reyes Peninsula, not *C. corymbosa* Herder. It is believed that *C. corymbosa* is an endemic restricted to the coastal dunes of Sonoma and Mendocino counties.

8. Mimulus (p. 241)

5. **M. nasutus** Greene. The large-flowered form of this monkeyflower is var. *insignis* (Greene) Grant, not var. *eximius*.

8. **M. Congdonii** Rob. Although this diminutive red-flowered monkeyflower is relatively rare, it is widespread in both Sierra and Coast Range hills. The plant from Mariposa County was named *M. Congdonii* by B. L. Robinson before Alice Eastwood described *M. modestus* from Marin County.

9. Limosella (p. 243)

1. **L. acaulis** Ses. & Moç. The diminutive mudwort of Marin County strands and wet flats belongs to this more austral species rather than to the more boreal *L. aquatica* L. In *L. acaulis* the style is more than 0.5 mm. long, in *L. aquatica* it is shorter.

17. Castilleja (p. 245)

The names of Indian paintbrushes in Marin County should be changed to agree with those proposed by F. W. Pennell:

1. **C. inflata** Penn.—*C. Wightii* Elmer ssp. *inflata* (Penn.) Munz.

1*a*. **C. Wightii** Elmer.—*C. latifolia* H. & A. var. *Wightii* (Elmer) Zeile.

1*b*. **C. Wightii** ssp. **rubra** Penn.—*C. latifolia* var. *rubra* (Penn.) J. T. Howell.

2. **C. affinis** H. & A.—*C. Douglasii* Benth.

4. **C. franciscana** Penn.—*C. affinis* of California authors, not of Hooker & Arnott.

RUBIACEAE. Madder Family
2. Galium (p. 252)

4*a*. **G. spurium** L. var. **spurium.** This is the "anomalous plant with smooth fruits" noted under *G. tinctorium* L. as a suspected hybrid (p. 253). It was correctly identified by Dr. Friedrich Ehrendorfer as the typical variety of this common European species.

7. **G. Nuttallii** Gray var. **ovalifolium** Dempster. The typical variety is found from Santa Barbara County to northern Baja California; var. *ovalifolium* is found throughout the Coast Ranges from Del Norte County to Baja California.

3. Asperula

1. **A. odorata** L. The sweet woodruff or waldmeister of the Old World has been found as a fugitive from cultivation in Ross and it tends to spread adventitiously wherever it is planted. It looks like a species of *Galium* (p. 252) but may be distinguished by its funnelform corolla.

VALERIANACEAE. Valerian Family
2. Plectritis (p. 256)

Variability of fruits in *Plectritis* should not be the basis for specific differentiation according to the studies of both L. T. Dempster and D. H. Morey. The three species and subspecies accepted here for Marin County are aligned as they are in Abrams and Ferris, Illustrated Flora of the Pacific States (4: 61–63, figures 5062-5064. 1960), a treatment based on Dr. Morey's work.

a. Corolla nearly regular, the lobes subequal. Spur of corolla stout and somewhat
   clavate, usually more than ⅓ length of corolla; dorsal keel of fruit obtusely
   angled, often with a median longitudinal groove.........................b
a. Corolla bilabiate, the lobes unequal...c
b. Fruit winged or wingless, if winged the wings thick with a broad grooved marginal border...1. *P. macrocera*
b. Fruit winged, the wings thin with a narrow scarcely grooved marginal border
   ....1a. *P. macrocera* ssp. *Grayii*
c. Spur of corolla shorter than the ovary, usually less than ⅓ length of corolla or
   obsolete; dorsal keel of fruit usually without median longitudinal groove...d
c. Spur of corolla longer than the ovary, more than ⅓ the length of the corolla;
   dorsal keel of fruit usually with a median longitudinal groove..............e
d. Keel of fruit sharply angled, acute..........2a. *P. congesta* ssp. *brachystemon*
d. Keel of fruit smoothly rounded.....................2b. *P. congesta* ssp. *nitida*
e. Corolla 5.5—8.5 mm. long......................................3. *P. ciliosa*
e. Corolla 1.5—3.5 mm. long.........................3a. *P. ciliosa* ssp. *insignis*

1. **P. macrocera** T. & G. San Rafael Hills (*Pollard*); Big Rock Ridge (*Howell*).

1a. **P. macrocera** ssp. **Grayii** (Suksdorf) Morey. Tiburon Peninsula; Mount
Tamalpais (between Pan Toll and Rock Spring); Bald Mountain west of San
Anselmo.—*P. macroptera* Rydb.

2a. **P. congesta** (Lindl.) DC. ssp. **brachystemon** (F. & M.) Morey. Sausalito, Tiburon, and Mount Tamalpais (Phoenix Lake, Lagunitas Meadows, Potrero
Meadow) to the San Rafael Hills, Carson country, and Point Reyes Peninsula.—
P. samolifolia (DC.) Hoeck.; *P. samolifolia* var. *involuta* (Suksdorf) Dyal; *P. aphanoptera* (Gray) Suksdorf; *P. magna* (Greene) Suksdorf.

2b. **P. congesta** ssp. **nitida** (Heller) Morey. Baltimore Park; Nicasio; Point Reyes
Station; Point Reyes Peninsula; Dillon Beach.

3. **P. ciliosa** (Greene) Jeps. San Rafael Hills; Bald Mountain west of San Anselmo; Fairfax Hills; Big Rock Ridge.—*P. californica* (Suksdorf) Dyal.

3a. **P. ciliosa** ssp. **insignis** (Suksdorf) Morey. Rare on Mount Tamalpais (*Eastwood 2507;* between Bootjack and Mountain Theater, *Howell in 1946*).—*P. californica* (Suksdorf) Dyal var. *rubens* (Suksdorf) Dyal.

3. Valerianella

1. **V. Locusta** (L.) Betcke. This annual European valerian, sometimes known as
corn-salad, grew as a weed in a Larkspur garden. In it the inflorescence is openly
branching; in *Plectritis* (p. 256) it is subcapitately congested.

CAMPANULACEAE. BELLFLOWER FAMILY
1. Campanula (p. 257)

1. **C. californica** (Kell.) Heller. The California bluebell reaches a southern
distributional limit in Santa Cruz County where it has been collected by Vesta
Hesse.

3. Triodanis

1. **T. biflora** (R. & P.) Greene. Venus looking-glass should have this name instead
of *Specularia biflora* (R. & P.) F. & M. (p. 258).

LOBELIACEAE. LOBELIA FAMILY

1. **L. Erinus** L. This much-used garden plant from South Africa has escaped to a sunny bank in Kent Woodlands (*B. Clear in 1960*). In this *Lobelia* the corolla surmounts an obconic ovary borne on a slender pedicel; in *Downingia* the elongate stalklike ovary is sessile.

COMPOSITAE. SUNFLOWER FAMILY
2. Grindelia (p. 265)

2. **G. stricta** DC. This coastal gumweed reaches a southern distributional limit in Monterey County.

3*a*. **G. procera** Greene. Local in San Anselmo, probably introduced. The tall gumweed, typical in the lower parts of the Great Valley, differs from *G. camporum* Greene in its achenes: in *G. camporum* they are topped by a low knobby corky crown while in *G. procera* this crown is lacking. Also the ray-flowers are usually more numerous in *G. procera*.

7. Haplopappus

The name of the genus containing mock heather [*H. ericoides* (Less.) H. & A.] and golden fleece [*H. arborescens* (Gray) Hall] was originally spelled *Aplopappus* (p. 268), but now *Haplopappus* is preferred.

8. Chaetopappa

The genus *Pentachaeta* (p. 268) has been combined with *Chaetopappa* with the following name-changes:

1. **C. bellidiflora** (Greene) Keck.—*Pentachaeta bellidiflora* Greene.

2. **C. alsinoides** (Greene) Keck.—*Pentachaeta alsinoides* Greene.

12. Aster (p. 269)

1*a*. **A. subspicatus** Nees. The *Aster* formerly known as *A. Douglasii* Lindl. has been collected in Mill Valley (*Howell 953* in 1925) and near Point Reyes lighthouse (*Rubtzoff 788* in 1951). It may be separated from *A. chilensis* Nees by its more foliaceous involucral bracts that are scarcely imbricated.

13. Erigeron (p. 270)

7*b*. **E. inornatus** (Gray) Gray var. **viscidulus** Gray. The eradiate leafy fleabane on Point Reyes Peninsula is referable to this variety with var. *Biolettii* (Greene) Jeps. as a synonym.

13*a*. Conyza

1. **C. bonariensis** (L.) Cronquist. This weedy South American fleabane has been known in the California floras as *Erigeron crispus* Pourret (p. 270) or *E. linifolius* Willd. In *Conyza* the outermost flowers in a head are pistillate, tubular, and nonligulate; in *Erigeron* they are either pistillate and ligulate or perfect and discoid.

19. Psilocarphus (p. 272)

1*a*. **P. tenellus** Nutt. var. **tenuis** (Eastw.) Cronquist. This is another example of a plant usually found in the interior parts of California that reaches a coastal habitat and thrives. It is locally common on a sandy flat near the radio station on Point Reyes Peninsula.

20. Gnaphalium (p. 272)

3. G. ramosissimum Nutt. The pink cudweed has been found as far north as Humboldt County.

4. G. beneolens A. Davidson. The name *G. microcephalum* Nutt., applicable to plants from Monterey County southward, was misapplied to Marin County plants.

20a. Helichrysum

1. H. petiolatum (L.) DC. Shrubby plants forming a dense thicket on the slope of Bolinas Ridge above Stinson Beach (*Arthur and Barbara Menzies in 1968*). This attractive white-woolly plant from South Africa may be readily separated from *Gnaphalium* (pp. 261, 272) by its ovate petiolate leaves.

24. Ambrosia (p. 274)

The genus *Franseria* (p. 274) is now referred to *Ambrosia* with the following change of names:

2. A. Chamissonis (Less.) Greene.—*Franseria Chamissonis* Less.

2a. A. Chamissonis var. **bipinnatisecta** (Less.) J. T. Howell.—*Franseria Chamissonis* var. *bipinnatisecta* Less.

25. Xanthium (p. 274)

2. X. strumarium L. The cockleburs of Marin County are referable to two varieties: var. *canadense* (Miller) T. & G. (*X. italicum* Mor.) in which the densely placed spines of the bur bear coarse hairs, and var. *glabratum* (DC.) Cronquist (*X. calvum* Millsp. & Sherff) in which the sparsely placed spines of the bur are glabrous or nearly so.

30a. Galinsoga

1. G. parviflora Cav. Garden weed in Larkspur. This plant, which is native from Mexico southward, may be separated from all members of the Sunflower tribe in Marin County (pp. 261, 262, 275) by its small heads with short white ray flowers.

31. Holocarpha (p. 275)

2. H. virgata (Gray) Keck. One of the common late-blooming plants on the foothills around the Great Valley, this tarweed was locally abundant in the Lagunitas Meadows in September, 1956. In it the flower-heads are racemose on elongate branches; in *H. macradenia* (DC.) Greene the heads are crowded at the ends of the branches.

32. Hemizonia (p. 275)

The five species now known to occur in Marin County may be distinguished by characters in the following key:

a. Leaves spine-tipped ....5. *H. pungens*

a. Leaves not spine-tipped ....b

b. Basal and lower cauline leaves pinnately divided; disk flowers not subtended
    by bracts ....4. *H. corymbosa*

b. Basal and lower cauline leaves entire or toothed; disk flowers subtended by
    bracts c

c. Rays white, veined or tinged with pink.......................3. *H. congesta*

c. Rays yellow ....d

d. Leaves deep green; plants of maritime bluffs and mesas ......2. *H. multicaulis*

d. Leaves silvery; plants of interior hills and valleys ....e

e. Herbage glandular above; spring blooming (April to June)
....2a. *H. multicaulis* ssp. *vernalis*

e. Herbage glandular throughout; summer and autumn blooming (July to
November) ....1. *H. lutescens*

1. **H. lutescens** (Greene) Keck. Common and widespread.—*H. luzulaefolia* var. *lutescens* Greene.

2. **H. multicaulis** H. & A. Olema; Dillons Beach.—*H. citrina* Greene.

2a. **H. multicaulis** ssp. **vernalis** Keck. Tiburon Peninsula (the type locality) to Carson Ridge and Point Reyes Station.

3. **H. congesta** DC.

4. **H. corymbosa** (DC.) T. & G.

5. **H. pungens** (H. & A.) T. & G. ssp. **maritima** (Greene) Keck. On roadside near Olema, undoubtedly introduced.—*Centromadia maritima* Greene.

33. Calycadenia (p. 276)

1. **C. multiglandulosa** DC. var. **cephalotes** (DC.) Jeps. This variety is now known to extend south of Marin County to Alameda and San Mateo counties.

2. **C. ciliosa** Greene. The white-flowered form (fma. *alba* Keck) of this rosinweed was detected on the serpentine near Bootjack on Mount Tamalpais in September, 1952. The species and form generally occur from Lake County northward. In *C. ciliosa* the heads are scattered along the stems; in *C. multiglandulosa* the heads are congested.

34. Madia (p. 276)

6. **M. anomala** Greene. This tarweed, believed to have been restricted to the North Coast Ranges, is reported from Mount Diablo and Sacramento County by Keck (Ill. Fl. Pac. States 4: 169).

41. Lasthenia (p. 280)

Ornduff has shown that the genus *Baeria* (p. 279) should be combined with *Lasthenia* (Univ. Calif. Publ. Bot. 40: 1–92. 1966). The following nomenclature is now in order:

1. **L. chrysostoma** (F. & M.) Greene.—*Baeria chrysostoma* F. & M.

2. **L. hirsutula** Greene. This distinctive plant of maritime slopes has been called *Baeria chrysostoma* ssp. *hirsutula* (Greene) Ferris but is referred to synonymy under *L. chrysostoma* by Ornduff. It deserves nomenclatural recognition.—*Baeria hirsutula* (Greene) Greene.

3. **L. macrantha** (Gray) Greene.—*Baeria macrantha* (Gray) Gray.

4. **L. minor** (DC.) Ornduff.—*Baeria minor* (DC.) Ferris; *B. uliginosa* (Nutt.) Gray.

5. **L. glabrata** Lindl.

6. **L. glaberrima** DC.

49. Matricaria (p. 282)

2. **M. occidentalis** Greene. Occurring as a casual roadside weed at the head of Drakes Estero on Point Reyes Peninsula (*H. G. Baker in 1965*). The western pineapple weed, which is native in the interior valleys of California, superficially re-

sembles the widespread *M. matricarioides* (Less.) Porter but that more common plant is not so robust and lacks a short 2-pointed pappus crown.

50. Chrysanthemum (p. 282)

4. C. Leucanthemum L. The widespread attractive ox-eye daisy is naturalized near Inverness and may be expected to occur in other parts of the county, especially near the coast.

5. C. maximum Ramond. The Shasta daisy has been observed as a roadside escape on the west side of the Bolinas Lagoon. Both this and the preceding differ from other species in the genus in Marin County in their large solitary flower heads with white rays. The Shasta daisy may be recognized by its larger heads (5 cm. or more across) and by its evenly toothed, longer stem leaves (10 cm. or more long).

52. Soliva (p. 283)

As characterized by their achenes, there are three species of *Soliva* in Marin County (cf. B. Crampton, Leafl. West. Bot. 7: 196–198, 1 figure. 1954). They are similar in appearance and may occur on grassy hillsides or as weedy plants in trodden ground along paths and roads.

a. Achenes with narrow callous-thickened wing, the wing developing near the top two short sharp teeth or the teeth obsolete ....1. *S. daucifolia*

a. Achenes with broad membranous wing, the wing projecting as 2 pointed teeth or lobes well above the achene body ....b

b. Achene wing entire ....2. *S. sessilis*

b. Achene wing conspicuously notched near the base ....3. *S. pterosperma*

1. S. daucifolia Nutt. Occasional: Sausalito Hills; Ignacio; Point Reyes Peninsula (Inverness, Heart's Desire Beach, near the lighthouse).

2. S. sessilis R. & P. Common and widespread: Tiburon (Old St. Hilary's), Mount Tamalpais (Fern Canyon), and San Rafael Hills to Ignacio, Nicasio, and Point Reyes Peinsula (Bolinas Mesa, Mud Lake, Ledum Swamp, road west of Drakes Beach). In some forms the wing is thicker and narrower but it is not certain whether this variation is genetic or edaphic.

3. S. pterosperma (Juss.) Less. Heretofore known in California only from the central Sierran foothills, this species has been found on Angel Island by Douglas Ripley.

53a. Santolina

1. S. Chamaecyparissus L. This dusty-miller-like shrub, which is commonly called lavender-cotton, is a fugitive from cultivation near Phoenix Lake and may be expected about old-fashioned gardens. *Santolina* belongs to the Mayweed Tribe (p. 263) and in Marin County most closely approaches *Tanacetum* (p. 283) from which it differs in the coriaceous bracts of its receptacle.

Tribe 9a. Calenduleae. MARIGOLD TRIBE

In the key to the tribes of *Compositae* (pages 258, 259), the marigolds key out with *Adenocaulon* as far as the achenes are characterized (p. 259). The leafy stems and showy flower heads of the marigolds, however, are distinctively different from the habit and heads of *Adenocaulon* (which is a member of the Cudweed Tribe, pages 260, 271).

58a. Calendula

a. Heads large and showy, the rays usually more than 1 cm. long. .1. *C. officinalis*

a. Heads smaller, the rays about 1 cm. long ....2. *C. arvensis*

1. **C. officinalis** L. The cultivated garden marigold tends to persist spontaneously. It has been reported from several stations on Tiburon Peninsula by Peñalosa and is to be expected elsewhere.

2. **C. arvensis** L. Reported as locally common in Stinson Beach in 1960, the field marigold has not spread in Marin County as it has in southern Sonoma County where it is a conspicuous floral feature in fields and orchards in early spring. Both this and the garden marigold are native in Europe.

59. Cirsium (p. 285)

2. **C. callilepis** (Greene) Jepson. This is the Marin County thistle in which the numerous spiny-fringed involucral bracts are closely appressed. The type selected to represent this species was collected in the Fairfax Hills near the Meadows Club. A rare form from Mount Tamalpais with the tips of the bracts spreading slightly was called *C. remotifolium* (Hook.) DC. var. *odontolepis* Petrak.—*C. remotifolium* of MARIN FLORA in part.

2a. **C. callilepis** var. **pseudocarlinoides** (Petrak) J. T. Howell. In this variety the lacerate-margined involucral bracts are narrower and loosely ascending. The type is from Mount Tamalpais where the plant is occasional on brushy or wooded slopes and where it hybridizes with *C. quercetorum* (Gray) Jeps. *Cirsium amblylepis* Petrak is the name of a Tamalpais plant that may be one of the suspected hybrid derivatives.—*C. remotifolium* (Hook.) DC. ssp. *pseudocarlinoides* Petrak.

4. **C. brevistylum** Cronquist. The name *C. edule* Nutt. is now applied to a thistle of Washington and British Columbia.

6. **C. proteanum** J. T. Howell. Since *C. Coulteri* Harv. & Gray was found to be the same as *C. occidentale* (Nutt.) Jeps., a new name has been given to the Venus thistle originally described as *Carduus venustus* by E. L. Greene (not *Cirsium venustum* Porta).

8. **C. Douglasii** DC.—*C. Breweri* (Gray) Jeps. var. *Wrangelii* Petrak.

9. **C. hydrophilum** (Greene) Jeps. var. **Vaseyi** (Gray) J. T. Howell. The thistle of wet places in serpentine areas of the Tamalpais region is only varietally distinct from the Suisun Marsh thistle.—*C. Vaseyi* (Gray) Jeps.

63. Centaurea (p. 288)

5. **C. repens** L. The Old World plant known as Russian knapweed occurred as a roadside weed between Santa Venetia and China Camp (*Frankel 213* in 1963). As in *C. Cyanus* L., the involucral bracts in *C. repens* are not spine-tipped, but the two are readily separated since *C. repens* is a perennial with all flowers in the head alike and fertile.

63a. Cnicus

1. **C. benedictus** L. In 1921 Alice Eastwood collected the blessed thistle of the Mediterranean region in Mill Valley but there is no record that it has been found there since. Among the thistles of Marin County, the achenes of *Centaurea* (p. 288), *Carthamus* (p. 288), and *Cnicus* are attached by their sides. *Cnicus* is readily

distinguished from *Centaurea* by its spiny leaves and from *Carthamus* by its pappus of 20 bristles that are arranged in a long outer series and a shorter inner series.

67. Microseris (p. 289)

The species are here realigned following the studies of K. L. Chambers who unites *Uropappus* (p. 289) with *Microseris* (Contrib. Dudley Herb. 4: 207–312. 1955; Leafl. West. Bot. 10: 106–108. 1964).

a. Pappus paleae gradually narrowed into the elongate bristle.................b

a. Pappus paleae notched or slightly erose at apex below the bristle............e

b. Plants perennial; involucre about 15 mm. long.................1. *M. paludosa*

b. Plants annual; involucre 12 mm. long or less c

c. Pappus paleae about 1 mm. long.................2a. *M. Douglasii* ssp. *tenella*

c. Pappus paleae 1.5–4 mm. long ....d

d. Achene slightly constricted just below the top; pappus bristles stout, subplumose
....2. *M. Douglasii*

d. Achene not constricted below the top; pappus bristles delicate, scabrous
....3. *M. Bigelovii*

e. Pappus paleae slightly erose at apex.........................4. *M. decipiens*

e. Pappus paleae sharply notched at apex ....f

f. Pappus white or whitish, deciduous; achenes narrowed into a stout beak
....5. *M. Lindleyi*

f. Pappus buff, persistent; achenes only slightly narrowed upward
....6. *M. heterocarpa*

1. **M. paludosa** (Greene) J. T. Howell.

2. **M. Douglasii** (DC.) Sch. Bip. Tiburon and Mount Tamalpais (Portrero Meadow, Lagunitas Meadows) to the Carson country, Hamilton Field, and Lucas Valley.

2a. **M. Douglasii** ssp. **tenella** (Gray) K. Chambers. In Marin County only known from the south end of Tiburon Peninsula (*Howell 18078* in 1943).—*M. aphantocarpha* (Gray) Gray.

3. **M. Bigelovii** (Gray) Sch. Bip. Angel Island; Sausalito; Kentfield; Mount Tamalpais (Portrero Meadow); Fairfax Hills; Point Reyes Peninsula (Mount Vision, Drakes Estero, Point Reyes lighthouse); Dillon Beach.

4. **M. decipiens** K. Chambers. A rare species occurring south to Monterey, in Marin County known only from coastal serpentine slope south of Stinson Beach (*Howell 22098* in 1946, *Chambers 668* in 1955).

5. **M. Lindleyi** (DC.) Gray. Tiburon; San Rafael Hills; Mount Tamalpais; Carson Ridge; Inverness Ridge.—*Uropappus linearifolius* (DC.) Nutt.

6. **M. heterocarpa** (Nutt.) K. Chambers. Mount Tamalpais (Summit Avenue, Phoenix Lake); Fairfax Hills.—*Uropappus Lindleyi* of MARIN FLORA, not *U. Lindleyi* (DC.) Nutt.

78. Sonchus (p. 291)

1. **S. oleraceus** L. A variant of the common sowthistle with leaves bipinnatifid occurs on Tiburon Peninsula and has been reported as var. *lacerus* (Willd.) Wallr. by J. Peñalosa.

79. Lactuca (p. 291)

1. **L. biennis** (Moench) Fernald. The name *L. spicata* (Lamk.) Hitch. was misapplied and belongs to a different plant.

80. Agoseris (p. 292)

In his study of the perennial species, Quentin Jones has realigned the native dandelions as follows:

1. **A. apargioides** (Less.) Greene. This is the widespread plant of grassy or brushy hills.—*A. hirsuta* (Hook.) Greene.

1*a*. **A. apargioides** var. **Eastwoodiae** (Fedde) Munz. The plant restricted to maritime bluffs and dunes.

2. **A. grandiflora** (Nutt.) Greene.—*A. plebeia* (Greene) Greene.

3. **A. retrorsa** (Benth.) Greene. The only record known from Marin County is from Bolinas Ridge where the plant was collected by Alice Eastwood. The achene in this species is abruptly truncate below the beak, not tapering into the beak as in the other species.

New Plant Names in the Supplement

Athyrium Filix-femina (L.) Roth var. **cyclosorum** (Ledeb.) T. Moore fma. **strictum** (Gilbert) J. T. Howell, comb. nov. *A. cyclosorum* var. *strictum* Gilbert, List of North American Pteridophytes 14, 32. 1901.

Eleocharis quinqueflora (F. X. Hartmann) O. Schwarz var. **Suksdorfiana** (Beauverd) J. T. Howell, comb. nov. *E. Suksdorfiana* Beauverd, Bull. Soc. Bot. Genève, ser. 2, 13: 267. 1922.

Corallorhiza maculata Raf. fma. **immaculata** (M. E. Peck) J. T. Howell, stat. nov. *C. maculata* var. *immaculata* M. E. Peck, Leafl. West. Bot. 7: 177. 1954.

Ambrosia Chamissonis (Less.) Greene var. **bipinnatisecta** (Less.) J. T. Howell, comb. nov. *Franseria Chamissonis* var. *bipinnatisecta* Less., Linnaea 6: 508. 1831.

Index to the Supplement